The Frontiers Collection

The books in this collection are devoted to challenging and open problems at the forefront of modern science and scholarship, including related philosophical debates. In contrast to typical research monographs, however, they strive to present their topics in a manner accessible also to scientifically literate non-specialists wishing to gain insight into the deeper implications and fascinating questions involved. Taken as a whole, the series reflects the need for a fundamental and interdisciplinary approach to modern science and research. Furthermore, it is intended to encourage active academics in all fields to ponder over important and perhaps controversial issues beyond their own speciality. Extending from quantum physics and relativity to entropy, consciousness, language and complex systems—the Frontiers Collection will inspire readers to push back the frontiers of their own knowledge.

More information about this series at https://link.springer.com/bookseries/5342

Bernd-Olaf Küppers

The Language of Living Matter

How Molecules Acquire Meaning

 Springer

Bernd-Olaf Küppers
University of Jena
Jena
Germany

ISSN 1612-3018   ISSN 2197-6619   (electronic)
The Frontiers Collection
ISBN 978-3-030-80321-6   ISBN 978-3-030-80319-3   (eBook)
https://doi.org/10.1007/978-3-030-80319-3

This Springer imprint is published by the registered company Springer Nature Switzerland AG
The registered company address is: Gewerbestrasse 11, 6330 Cham, Switzerland

Thus, Nature speaks down to other senses,
to known, misjudged, unknown senses;
thus, she speaks to herself and to us
through thousands of appearances.

—*Johann Wolfgang von Goethe (Theory of Colors)*

Acknowledgements

The ideas presented here emerged from discussions with scientists from quite different areas of research. Above all, I acknowledge a lasting debt of gratitude to Manfred Eigen, who in the early 1970s was my doctoral supervisor at the Max Planck Institute for Biophysical Chemistry in Göttingen. Manfred Eigen not only paved my way from physics to molecular biology, but also supported, over many years, my increasing interest in the information-theoretical foundations of biology.

Research at the interface of physics and biology inevitably leads to complex philosophical issues of science. In this respect, Erhard Scheibe and Carl Friedrich von Weizsäcker were my closest interlocutors in my early years as a scientist. Over time, many ideas in this book were also shaped in discussions with Gregory Chaitin, Günter Hotz, Jean-Marie Lehn, Juan Roederer and Peter Schuster, to mention but a few. My thanks go to all of them.

I also remember with gratitude a series of interdisciplinary meetings on frontier issues of science initiated by the John Templeton Foundation. Exchanges of ideas with Paul Davies, Christian de Duve, Freeman Dyson, George Ellis, John Lucas, Arthur Peacocke, John Polkinghorne, Martin Rees and others all influenced in one way or another my view on the scope and reach of science.

A long-lasting scientific friendship connects me with Koichiro Matsuno and Paul Woolley. Beside our scientific discussions, Koichiro sharpened my view on the differences between Far Eastern and Western culture, which are also reflected in these cultures' different attitudes to scientific and technological progress. Paul I thank for many valuable suggestions regarding the manuscript. His alert eye and constructive criticism were an enormous help for me. Finally, I would like to thank Springer Nature, and in particular Angela Lahee, for their support in publishing this book.

Introduction: Bridging the Gap between Matter and Meaning

Until the middle of the twentieth century, it was a widely held view among leading physicists that life's phenomena elude a complete physical description. The characteristic features of living matter, such as self-maintenance, self-control and self-reproduction, seemed to have no explanation in physical terms. Moreover, living systems have a degree of ordered complexity that from the perspective of traditional physics is highly improbable. Altogether, this gave rise to a strong impression that living matter obeys its own laws.

An epoch-making turning point in our understanding of life was the elucidation of the molecular structure of DNA by Francis Crick and James D. Watson in the 1950s. In the years that followed, it became evident that living matter is not governed by life-specific laws, but only by the physical and chemical properties of biological molecules, among these above all two classes of macromolecules: the nucleic acids and the proteins. In living matter, these molecules operate together like the legislative and executive sides of government. The nucleic acids hold instructions for the formation of proteins, which for their part execute all the life functions that are encoded in the genetic script. This highly coordinated interplay is based exclusively on the known laws of physics and chemistry.

From the genetic instructions, everything about living matter can be explained, at least in principle, by the genome's interactions with its physical and chemical environment. The translation of the genetic script into proteins, which are the carriers of biological function, is mediated by the genetic code. It establishes the link between living matter's genotype and phenotype. To describe this interplay adequately, the concept of information has been introduced into molecular biology. Through this, living matter has become accessible to a theoretical understanding based on the models of storage and processing of information.

Moreover, exploring the molecular basis of life has uncovered some striking parallels between the genetic script and a text written in human language. Thus, the genome's molecular building blocks ("nucleotides") have the property of letters that are organized hierarchically into functional units corresponding to words, sentences and paragraphs. Like a written text of human language, the genetic script has punctuation marks and a defined reading direction.

In the phase of molecular biology, theoreticians used the information concept of the then nascent communication technology. For example, this allowed one to measure the amount of information stored in the genetic script by the number of binary digits (bits) necessary to specify the precise sequence of the genome's building blocks ("letters"). This number, however, is only a measure of the complexity of the program resident in the genome, and it tells us nothing about the program's actual meaning, i.e., the operating instructions that it carries.

In fact, the communication engineer's concept of information is entirely detached from the meaning of a message. From a technical point of view, this is quite understandable. The communication engineer's task is merely to transfer a sequence of signals, symbols or binary digits with as few errors as possible to a receiver, so that the message's content, whatever its meaning may be, is not altered. From the point of view of technological communication, a randomly jumbled sequence of characters has the same information measure as a meaningful message of the same length. At the other extreme, it is precisely the meaning content of genetic information that constitutes the difference between living and non-living matter.

Theoretical biology, therefore, requires a comprehensive concept of information that also includes the semantic dimension of genetic information. However, to practitioners of the exact sciences, whose work is based on observation, measurement and mathematical formalization, semantic elements must seem highly foreign. The semantic aspects of reality would seem to be, at best, accessible to the humanities and their methods of interpretation, which in themselves are largely subjective. So, the question arises as to whether there is some hitherto unknown pathway between these two hemispheres of scholarly thinking that can guide us to grasping the semantic dimension of information in an objective and precise manner too.

In the 1980s, this question moved increasingly into the focus of theoretical biology, when the physical and chemical theory of molecular self-organization and evolution of life took shape (cf. my *Molecular Theory of Evolution* [1]). At that time, it was becoming clear that an understanding of the generation of information in Nature will be the key to a deeper comprehension of living matter. I outlined this issue in *Information and the Origin of Life* [2]. In his foreword to that book, the physicist Carl Friedrich von Weizsäcker wrote: "Scientifically, this theory seems to me to close a gap that is perhaps comparable to the geographical discovery of the North-West Passage north of America: no one had reason to doubt that these waters existed, but it was uncertain whether our ships would be able to navigate them." [2, p. xiv]

Where are we now? Has our voyage of discovery taken us to uncharted scientific territory? During the past three decades, rapid progress has been made in molecular biology (Chap. 6). The development of modern sequence analysis has led us to new and profound insights into the fine structure of genetic information (Sect. 6.2). The discovery that a particular class of nucleic acids (RNA) can catalyze their own reproduction, without the help of proteins, has jolted the RNA world into the center of research interest, opening new paths toward an experimental and theoretical understanding of life's origin (Sect. 6.9). Moreover, within the frame of biological information theory, a "royal road" is today opening up that may lead us to modeling the semantics of genetic information (Sect. 6.8).

Given the huge amount of information that has accumulated from genetic research and which is deposited in more than a thousand databases worldwide, it is becoming clearer almost by the day that biology urgently needs a systematization frame that goes beyond the classical Darwinian theory of evolution. The theoretical concept we are looking for must be based on a natural principle that grasps the peculiarities of life's processes within the framework of the known laws of physics and chemistry, without ascribing a special status to living matter. In this book, I posit that all molecular biological findings support the hypothesis that the principle sought is a molecular language.

It is evident that this idea not only has far-reaching consequences for the theoretical foundation of biology; it also bundles numerous issues of our scientific understanding of the world and brings them to a focus. This is the reason why the fundamentals of science occupy a large part of this book. Let me highlight here some points that lead directly to the core of the book. The conjecture that language might be relevant for understanding living matter, its origin and evolution, emerged long ago. No less a visionary than Charles Darwin wrote in his *Origin of Species* (1859): "If we possessed a perfect pedigree of mankind, a genealogical arrangement of the races of man would afford the best classification of the various languages now spoken throughout the world; and if all extinct languages, and all intermediate and slowly changing dialects, were to be included, such an arrangement would be the only possible one." [3, p. 410]

The physiologist Friedrich Miescher, who discovered nucleic acids at the end of the nineteenth century, saw no other way to explain the diversity of genetic dispositions than by comparing it to the unlimited richness of "words and expressions of all languages" [4, p. 117; author's transl.]. In *Laws of the Game* [5] and other writings, Manfred Eigen also pointed to the obvious parallels between the organization of genetic information and that of human language. In his Nobel lecture on *The generative grammar of the immune system* [6], Niels Jerne compared the immune system's "immense repertoire" to "a lexicon of sentences which is capable of responding to any sentence expressed by the multitude of antigens which the immune system may encounter" [6, p. 220]. He found it "astonishing that the immune system embodies a degree of complexity which suggests some more or less superficial though striking analogies with human language, and that this cognitive system has evolved and functions without assistance of the brain" [6, p. 223].

In this book, I will go a step further and claim that language is not merely a helpful analogy to describe the organization of living matter, but a principle of Nature that has its roots in the laws of physics and chemistry. This hypothesis breaks with our traditional view according to which language is a unique property of humans. Since language is mankind's gateway to the world (Chap. 1), it is in the nature of things that we have, first and foremost, an anthropocentric idea of language. From this perspective, any talk about a language of living matter must inevitably seem to be metaphorical. To escape the constraints of a superficial analogy, one has to deepen the idea of language by abstracting from the complex and specific peculiarities of human language and uncovering its structural features. Afterward, one must demonstrate that these structures are already present at the molecular level of Nature. With this task, we are undoubtedly breaking new ground in the exact sciences.

Up to now, the most advanced approach to the structural aspects of human language has been developed by the linguist Noam Chomsky, in his book on *Syntactic Structures* [7]. His investigations uncovered an overarching aspect of human languages, termed "universal grammar," comprising the rules according to which words and sentences are formed in all natural languages. The studies furthermore suggest that grammatical rules are recorded in innate structures of the brain. This, in turn, would mean that the universal grammar is genetically anchored.

A prerequisite for Chomsky's analysis is the assumption that grammatical structures can be justified exclusively at the syntactic level of language, i.e., without reference to its semantic dimension. We will follow Chomsky's argument by reconstructing the nucleation of language at the level of prebiotic macromolecules (Sect. 6.9). The language itself we will denote as "molecular" language and its developed form as "genetic" language. Correspondingly, this book leads from the structures of human language to the language of living structures.

To justify our assertion of the existence of molecular language, we start from the well-grounded working hypothesis that all life processes are based exclusively on physical and chemical laws. These laws, however, may act together in a particular manner that we usually describe as a principle. An example of this is the principle of natural selection, which takes effect not only among living beings but also among molecules in non-living systems. The only requirements for this are self-reproduction and an overall growth limitation placed upon the population.

Self-reproducing nucleic acids are paradigmatic for selection in the Darwinian sense (Sect. 6.5). However, in the absence of any biosynthetic machinery, the molecules' structural properties themselves are the target of selection. They determine the molecules' reproduction dynamics and thereby their selection value. The greater a molecule's reproduction rate, reproduction accuracy and inherent stability are, the higher is its selection value. Since nucleic acids' structural properties depend on their nucleotide sequence, selection automatically favors sequence patterns that contribute to efficient reproduction. Moreover, in the interplay between random mutation and selection, these patterns will be strongly conserved. They can be considered as "proto-words" or "proto-sentences" of a molecular language, stabilizing the advantage acquired by the molecule's structure (Sect. 6.8).

The folding of a nucleotide sequence to produce a three-dimensional autocatalytic structure is determined by physical and chemical forces only. At the same time, this is the most elementary relationship between structure and function in prebiotic molecules that one can imagine. At this level, the physical and chemical origin of molecular language lies before us, as it were, in a nutshell. It is fascinating that a relatively simple autocatalytic mechanism, combined with random mutation and natural selection, already leads to the formation of syntactic structures in non-living matter. This is the starting point for the development of a language that finally passes through numerous stages of evolution, from molecular language to genetic language up to the sophisticated forms of human language.

Detailed analysis verifies that living matter's language shows all the structural features that we also associate with human language: Molecular language is based on a finite alphabet. Its words are hierarchically organized into sentences, paragraphs and so forth (Sect. 1.5). Its syntax is aperiodic (Sect. 5.4) and has a grammatical structure (Sect. 6.9). Finally, genetic information expression leads to a dynamization of information that shows all the features of linguistic communication (Sect. 6.3). It breathes life, in the truest sense of the expression, into the abstract formula "Life = matter + information" (Sect. 5.7).

The concept of molecular language will significantly alter our traditional view of the origin and evolution of life. Beside natural selection, language must be considered the second decisive driving force of biological development. Thus, evolution theory becomes a matter of linguistic theorizing (Sect. 6.8). This shift in perspective suggests that life's evolution should be viewed dualistically, i.e., as the evolution of the genetic language's syntax and the evolution of its semantics. Both processes are based on natural selection, but they refer to different evolution processes, namely the non-Darwinian development of living matter's genotype and the Darwinian development of its phenotype (Sect. 6.10).

The non-Darwinian development of syntax is non-adaptive. It is entirely restricted to the structural properties of the self-reproducing information carriers and their possible interactions. The later development of semantics is superimposed upon this process. It is the step at which genetic information is translated into proteins and obtains its relevance, i.e., its meaning regarding the outer world. The molecular language of living matter now becomes context-dependent, and Darwinian evolution by adaptation comes into play. This is also the point at which molecular language goes over into genetic language.

Language-driven evolution has several distinctive features. As described, the nucleation of molecular language's grammar can be reduced entirely to physical and chemical processes combined with mutation and selection. At this level, nucleic acid molecules can already develop syntactic structures corresponding to words and sentences. However, by cooperative interactions, nucleic acid molecules can also form reaction cycles stabilizing and enlarging their structural information (Sect. 6.10). In this way, a reservoir of genetic "words" and "sentences" can build up, filling living matter's linguistic toolbox. At this level, which is still the level of syntactic structures, a form of molecular organization begins to emerge. It is characterized by cooperation, compartmentation, self-regulation, hierarchy formation and other functional elements. Together, these constitute the proto-semantics of genetic information (Sect. 6.10) and thus represent a case study for the model of a semantic code (Sect. 6.8) that describes the emergence of the meaning of items of information.

By mutation and selection, numerous forms of organization may evolve, differing in both the kind and the weighting of their functional features. One can compare them with the myriads of possible ice crystals, each of which has an individual and unique form although all result from the same physical mechanism (Sect. 5.7). Molecular organizations based on the chemical rules of living matter's proto-grammar show the same combinatorial richness of forms that is given by the unlimited diversity of linguistic expressions in human language. At first, however, molecular organizations were nothing more than linguistic pre-structures—blank forms for the further evolution of life. They were syntactic structures without semantics.

Language-driven evolution provides a plausible explanation for the genome's noncoding regions. These must be interpreted as the information structure on which the genetic organization of an organism is based. Obviously, only a minor part of the genome's information has a relation to the outside world at all, as expressed by the organism's phenotype. This information is located in the genes that are translated into proteins. The major part of the genotype, in contrast, seems to function only within the internal context, manifested in the organism's organization. In other words, the genome's noncoding information must be expected to serve the language mechanism, which causes and controls the genes' dynamization and establishes the genome's relation to the outside world. The same applies, by the way, to human language. In a written text, for example, many words do not relate to the actual subject matter at all. Instead, they are necessary to structure a sentence grammatically and logically. They only constitute the framework into which the minor fraction of words that are subject-matter-related is embedded (Sect. 6.9).

According to the Darwinian understanding of evolution, the semantic dimension of genetic information originates from organisms' evolutionary adaptation to their environment. Figuratively speaking, organisms gain information about their environment by mutation and selection, which becomes fixed in their genes. In this sense, genes are thought to map information about the organism's external world. They determine the organism's phenotype on which Darwinian selection operates. Given a sufficiently rich and varied environment, the evolution of the phenotype can be justified on the basis of Darwin's theory.

However, in the earliest stages of evolution, there was no information-rich environment (or context) that could serve as the reference frame for molecular evolution, directing the evolution of information toward increasing complexity. The generation of a sophisticated program complexity itself presupposes an external source of sufficiently complex information. Without it, evolution would be a kind of perpetual motion machine, creating information out of nothing—an idea, however, that can be shown to be impossible (Sect. 5.6).

Evidently, there is a blind spot in Darwin's theory of adaptation. The only way to put his idea onto a solid basis is provided by recourse to the language of living matter. This leads directly to a highly significant aspect of language, namely the context-dependence of linguistic expressions (Sect. 1.8). To justify this significance, one has to reconsider the process of gene expression by which the relationship between genetic information and its environment, the "external" world, is established. This process requires a machinery for biosynthesis, to translate genetic information into the immense variety of biomolecules from which life processes emerge.

Moreover, gene expression also needs perpetual feedback between the genome and its gene products, to coordinate the myriads of molecular processes. This feedback takes the form of communication, even though molecules do not "talk" to each other in the literal sense. Communication does not presuppose consciousness; rather, it (only) requires the exchange of information between sender and receiver, in this case the genome and its physical and chemical environment. The environment suffices to give a meaning to the —a priori meaningless—nucleotide sequence of the genome.

The contextuality of information modifies the classical idea of genetic determinism without, however, abolishing it. Genetic determinism used to be based on the assumption that genetic information is necessary and sufficient for constructing the living organism. This idea must today be reinterpreted as a "generative" determinism (Sect. 6.3). According to this, the syntax of genetic information is still necessary and sufficient for the organism's self-reproductive maintenance, but its semantics are constituted solely by the genome's expression, i.e., through its "communicative" interaction with its molecular context. We know this very well from human language, where the meaning of linguistic expressions is sharpened in the dialog between communication partners.

The idea of genetic language leads to a new interpretation of Darwinian evolution. From the perspective of language-driven evolution, the source of evolutionary progress is not the environment, but the nearly unlimited number of linguistic expressions that can be generated at the genotypic level by molecular language. These expressions represent possible forms of functional organization. When translated into a phenotype, they are tested for fitness by natural selection. In other words, the actual motor of biological evolution is not the environment itself, but the change in use of living matter's language in a continuously changing environment. Those forms of expression will survive that turn out to be meaningful under the prevailing conditions. By this mechanism, the biosphere emerged over time. Without the existence of genetic language, the genome would lose all its information when the environment changes. It would eliminate large parts as "junk". Darwinian test-tube experiments demonstrate this clearly (Sect. 6.5).

The language paradigm can be expected to open up new questions and pathways in the exploration of the genome's structure and function. It may also have an impact on biotechnology and medical research. From the epistemological perspective, the language paradigm provides a systematic framework for biological theory. Quite different ideas on the mechanisms of evolution ("neutral selection", "selfish genes", "convergent evolution", "punctuated equilibria", "tinkering") no longer appear to contradict one another, but rather to be complementary features of a language-driven evolution.

The crux of the matter is the interface between physics and biology, where the language of living matter has its roots. However, this book goes far beyond that. It takes up the problem of how semantic information could arise in living matter—a problem intertwined with that of the genesis of meaning and with ramifications reaching into all areas of science. This broad issue is also reflected in the structure of this book. To justify the idea of living matter's language, one has to look in depth at our scientific and philosophical thinking, at language as such (Chap. 1), at science's claim to truth (Chap. 2) and its methods (Chap. 3), at the unity of science (Chap. 4), its limits (Chap. 5) and perspectives (Chap. 6). An epilog (Chap. 7) introduces Nature's semantics and considers some implications of this for our view of Nature. Accordingly, this book is also an account of how progress in the life sciences is transforming the whole edifice of science, from physics to biology and beyond.

References

1. Küppers B-O (1983) Molecular Theory of Evolution: Outline of a Physico-Chemical Theory of the Origin of Life. Springer, Berlin/Heidelberg
2. Küppers B-O (1990) Information and the Origin of Life (transl: Woolley P) MIT Press. Cambridge/Mass [Original: Der Ursprung biologischer Information, 1986]
3. Darwin C (1859) On the Origin of Species by Means of Natural Selection, or the Preservation of Favoured Races in the Struggle for Life. Murray, London
4. Miescher F (1897) Letter to Wilhelm His, 17 Dec 1892. In: Histochemische und physiologische Arbeiten, Bd 1. Vogel, Leipzig
5. Eigen M, Winkler R (1993) Laws of the Game: How the Principles of Nature Govern Chance. Princeton University Press, Princeton [Original: Das Spiel, 1975]

6. Jerne NK (1984) The generative grammar of the immune system. Nobel lecture. Nobel Foundation
7. Chomsky N (1957) Syntactic Structures. Mouton, The Hague

Contents

1

Language: Gateway to the World

1.1 Forms of Knowledge

"All men naturally desire knowledge. An indication of this is our esteem for the senses; for apart from their use we esteem them for their own sake, and most of all the sense of sight. Not only with a view to action, but even when no action is contemplated, we prefer sight, generally speaking, to all the other senses. The reason of this is that of all the senses sight best helps us to know things, and reveals many distinctions."

With these words Aristotle's *Metaphysics* begins [1, book 1, 980a]. In fact, the cultural history of man is marked by a steady increase in knowledge. However, only in our time has this knowledge grown to such a vast extent that it can hardly be surveyed anymore. An example of this is the enormous increase in knowledge that has accompanied recent developments in science. It has built up a virtually impenetrable jungle of information around us, in which only a few people are still able to find their way.

The almost explosive development of knowledge is undoubtedly a primary reason for the apprehension with which many people view science: when the complexity of scientific discoveries is no longer transparent and under-standable, it becomes eerie and feels threatening. It is therefore not surprising that many people regard science as a destructive rather than a constructive element of our world. The increase in knowledge has been so great that, even among scientists, mutual understanding is often scarcely possible. More than that: our experience must be updated almost daily, which in turn requires a perpetual rearrangement of our stock of knowledge. The so-called expert controversy, in which everyone claims to possess real insight and to have the

© Springer Nature Switzerland AG 2022
B.-O. Küppers, *The Language of Living Matter*, The Frontiers Collection,
https://doi.org/10.1007/978-3-030-80319-3_1

latest state of knowledge on his side, is the most clearly visible expression of this development.

Nevertheless, the widespread talk about a "flood of knowledge" is somewhat misleading. It conveys the impression that understanding in science accumulates continually, as new insights are found and integrated into our scientific view of the world. Yet is the mere increase in printed or electronically stored information really a significant indicator of progress in knowledge? Can knowledge be measured exclusively in digital bits? Are not, instead, content and quality the decisive hallmarks of advanced knowledge? How can we evaluate scientific progress in these terms? Such questions lead directly to the search for the essence of the human culture of knowledge. Before we set out on this path, however, let us briefly clarify some terms related to "knowledge".

The most elementary expression of cognition has always been sought in direct intuition and perception, in short: in the obviousness of things. Aristotle described it this way when, at the beginning of his *Metaphysics*, he emphasized the importance of our senses in satisfying man's thirst for knowledge. However, the bare experience of things is only a preliminary stage to a conceptually and methodically elaborated knowledge in which cognition is linked and theoretically substantiated by definitions, causal explanations and proofs.

Knowledge in the real sense is cognition of complex issues. For example, when one says that someone has particular cognition one is stating that the person concerned has not only perceived something, but also knows how the observed "something" is constituted and how it is related to other observations. In other words: It is only through insight into the causal interrelationships of reality that the bare experience of reality becomes a cognition of reality and thus constitutes theoretical knowledge.

There are other forms of knowledge. For example, if we say that we know how to achieve what we want to do, then we are talking about a kind of knowledge that guides our actions, and that is in the broadest sense relevant to our life pursuits. This knowledge is primarily a practical knowledge of specific skills by which we can cause or produce something.

Entirely different answers, on the other hand, are required by the questions of what we should do at all and whether what we intend to do is also morally justifiable. Answers to these questions need orientation guides that determine the goals, and the value yardsticks, of our actions. Such orientational knowledge is difficult to justify. It is mostly based on metaphysical or transcendental reasoning, including, last but not least, religious convictions and values.

The knowledge that can claim maximum validity is undoubtedly the knowledge of how to effect or to produce something, because this enables us to solve practical problems. On the other hand, practical knowledge is only possible because it is based on theoretical knowledge, which yields information about the interrelationships between causes and effects. These two forms of knowledge are, therefore, inseparably linked and document our rational approach to reality.

In current philosophy, there is much debate about the question of which knowledge form has priority in the development of science and technology. Does theoretical knowledge result from the understanding of life's practicalities, or is theoretical knowledge—conversely—an ultimate prerequisite for the development of practical knowledge? Would it have been possible to invent the scales without any concept of what weight is? Or, the other way around, could we have a notion of weight without ever having built a set of scales? We can ask the same questions in connection with other practical inventions, for example the technique of leverage and the lever laws.

Asked in the most general way: Are our scientific concepts and theories first and foremost cultural constructs that have emerged from dealing with the experimental technical issues of our living environment, or are they a distillate of objectively existing relationships that make it possible to deal with reality technically? The cultural history of mankind offers plenty of indications to support the one or the other explanatory variant. However, as we shall see later on, an absolute justification of human cognition is excluded, for a fundamental reason. This means, in turn, that the epistemological question, asking which of the two forms of knowledge has priority, cannot be answered finally.

The distinction between theoretical and practical knowledge goes back to Aristotle. He distinguished for the first time between an understanding of the universal principles of reality and knowledge focused on the demands of everyday life. This distinction was a remarkable step, in so far as in early antiquity the Greek word *theōria* had an ethical and religious connotation and thus a normative aspect.

The prescriptive function of *theōria* concerned both the understanding of Nature and the social organization of the human community. In other words: *theōria* was assumed to determine not only the movement of the stars but also the forms of reasonable coexistence in human society. Accordingly, theoretical and practical knowledge made up a unified whole. The related idea of *lógos* expresses clearly this line of thinking. In the antique understanding, *lógos* is the reasonable world law, which directs everything and mediates the unity of reality in the multiplicity, diversity and contradictoriness of its appearances.

Concerning practical knowledge, Aristotle further distinguished between knowledge that is focused explicitly on goodness and happiness, and knowledge that is aimed at the work to be done (*poiesis*). The latter is to be understood in the sense of skill (*téchne*) or creative production. Poietic actions include matters of everyday life as well as those of the world of arts. The latter aspect of *poiesis* is still reverberating today in the word "poetry".

Aristotle also argued that the "poietic" knowledge was the very basis of all knowledge, including in particular cognition about Nature. In fact, Nature appeared to him as an active subject that pursues goals and purposes. Through her "poietic" actions, Aristotle concluded, Nature creates and organizes herself.

In this connection, Aristotle's doctrine of the four causes played a central role. A simple example explains the basic idea behind this doctrine: If one wants to build a house, one needs specific materials ("material cause"); a building plan that determines the shape of the house ("formal cause"); craftsmen who fit together the material components ("efficient cause"); and a goal that describes the purpose which the building is ultimately to serve ("final cause"). In other words, Aristotle was saying: All things have been created, they consist of matter, they have a shape and they have a purpose. This applies not only to the things that have emerged from human activity but also to things of natural origin.

According to Aristotle, the four causes give a sufficient answer to the central question of why things exist and why things are the way they are. Aristotle furthermore assumed that the four causes are interrelated pairwise: the material cause with the formal cause and the efficient cause with the final cause. They behave as the determinable does to the determining: The material of a thing is in itself indeterminate and is given its characteristic shape only by the formal cause. The per se indefinite movement receives its direction only through the final cause.

Aristotle's notion of movement always raises the question of "where to". Since there is no motion without direction, every movement points inherently to a target (*télos*). However, in the antique understanding, "movement" means not only change in spatial position, but also any change in quantity or quality. Thus, the theory of movement, which Aristotle places at the center of his natural philosophy, inevitably leads to a teleological understanding of reality. Every process, whether artificial or natural, must be seen as a purposeful formation of things because the Aristotelian concept of movement necessarily implies a direction toward a goal.

For example, a seed becomes a plant because the state of being a plant is the seed's inherent target. Natural things differ from artificial things only in that they have the movement within themselves, while artificial things are moved from the outside. Since each movement, as Aristotle postulates, is induced by another movement, this raises the question of the "prime mover" of the world, who is himself unmoved. With the problem of the "unmoved mover" as the first cause for all things that exist, Aristotle finally enters the realm of speculative metaphysics.

It seems that Aristotle did not want to understand his theory of motion in the sense of cosmic teleology, in which the world is heading toward a universal goal or fulfilling a universal purpose. Although Aristotle regarded every natural process as targeted, he would have denied the existence of an overarching goal. Following this interpretation, one can say that Aristotle considered the dynamics of the world to be a network of purposeful processes, but not a goal-directed network serving an overall purpose. The latter interpretation is more likely to have stemmed from the over-interpretation of the Aristotelian world-view by medieval theology, which set out to justify the wisdom and perfection of a Creator God. For this reason, one must carefully distinguish between the Aristotelianism of medieval theology and the authentic ideas of Aristotle himself.

With the view that every movement, i.e. every change, is goal-directed, Aristotle gave the concept of Nature an interpretation that was definitive for more than two thousand years in the Western understanding of Nature. It was not until the beginning of the seventeenth century that the Aristotelian world-view collapsed when Galileo Galilei succeeded in proving the inconsistency, and thus the untenability, of the Aristotelian doctrine of motion. However, the path that ultimately led to a systematic reorientation and renewal of the sciences and thus to a modern, mechanistic concept of Nature was an arduous one, because the Aristotelian world-view had burned deep into occidental thinking, leaving many traces.

1.2   Toward a New Atlantis

Aristotle had dealt with almost all the philosophical and scientific questions accessible to the thinking of his time. They concerned, beside metaphysics, the philosophy of science and nature, physics, biology, ethics, political theory, logic and language. His scientific heritage is correspondingly extensive, so that Aristotle can rightly be considered the first universal scholar in the modern sense of the word. The influence of Aristotelian thinking on science was so strong that it took two thousand years before a new era of science could begin.

At first, the methodological redefinition of science became the focus of public interest. This step has been prepared by Francis Bacon and René Descartes. At the beginning of the seventeenth century, when science and technology were emerging, both had rediscovered the importance of practical knowledge, to which Aristotle had already turned his attention. Bacon and Descartes were inspired by the idea that the living conditions of humans can be improved with the help of science and technology. As a result, they focused on the question of which methods are best suited to increase scientific and technological knowledge.

While Descartes also had an epistemological interest in science, Bacon considered science and technology exclusively in terms of their usefulness to society. Bacon believed that social and political peace in a community could only be guaranteed if all people could live free of worry and conflict. According to Bacon, however, this is only possible if human society gains— with the help of science and technology—unlimited power over Nature.

Driven by his maxim "knowledge itself is power", Bacon was interested in the conditions under which scientific and technical progress can be organized as effectively as possible. This aim was served by his methodology, as he described it in his programmatic *Novum Organum Scientiarum* (Fig. 1.1). The title of this treatise, which announces the "renewal of science", alludes to the famous *Organon*, a collection of theoretical and logical writings by Aristotle that had influenced Western thought for centuries.

The Greek word *organon* means "tool". The new tools with which Bacon wanted to tackle Nature to elicit its secrets were observation and experiment. Bacon believed that from the generalization of observations he could finally derive the rules that Nature obeys and which, he thought, were the key to gaining power over Nature.

All in all, Bacon described the scientific method as the art of discovery (*inventio*) from experience. This approach, in which observation and experiment play a central role, seems to be so close to the idea of empirical science that one is immediately inclined to acknowledge Bacon as a mastermind of modern

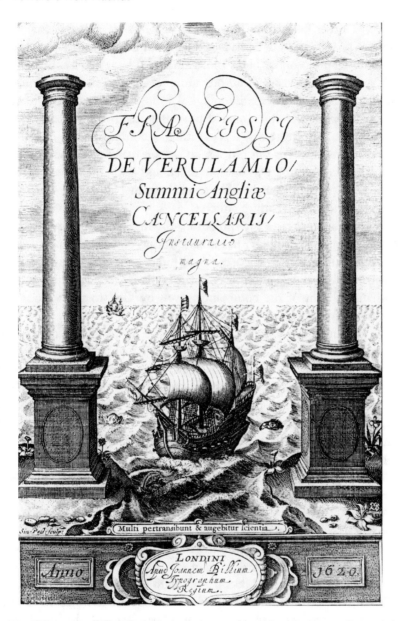

Fig. 1.1 Title page of Francis Bacon's major work *Instauratio Magna*. Bacon's book, published in 1620, is to a large extent a collection of aphorisms. Only the second part of the book, entitled *Novum Organum*, was completely worked out by Bacon. For this reason, later editions of the book were given the title of the dominant part, *Novum Organum Scientarium* [2]. Bacon's reflections have become eminently influential in modern scientific thinking, as they initiated the cultural transition from medieval science to modern science, even though Bacon's own thinking still seemed to be largely medieval. His program, which propagates the awakening of science and technology for the benefit of humanity, remains the guiding motive of contemporary science. [Image: Wikimedia Commons]

science. On closer inspection, however, it becomes clear that Bacon's understanding of science was not as revolutionary as it might seem at first glance.

The cultural philosopher Hans Blumenberg [3] has given a subtle and, as it
would appear, conclusive interpretation of Bacon's actual intentions.
According to this, all the methodological innovations that seem to emerge
from Bacon's work are only aimed at providing man with the tools to enable
him to transfer his creative will to Nature, according to the model of the
divine act of creation. For this purpose, however, man must observe Nature as
the work of the Creator without prejudice and with the utmost precision.
Only in this way will man be able to elicit from Nature the rules—divine
instructions—to which Nature adheres.

Following Blumenberg's interpretation, the method of observation, which
Bacon emphasizes so strongly, has a background entirely different from that
of the modern empirical sciences. As Blumenberg elaborates, Bacon was
obviously recalling the creation story in the book of Genesis, which is nothing
more than the sum of the commands given by God to the various beings
when they were called by their names: to be.

Thus, the human in Paradise just repeats the names that had appeared in
the creation commands of God. These are the actual names of the things. If
these names are called out, the things obey in the same way as they did when
they followed the divine act of creation, namely to come forth from nothing.
According to Blumenberg, the task of Bacon's new science would have been
"to extract from Nature the means by which power is exercised over her, just
as Nature had been a product of power from the very beginning" [3, p. 87;
author's transl.].

Blumenberg is probably right when he says that Bacon's conception of
science was closer to the black art of magic than to the then nascent methods
of modern science, which are devoid of any magic power. Bacon's idea that
Nature will disclose its secrets if one only encounters it without prejudice, i.e.,
by merely observing, is in fact nothing more than a magical incantation, one
in which mankind is supposed to gain control over Nature just by calling
things by their names. Thus, Bacon's program proves to be deeply rooted in
the biblical creation myth: science and technology are destined to regain the
sovereignty and power over Nature that had been lost with the banishment of
humans from Paradise. This thought is also supported by Bacon's utopia *New
Atlantis* [4].

At the same time, it becomes clear that a renewal of science by critical
thinking could not arise from an understanding of Nature that was still

attached to the magical world of the Middle Ages. On the contrary, Bacon even believed that one has to put the creative mind into chains. Since he was convinced that the secrets of Nature could only be disclosed by encountering Nature without prejudice, he demanded that we temporarily prevent our restless intellect from jumping to premature conclusions. Therefore, Bacon advised that "understanding must not […] be supplied with wings, but rather hung with weights, to keep it from leaping and flying" [2, p. 97].

Following on from Bacon's "power of knowledge", none other than Descartes made a large-scale attempt to renew science. Descartes also advocated an ideal according to which he considered scientific knowledge would enable humans to appropriate Nature and to become her ruler for the good of humanity. In the foreword to his *Principles of Philosophy* published in 1644, Descartes compares science and its various disciplines to a tree [5]. Its fruits, however, as he points out, cannot be found at the roots (in metaphysics) or on the trunk (in theoretical physics), but rather in its branches, i.e., in mechanics, anatomy, physiology and psychology. According to Descartes, mechanics serves for technology, while anatomy and physiology are essential for medicine; finally, psychology allows the mental control of emotion, feelings and passions.

Like Bacon, Descartes emphasized the importance of science for the improvement of human living conditions. In this connection, he thought primarily of working conditions, the maintenance of health and the prolongation of life. However, while Bacon misconstrued the task of scientific mastery over Nature as "an excessive implementation of the basic idea of magic" [3, p. 88; author's transl.], Descartes had a completely different vision of science. He concluded that control of man's outer and inner nature could only be achieved on the basis of knowledge that was methodologically assured and subject to constant monitoring. Unlike Bacon, for whom scientific knowledge was just the means to an end, Descartes regarded scientific knowledge as a goal to be pursued for its own sake. In short, the essential difference between Descartes and Bacon is their different perspective on knowledge, which makes Descartes appear to us as the genuine pioneer of modern science.

In his *Discourse on Method* [6] of 1637, Descartes established four rules for the proper use of reason. First of all: one should never acknowledge something as being right that is not obviously true and does not appear indisputable. Secondly: one should divide every problem into partial problems as far as possible, in order to be better able to solve it. Thirdly: one should always start with the simplest and most easily understandable issues and then move step by step to the compound, complex ones. Finally: one should in all cases strive for generality and completeness of knowledge.

The first rule puts systematic doubt at the head of the process of scientific cognition. According to this rule, only knowledge that can withstand rigorous doubt can claim to be true. The idea of critical knowledge, which is formulated by this rule, is of such fundamental importance that it can rightly be described as the key principle of modern science.

The second and third rules set out the starting-direction for gaining knowledge in science: the problem to be investigated should first be broken down into sub-problems. Then one should proceed from simple issues to complex ones. This method, based as it is on the dissection of the object or problem under investigation, has been called since the eighteenth century the "analytical" research strategy.

The fourth rule takes account of the idea that only a general knowledge can provide information about the law-like character of the world. The far-reaching demand for the completeness of cognition is, however, a mere desideratum. It aims at the inner consistency and absence of contradictions of the system of cognition, and is intended to justify its claims to truth.

To sum up: Descartes' methodology represents the basis of modern science, the guiding principles of which are, beside systematic doubt: analytical access to reality, and the objectification and generalization of cognition. Only such a methodologically secured knowledge, placed under unrestricted testing by experience, can ensure the rationality of our world-cognition.

At the transition to modern times, it also became clear that man can only intervene in Nature if he has precise knowledge of its fabric of causes and effects. Only this causal knowledge allows man to make reliable predictions about natural events and thus to intervene in a direct manner in Nature. For this purpose, the search must be made for facts that can be represented as cases of mathematically formulated laws. The implementation of this idea by Johannes Kepler, Galileo Galilei, Christiaan Huygens, Robert Boyle, Isaac Newton and other natural scientists of the sixteenth and seventeenth centuries was the most crucial step on the way to today's exact science.

1.3   Language and the World

The widening of causal knowledge in modern times has enabled mankind to become more and more independent of its natural living conditions. Consequently, we have gained increasing control over our outer and inner nature. It is evident that this development inevitably heralded the end of the Aristotelian world-view. If man can appropriate and transform Nature for his purposes, then Nature cannot at the same time obey a causal law that gives it a final destination. The Aristotelian view of Nature, which asserted the intrinsic purposefulness of natural processes, was displaced by the mechanistic conception, according to which natural events are fundamentally undirected and purpose-free.

Indeed, the mechanistic view of the world leaves no room for teleological explanations. Nevertheless, the idea that Nature as a whole arose from the goal- and purpose-orientated rationality of the Divine has persisted down to the present day. It finds its expression in the admiration of the perfect harmony with which the laws of Nature act in the universe in general and in the living organism in particular. This kind of spiritual view is also widespread among scientists. In fact, over the centuries a picture of Nature has solidified that suggests a harmonious order of Nature with a textbook-like character to help us understand reality. In this sense, the scientific progress of modern times reflects man's continued efforts to read in the "Book of Nature", to check by experience what has been read, and to transfer it into technical innovations.

However, as experience itself is burdened with all kinds of prejudices, it could only develop its correcting power in step with improvements in accurate observation and controlled experimentation. Thus, over the centuries, there have always been new interpretations of what has been read in the book of Nature. In the light of these, our modern and sophisticated understanding of reality has emerged step by step. Even today, the image of the readability of Nature is still present. When some years ago the human genome was the successfully deciphered, this event was enthusiastically celebrated with the words that the "Book of Life" now lies open before us and we only have to read it in order to make its contents useful for us ([7], see also [8]).

Blumenberg, to whom we owe a comprehensive historical analysis of the book metaphor, has even gone one step further. He has coined the phrase "readability of the world" to characterize the epistemic approach which underlies our understanding of the world. Following Blumenberg, our world cognition becomes more extensive and more discerning with each new reading of the world. "Only in the course of time and seen in historical perspective", he explains, "will we realize what cannot both be and be had at the same moment." [3, p. 21; author's transl.]

In other words, the meaning of a text is not determined by one of its many possible interpretations. Rather, it emerges gradually as the sum of all readings. From this point of view, the understanding of the historical depth of the readability is given priority over any time-bound reading of the world. However, the historical diversity of world interpretations leads to a disrupted understanding of the world that ultimately displays only its unreadability. In contrast, science aims to engender an ever more precise and clear understanding of the world. It is not the ambiguity of what has been read that paves the way to knowledge, but rather the reduction of ambiguity and uncertainty by the exclusion of untenable or false notions about the world.

This also raises the question of what role language plays in our understanding of the world. Obviously, something can only be "readable" if it has the structure of a language. Otherwise, the idea of the readability of the world would only be a suggestive, and ultimately a meaningless, metaphor. With this kind of consideration, we are approaching an epistemological thought that the philosopher Hans-Georg Gadamer once encapsulated in the mysterious-sounding words: "Being that can be understood is language." [9, p. 490]

This sentence allows two interpretations. On the one hand, it could mean that human language is the ultimate prerequisite for understanding "being". Accordingly, man's language would make up the unassailable point of man's reference to his world, i.e., man's world understanding would come to be only in and through his language. This view has met with a broad consensus in the contemporary philosophy of the humanities. Gadamer himself also held it.

However, Gadamer's statement permits further interpretation. It could alternatively mean that only being can be understood which itself already has the structure of a language. This view would turn the first interpretation upside down, stating that being has the structure of a language from the outset, through which it speaks to us and becomes recognizable.

In both interpretations, language is considered to be the gateway to the world. The first interpretation highlights human language as the reference point for the knowledge of the world. In contrast, the second interpretation generalizes the concept of language and extends it beyond that of human language. It declares the linguistic structure of the world itself to be the starting point for our knowledge of the world. We shall see that there are good reasons to follow the second interpretation, because it places the exact sciences on a firm footing and secures their claim to objectivity. Only in this case can the "readability of the world" be free from the time-bound and subjective character of a world understanding that is fettered to the historically developed language of humans.

The universal language that we are looking for is one of abstract structures. It is the language of mathematics. Precisely this thought was expressed some hundreds of years ago by Galileo when he wrote: "Philosophy is written in this grand book—I mean the Universe—which stands continually open to our gaze. But it cannot be understood unless one first learns to comprehend the language and interpret the characters in which it is written. It is written in the language of mathematics, and its characters are triangles, circles and other geometric figures, without which it is humanly impossible to understand a single word of it; without these one is wandering about in a dark labyrinth." [10, p. 183 f.]

Since it is evident that mathematics represents the structures of our reality in a correct way, its language must be congruent with the real background of our world. Consequently, the language of mathematics is the language of the structures of the world. Only in the light of this deduction do the conclusiveness and truth of the mathematical world description become comprehensible. If mathematics were nothing more than a culture-bound, artificial language that at some point in time was invented by man, one would not be able to explain the overwhelming success that mathematics has had in the solution of theoretical and practical problems. Without the language of mathematics, the architecture of our world could be neither appropriately described nor understood.

Admittedly, the term "language" has so far remained mostly undefined. On the one hand, we associate the notion of language with human language. On the other, we give the concept of language an interpretation that goes far beyond human language. If we do not clarify the concept of language further, we shall have to face the accusation that the idea of a "language of structures" is based on empty words. It is therefore essential to generalize the concept of language in such a way that it can be meaningfully applied to the structures of the world.

We may even be confronted here with a new "Copernican revolution". The thesis that the world has the structure of a language could lead to a revolution in our world understanding, one comparable to the change from the geocentric to the heliocentric world-view. As is well known, the latter was introduced at the beginning of the sixteenth century by Nicolaus Copernicus, who asserted that the earth revolved around the sun, and not vice versa. This new world-view led to a dramatic change in the self-perception of man, who suddenly had lost his place at the center of the universe. In this sense, it would indeed be a similar turning point if the center of our understanding of the world were no longer human language, but rather an overarching language that in human language has just assumed a particular form.

This turn in no way ignores the extraordinary complexity and uniqueness of human language. Neither the specific autonomy of human language nor the particular evolutionary characteristics of man are called into question if one extends the concept of language beyond that of human language. Here, the same applies as for the Darwinian revolution in our view of man: By providing evidence for the evolutionary relationship of all living beings, the especial status of mankind has indeed been abolished, but not the fact of mankind's unique manifestation. Instead, only the coordinates of our world-understanding have shifted.

Comparable to the evolution of man, the origin and evolution of human language must also be seen as a result of an adaptation, namely its adaptation to the life-world of man. The specific adaptation of language to the life-world of man would also explain the fact that the human language is only of limited use for representing physical reality outside our immediate area of experience and perception. Thus, for example, an adequate description of the submicroscopic world is only possible when one employs the language of mathematics, which is free from the specific forms inherent in human language. Mathematics is neither situation-dependent nor is it limited to particular areas of experience or perception by human beings.

Since there is no Archimedean point outside human language from which one can speak about language, it appears that we have here run into a circular argument. Can one really extend the concept of language beyond that of human language, even though the idea of language is itself only accessible through human language? This apparent problem is resolved by the fact that we already have an abstract language in the form of mathematics, and this indeed stands outside the common language through which humans communicate. Mathematical language goes beyond human language because it also represents facets of our reality that cannot be grasped by a language adapted to the objective areas of human experience.

Numerous models have been put forward in the endeavor to explain the evolution of human language from proto-languages. They combine—under the umbrella of evolutionary biology—research results from linguistics, speech science, genetics, neuroscience and anthropology. The different approaches to language evolution range (to give just two examples) from game-theoretical models [11, 12] to the comparative study of primitive communication systems of animals [13]. An introduction into the various concepts of language evolution can be found elsewhere [14, 15].

Enlightening insights into the basics of language capability, we owe, in particular, to neurobiology. An area of the human brain, known as Broca's area, is in focus here. It is believed to play a central role in the linguistic capability of humans (Fig. 1.2). More astonishing was the discovery of an area in the brain of great ape species that is anatomically similar to Broca's area of the human brain [16]. Moreover, apes use highly-developed elements of sign and body language, which is quite obviously an expression of a world

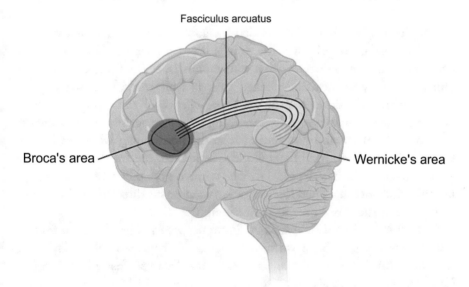

Fig. 1.2 Speech areas of the human brain. Both areas of the cerebral cortex, named after their discoverers, are known to control speech. According to the classical model of language neurobiology, Broca's area is responsible for speech production, and Wernicke's area apparently controls speech-comprehension. However, there is evidence that Broca's area is also involved in language comprehension. Above all, Broca's area seems to be responsible for speech motor skills, sound and sound analysis, articulation, and the formation of abstract words, whereas Wernicke's area seems to be relevant for auditory sensory perception and logical speech-processing. These two areas are connected to one other by a nerve pathway, the *fasciculus arcuatus* ("curved bundle"). It is most probable, however, that other areas of the brain are also involved in the ability to speak. Maybe the neural organization of the ability to speak is even more highly dispersed over the various brain areas than is suggested by the traditional anatomical definition of Broca's and Wernicke's areas [17]. [Adapted from OpenStax College, Wikimedia Commons]

relation based on communication and mutual understanding. Such findings, suggesting anatomically comparable speech areas in the brains of primates and the human brain, support the thesis that human language evolved from primitive language structures.

In the following paragraphs, however, we shall go one step further, asking whether language is a general principle of Nature that exists independently of humans and already becomes manifest in the structure of living matter. Language must indeed be seen as the essence of living matter per se. We will give this thought a lot of space in this book. A strong indication of the scope of this working hypothesis found in the astonishing parallels between human language and the language-analog structure of genes, which we will examine in Sect. 1.5.

The frequently raised objection that the application of linguistic categories in biology has no cognition value as it is based on a metaphor is a misunderstanding of the essence of what a metaphor is. It is admittedly true that human language is intimately linked to man's life-world and that the linguistic terms have developed in man's experience of real things and phenomena. *A priori*, however, there is nothing to be said against generalizing the concept of language and transferring it to the non-human world, as human language is already in itself metaphorical. In speech, the words of a language always refer to other words, which is precisely the basic pattern of metaphorical speech. Therefore, criticism cannot be directed at the use of metaphors as such, but only at their wrong use.

Metaphors are linguistic means of "improper speech", in which a word that is meant is replaced by a word with a factual, conceptional or figurative similarity: for example, if one uses the term "source" instead of "cause". Often, metaphors are also used to name things for which everyday language does not possess any proper terms, such as "arm of a river" or "eyeball". Scientific language likewise uses metaphors. Here, metaphors are relevant in so far as they give the abstract terms of science a descriptive content. Mixed concepts, such as "crystallization nucleus", "atomic nucleus", "evolutionary tree" and the like, form a linguistic bridge that connects the abstract concepts and realms of science with the familiar world of human experience. This procedure sometimes reaches its limits, as shown by the discussion about the intuition content of quantum physics (Sect. 1.10). Here, everyday language fails to describe the physical world adequately—a fact that cannot be compensated for by the use of metaphors.

Moreover, metaphors can suggest ideas that are only partially consistent with the actual findings of science. The metaphor "evolutionary tree" is an excellent example of this. It gives the impression that the lines of descent of all living beings can be represented in a tree-like manner, even though the actual findings of evolution theory suggest a branching pattern that more closely resembles that of a shrub or a piece of coral. Clearly, these metaphors have not been adapted to the actual facts after their introduction. Beside, the tree metaphor implies more strongly the idea of an ever-higher development of life by evolution than the shrub metaphor does.

It is part of the essence of metaphoric speech that language jumps from one image to another. From the epistemological point of view, the use of metaphors is, therefore, nothing other than the creation of analogies. This technique has enormous heuristic value. However, if there is no control through experience, analogies may lead to wrong conclusions and subsequently to unjustifiable and even wild speculations. Metaphysics is particularly susceptible in this respect, as it often tries to employ analogies to open up areas of reality that are beyond all possible experience. In epistemology, for example, the technique of analogizing has repeatedly led to violent controversies. The reason is that there are no binding criteria that can be used to differentiate between meaningful and meaningless (or even absurd) analogies.

Finally, one may ask whether the concept of the "language-analog" structure of reality also falls into the realm of inappropriate metaphors. In fact, at first glance, the idea that reality has the structure of a language seems to involve circular logic: using the terms and means of human language, we argue that there is something like a language beyond human language. However, the suspicion of circular analogy-formation is invalidated by language itself. As already outlined above, language is always based upon terms that acquire their meaning only in the context of other terms. This enforces the use of metaphors, not least when one speaks about language itself. Thus, the existence of metaphorical speech is no reason to criticize or exclude the transfer of linguistic categories to the non-human spheres of the world. The only check for the quality of a metaphor is the sustainability of its explanatory power. In this sense, we will consider the idea of the world's language-like structure as a working hypothesis that we are going to examine on the basis of a variety of scientific concepts.

1.4 What is Information?

The background of our discussion is the idea that language is a general property of Nature. To deepen this idea, one must break away from the specific characteristics of human language. For this purpose, however, one needs a reference term that is sufficiently abstract and to which all forms of language can be related. Since language in the broadest sense carries and transports information, the idea arises that the concept of information could be the sought-after reference frame. This, in turn, leads to the question: What is information?

The first answer to this question is provided by language itself, more precisely: by the linguistic root of the concept of information. The word "information" derives from the Latin word "informare", which originally means "to form" or "to shape". Of course, the notion of form alone is unable to describe exhaustively the essence of information and language. Nevertheless, "form" is a primary element upon which information and all languages are based in equal measure; both information and language use signs or strings of characters, which at first are nothing more than formed matter.

It is essential to give priority to this crucial aspect from the outset, because that allows us to counteract the prevalent opinion according to which information serves only interpersonal understanding and is thus equivalent to the idea of communication. This prejudice may have to do with a publication by the pioneers of information theory, Claude Shannon and Warren Weaver. They called the nascent theory of information, somewhat misleadingly, *A Mathematical Theory of Communication* [18, 19] so that the impression inevitably arose that information and communication were one and the same.

In reality, however, Shannon and Weaver were only concerned with the technical question of how a string of signs can be transmitted from a sender to a receiver as efficiently and as error-free as possible. Typically of communications engineers, they were not interested in the content or meaning of a string of signs, but only in the technical problems of its transmission. Therefore, Weaver expressly emphasised that the idea of information is used in information theory in a special sense and must not be confused with the idea of meaning. "Two messages", he explained, "one of which is heavily loaded with meaning and the other of which is pure nonsense, can be exactly equivalent, from the present viewpoint, as regards information." [19, p. 8]

In contrast, communication is more than the mere transmission of signs between sender and receiver. Usually, we associate the word communication with the exchange of meaningful information, news, instructions and the like. Even the bare recognition of signals as signals (or symbols) is already a preliminary stage of communication and understanding. Nevertheless, the word "information" does not necessarily mean "communication". As mentioned above, information can also be understood as a measure of the form of a structure. Aristotle already pointed out, that there is no matter without form. Consequently, information is a property of every material object.

With direct reference to Aristotle, physicist Carl Friedrich von Weizsäcker has described the inextricable interrelation of form and matter as follows: "Historically speaking, matter is, to begin with, the conceptual opposite of form. A cupboard, a tree are made of wood. Wood is their 'matter.' The name of the term 'matter' is in fact taken from this example: *materia = hyle*, which means wood. But the cupboard isn't simply wood, it is a wooden cupboard. 'Cupboard' is what it is intrinsically; cupboard is its *eidos*, its essence, its form. But a cupboard must be made of something; a cupboard without matter is a mere thought abstracted from reality. On the contrary, this cupboard made of wood is a real whole of form and matter, a *synholon*; form and matter are 'grown together' in it, it is something concrete." [20, p. 274 f.]

Weizsäcker's remarks refer to the information-theoretical roots of our conceptional thinking. According to this view, objects differ in their form content, whereby the term "form" here refers not only to the particular shape but also to the quality of an object. If we want to delimit an object conceptually, we have to decide between a finite number of possible alternative forms. We have to decide whether the object is round or angular, solid or hollow, coloured or not, light or dark and so on.

It is evident that the number of alternatives that must be decided between depends on the questions of in which general category, and in which context, the object is considered. This aspect is also referred to as the "context-dependence" of information (Sect. 1.8). It demonstrates at the same time the relativity of all information. In the general category "physical body", for example, Weizsäcker's "cupboard" has a form content entirely different from what it has in the category "item of furniture". In other words, the information content of an object is given by the number of yes/no decisions required to narrow down the essence of the object. In Sect. 1.10 we shall deepen this down to the physical level of quantum objects.

The idea that the form content of a structure is determined by a finite number of yes/no decisions applies also to signs and sequences of signs. The communications engineer Ralph Hartley [21] was the first who made this aspect the basis of a definition, equating the structural information of a message with its binary decision content. Accordingly, the structural information—i.e., the form content—of a string of signs is given by the set of yes/no decisions required to select a particular binary sequence from the set of all conceivable sequences of equal length. Shannon finally generalized Hartley's definition by defining, in addition, a probability distribution for the set of strings, thus giving each string statistical weight.

With the expression "content", we have so far denoted the structural information of a message. In common parlance, however, we understand the content of information as the meaning that it has for a recipient. The meaning of a piece of information can, in several ways, take effect by prompting an action or reaction on the recipient's part or in the recipient's surroundings. Consequently, further dimensions of information come to the fore, which go beyond the simple structural aspect of the sequences of signs that communication engineers deal with.

In linguistics, for example, the different dimensions of a sequence of letters are denoted as the syntactic, the semantic and the pragmatic aspects of language. The "syntactic" aspect refers to the arrangement of the letters, considering the rules according to which the letters are joined together to form words and sentences. The "semantic" aspect includes the relationships among the letters and what they mean. The "pragmatic" aspect includes the relationships between the letters, what they mean and the effect that they engender with the (human or other, e.g. mechanical) receiver. One can generally adopt this classification for the concept of information. Through this, the concepts of information, communication and language are closely linked to each other, and this also becomes the basis for a deeper understanding of living matter ([22] and Chap. 6).

However, the dissection of the notion of information into syntax, semantics and pragmatics can only be justified within the framework of scientific systematization. Strictly speaking, the three dimensions form an indissoluble unit, because one cannot speak of syntactic, semantic or pragmatic information without in each case co-thinking the other two dimensions. Nevertheless, the three dimensions of information allow the ordering of the different problems that are encountered in the scientific analysis of life phenomena.

Let us take a closer look at the three dimensions of information, considering first the syntactic dimension. The syntax only attains any meaning when the semantics of the symbols as such, i.e., their symbolic character and the reciprocal relationships between them, are already defined. Semantic information, in turn, presupposes a syntactic structure that can function as a carrier of the semantics.

Semantics, on the other hand, goes beyond bare syntax. The reason for this is that the meaning of an item of information only becomes effective through the relationship of its symbols to the external world. In other words, a piece of information which first represents only potential information acquires its semantics only through its pragmatic relevance, i.e. by the effect that the item of information engenders upon the recipient and the recipient's surroundings. The latter defines the pragmatic dimension of information. Because of their intimate interrelationship, semantics and pragmatics are frequently lumped together into the so-called semantic-pragmatic aspect of information.

Semantics represents the hinge between syntax and pragmatics; one might also say: between the structural and the functional aspects of information. Even though functions are always bound to structures, they cannot be reduced to them. Thus, obviously, there is no unambiguous connection between the structural and the functional level. An example of this is human language, in which the meaning and thus the pragmatic (or functional) relevance of words and sentences are determined by the syntax, but not reducible to it. This kind of non-causal determination is also called "supervenience" (for details see Sect. 3.8).

Although there exists a relationship between the level of meaning and the world of structures, there is obviously no law-like link between the two domains. Thus, the exceedingly difficult question arises of whether the semantic dimension of information can ever become the subject of an exact science based upon abstraction, idealization and generalization. Moreover, in human communication, the exchange of meaningful information is always an expression of human beings' mental activity, with all its individual and unique facets. Thus, this in turn raises the question of whether the specific properties of human communication can be blanked out so that the general characteristics of semantic information—as far as any such exist—emerge.

This problem seems to overstep the boundaries of exact science. Therefore, it is easy to understand that the founding fathers of information theory initially sought a quantitative measure for information that is independent of the content of the information. On the other hand, however, the semantic aspect of information is crucial for a deeper understanding of information-based systems in Nature. The most notable example is provided by biology. Without the concept of information, a deeper understanding of living matter is not possible.

However, to apply fully the concept of information in biology one also needs access to the semantic dimension of information. We will analyze this problem in Sects. 6.8 and 6.9 and propose a general solution to the problem. For now, let us just note that syntax and semantics are intimately linked

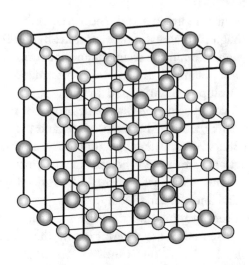

Fig. 1.3 Structural order of a crystal. The figure shows a schematic representation of the crystal structure of common salt. In the lattice structure the building blocks—chloride ions (red spheres) and sodium ions (green spheres)—are arranged in a regular and strictly alternating order. Such systems, in which the components are firmly fixed, are unable to develop independent functions. At best they can be a subordinate part of a functional system as is, for example, the case for the strictly ordered facets of an insect eye. [Image: Lanzi, Wikimedia Commons]

together and that, in biology, both dimensions of information become manifest in the relationship between the structure and the function of genetic information.

Let us be more specific about the idea of functionality, because we shall come across it time and again in this book. Functionality is characterized by order in time. This order is, by its nature, different from order in space, such as that of a crystal (Fig. 1.3). Spatial order can be adequately described by stating the coordinates of all components of the system.

In contrast to spatial order, there is no comparable representation of temporal order on which functionality is based. We can measure physical time with the help of clocks, but we cannot grasp in this way the meaningful —i.e., coordinated and self-regulated—interactions among the system's components. In the living cell, for example, there are numerous biochemical reaction cycles that are enmeshed into a complex metabolic reaction system that is synchronized and controlled by genetic information. Such a system cannot be represented by a rigid structural order, but only by a flow diagram of the different operation sequences that maintain the system's functionality (Fig. 1.4).

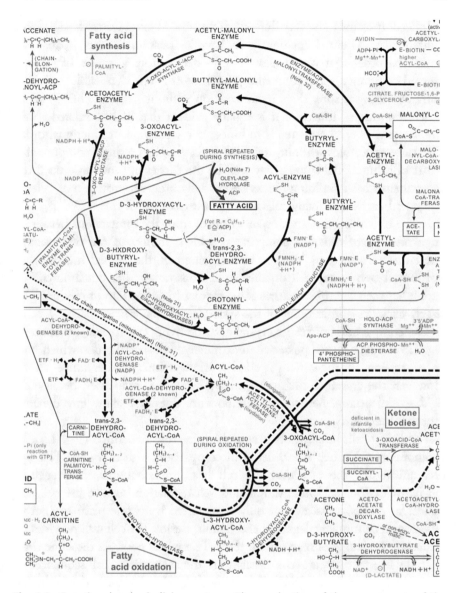

Fig. 1.4 Functional order in living systems. The production of the components of the cell is regulated in biochemical reaction cycles. This diagram shows the chemical pathways of fatty-acid synthesis and fatty-acid oxidation. Yet these represent only a tiny fraction of the exceedingly complex network of metabolic pathways in the cell. The interplay and working order of the numerous reaction cycles is represented by flow diagrams of this kind. There is no other way to visualize the enormous complexity of molecular life processes. [Excerpt from Roche Biochemical Pathways (poster)]

Indeed, it seems to be very difficult to describe in exact terms the essence of functionality. There is apparently no law-like relation between spatial and functional order that would allow one to derive from a system of spatially ordered components any compelling conclusions about its possible functions. Only one thing is sure: there is no function without structure, but there are structures without function. It is equally evident that highly ordered spatial arrangement of structures tends to hinder the formation of functional order. This statement holds without prejudice to the fact that a single component as a part of a functional system may well be the carrier of function even if this component's basic building blocks are spatially ordered. Protein molecules as functional carriers of living cells are the best example of this.

In the organism, the working order arises from the interactions between a huge number of different components which, moreover, are organized in numerous substructures. Already in the living cell, the cellular functions are distributed over many cell organelles, demonstrating that dynamic coordination between the substructures must require high freedom of motion. This largely excludes a rigid order of the system's components in space.

The comparison between a crystal and the living cell makes this clear. As Fig. 1.3 shows, a crystal possesses an extremely high spatial order, which finds its expression in the strictly periodic arrangement of the lattice atoms. However, the crystal as such cannot have functional properties, because the rigid spatial arrangement of its components does not allow any dynamic interplay among its parts.

The living cell, in contrast, is entirely differently structured (Fig. 1.5). Its components are much less ordered in comparison with a crystal. This is a prerequisite for the fact that the cell has functional properties such as self-reproduction, self-preservation and self-regulation. Inside the cell, the molecular components are not only transported from one place to another but are also continually degraded and rebuilt. Beside, the cell wall provides for a constant exchange of matter with the cell's environment. At the same time, all cellular processes are coordinated in their chronological sequence, to ensure the preservation of the cell. Here not spatial order, but order in time, is of primary importance.

The example of the living cell makes it clear that only the acquisition of freedom to move in space makes the constitution of complex temporal orders as the basis for functional systems possible. Nevertheless, the living cell also shows a specific spatial order; this results from the clustering of its components in operational centers, which in turn contributes to the hierarchical order of the life processes. Thus, in living matter, the antagonistic principles of spatial and temporal order do not exclude one another; rather, they act

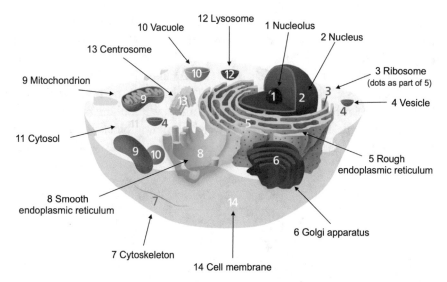

Fig. 1.5 Cutaway drawing of a eukaryotic cell. In contrast to crystals, the living cell is *functionally* ordered. The cell's numerous components are clustered in such a way that they can interact optimally with each other. Thus, the cell is structured into compartments, the so-called organelles, which take on particular tasks to support reproduction and metabolism. The largest organelle of the cell is the nucleus, which functions as the control center. The cell nucleus mainly contains the genetic material of the cell. The other organelles, such as the endoplasmic reticulum, the Golgi apparatus, the lysosomes and others, are responsible for the cell's metabolism. All activities within the cell are meticulously tuned to each other to ensure the maintenance of its various life processes. It is characteristic of these processes that large molecular components are continuously broken down, and new ones are built. Thus, working order cannot be based on a rigid spatial order such as one sees in a crystal. [Adapted from Kelvingson, Wikimedia Commons]

together in a well-balanced relationship. Other examples from biology also demonstrate that the balanced action of antagonistic principles has been a decisive factor in driving the evolution of life [23].

1.5 The Genetic Script

All fundamental life processes, from inheritance to metabolism, are instructed and controlled by information. So it is bordering on a miracle that the genetic information of a living organism is stored in a single molecule (or a few such molecules) which is passed on from generation to generation and thus ensures the continued existence of life.

The molecules of heredity belong to a particular class of biological macromolecules which were first discovered in the cell nucleus and accordingly received the name "nucleic acids" (Fig. 1.6). There are two chemically similar variants: deoxyribonucleic acid (DNA) and ribonucleic acid (RNA). Nucleic acids are chain-like molecules ("polymers") that are formed by linking smaller molecules ("monomers"). In this way polymers with millions of links can form. The low-molecular-weight building blocks of nucleic acids are also known as "nucleotides". Although the nucleic acids may have

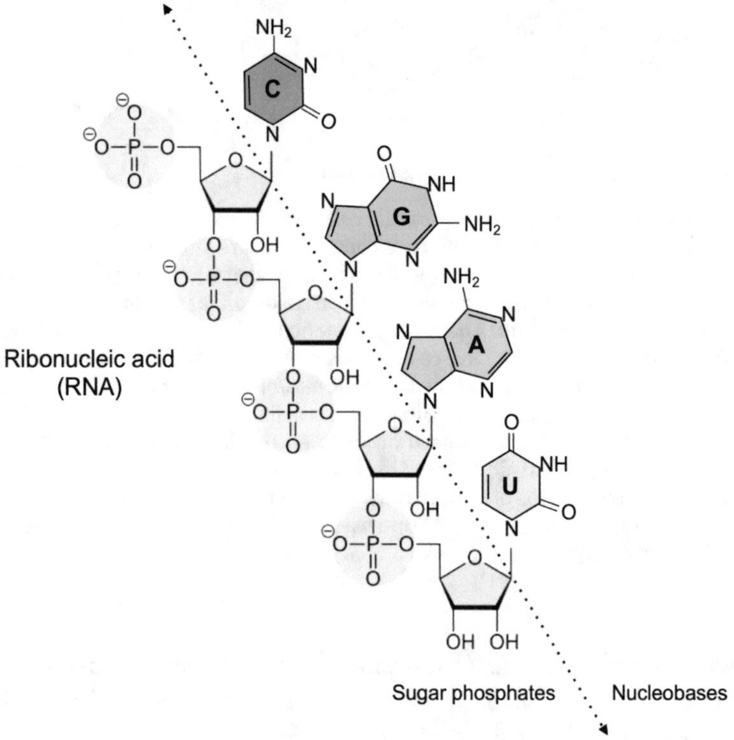

Fig. 1.6 Chemical structure of ribonucleic acids. The nucleic acids are chain-like molecules and comprise only four classes of building blocks ("nucleotides"). Each nucleotide, in turn, is composed of three small molecules: an organic base ("nucleobase"), a sugar molecule and a molecule of phosphoric acid. The sugar comes in two forms, ribose and deoxyribose; accordingly, there are two chemically closely related forms of nucleic acid, ribonucleic acid (RNA) and deoxyribonucleic acid (DNA). A further minor difference between the two is that RNA uses the nucleobase uracil whereas DNA uses the base thymine. The nucleobases are usually referred to by the initial letters of their chemical names: A(denosine), G(uanine), C(ytosine) and U(racil)/T(hymine)

considerable chain lengths, their chemical structure is relatively simple, because they are made up of only four classes of nucleotides, which are arranged like beads on a necklace (Fig. 1.7).

One of the fascinating properties of the hereditary molecules is that their chemical building blocks act like the letters of a written text. Strictly speaking, the genetic text is a script, i.e., a text specifying options for actions: in this case, possible process sequences. In fact, the entire blueprint of the organism is already encoded in the simple sequence of nucleotides. Just as a misprint can distort the meaning of a text, even the exchange of a single nucleotide can lead to the loss of relevant information and thus, under certain circumstances, to the death and decay of the organism.

Considering the enormous complexity of living beings, it may come as a surprise that the genetic alphabet consists of only four "letters". However, even with such a small alphabet, a huge number of different sequences can be generated, and the number of possible combinations increases immeasurably with increasing chain length n. It can be calculated according to the simple formula: $4 \times 4 \times 4 \times \ldots \times 4 = 4^n \approx 10^{0.6n}$.

To give some numbers: For example, the DNA of the bacterium *Escherichia coli* consists of about 4×10^6 nucleotides. Even at this relatively low level of life there already exist $10^{2,400,000}$ sequence alternatives to the genetic text that encodes the genetic information of a bacterium. For the human genome, the number of combinatorically possible sequence alternatives exceeds the power

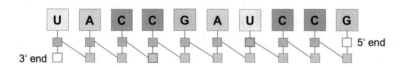

Fig. 1.7 Language-like nucleotide sequence of RNA. The nucleotides function in the heredity molecules like the letters in human language. Just like human language, genetic information is organized hierarchically. It contains words, simple and complex sentences, paragraphs, cross-references etc. (for details see Table 1.1). Moreover, the genetic information also comprises punctuation marks, which denote the beginning and end of a unit to be read. Last but not least, the chemical structure of the nucleic acids defines the direction in which the genetic information is to be read. Thus, one end of the nucleotide chain has a free binding site at the sugar molecule ("hydroxyl" or 3' end), the other a free binding site at the phosphor molecule ("phosphoryl" or 5' end). In vivo, nucleic acids are synthesized in the 5'-to-3' direction (see Fig. 6.7). The wealth and diversity of living beings have their origin in the vast number of possible manners in which a given number of nucleotides can be arranged in sequences

of the imagination. With about 10^9 nucleotides the human genome has a capacity of $10^{600,000,000}$ sequence alternatives! Thus, it borders on a miracle that the entire diversity and uniqueness of the life forms can be traced back to the simple combinatorics of just four nucleotides. At the same time, such numbers show what tremendous scope is at evolution's disposal to let its inventiveness play out (Sect. 6.10).

According to the doctrine of genetic determinism, all the characteristics of the living organism are determined by the sum of its genes, the so-called "genome". However, genetic information only becomes operative when it interacts with its surroundings under the physiological conditions of the living cell. The genome and its environment behave like the sender and receiver of information. Moreover, close feedback occurs between the two, insofar as the genetic text—which in itself is ambiguous—acquires an unambiguous meaning only through its surroundings.

From a physiological point of view, the genetic information is "expressed", i.e., translated step by step into the material organization of the growing organism. However, the physiological surroundings of the genome, and thus the context of genetic information, changes with every step of development, so that the information content of the genome is continuously reassessed during its expression. The changed information content, in turn, affects the surroundings, so that throughout its expression the growing organism is continually re-organized.

The feedback between the genome and its surrounding conditions has the consequence that the final form of the organism is built up by continuous reshaping and transformation of each step during its development. This is typical of self-organizing systems such as the living organism. Externally organized systems are fundamentally different. For example, in the case of a machine, the parts of the system are assembled according to a given plan. In contrast to self-organizing systems, the forces that fit the machine parts together are not an integral part of the system itself. Instead, they are usually external forces which act in a goal-directed way. However, the borders between "natural" and "artificial" are fluid. Since there are no forces in Nature that are goal- and purpose-directed, steering self-organization to its target, self-organizing systems can only build up by successively employing optimization processes based on a feedback mechanism. In this sense, one could also imagine automatons capable of reproducing themselves (Sect. 6.6).

The fact that the content of genetic information is highly context-dependent lends weight to the assertion that information in an absolute sense does not exist, but that it exists only in a relative way, depending upon its sender and the receiver. At the same time, it becomes clear that the content of the information encoded in a given set of genes must always depend on the specific stage of the genes' expression.

At first glance, it seems that the idea of the genetic determinacy of the organism has to be abandoned. However, this impression is deceptive. Even for the highly complex processes of gene expression, the information stored in the genome is (in the context of its environment) sufficient to guarantee the reproductive preservation of the organism. Seen in this light, the core statement of genetic determinism, according to which the genome specifies all the characteristics of the living organism, remains correct. In reality, it is not the concept of genetic determinism that proves problematic, but the concept of the gene. Contrary to the traditional view, genes can no longer be regarded as containing independent, quasi-absolute information. Instead, they acquire their trait-determining properties only in the context of other genes and their expression.

It is remarkable that gene expression appears to be a dialog between the genome and its surroundings. Even if this kind of dialog—unlike interpersonal communication—does not involve any kind of reflection on the subject of the communication, it is undoubtedly a form of meaningful information exchange. Here, communication serves to build up and reproductively preserve the living organism. To this end, all basic processes taking place in the living organism must be coordinated and aligned coherently. This in turn presupposes a continual exchange and evaluation of information—in fact, precisely what we previously described as "meaningful" communication.

From this point of view, it is not surprising that genetic information is based on principles comparable to those of human language. This analogy does not only refer to the fact that the nucleotides in the hereditary molecule act like the letters of a language. More than that, like human language, genetic information is also organized hierarchically (Table 1.1): Three of the four letters of the "genetic alphabet" build a code-word, comparable to a word in human language. Such code words are joined up into functional units, the genes. These have a similar meaning to a sentence in human language. The genes, in turn, are connected in higher-order functional groups, the chromosomes. These are each comparable to a longer text. Some code words even take over the function of punctuation marks, with which segments of the genetic text are delimited from each other.

Moreover, the hereditary molecules have other characteristics of language, such as the unambiguous succession of (genetic) letters and the provision of a reading direction (Fig. 1.7). If the genome were branched or had no reading

Table 1.1 Hierarchical organization of genetic information demonstrating the language-analog structure of the genetic script (based on [22, 24, 25])

Syntactic unit	Numbers[a]	Structural demarcation[b]	Function	Human language
Nucleotide	Genetic letter	4	Chemical structure	Primary coding symbol	Letter
Codon	Nucleotid triplet	64	Nucleotide	Translational unit for amino acid	Phoneme/morpheme
Gene	Up to 1000 codons	>1000	Codons	Coding unit for protein[c]	Simple sentence
Scripton (operon)	Up to 15 genes[d]	>1000	Promotor, Terminator	Transcription unit (mRNA)	Complex sentence
Replicon	Up to several hundred scriptons[d]	1 or several	Replicator, Terminator	Reproduction unit	Paragraph
Segregon (chromosome)	Serveral replicons	1 or a few	Centromere, Telomere	Meiotic unit	Chapter
Genome	A few segregons	1	Cell's nuclear membrane	Mitotic unit	Complete text

[a] The numbers in the first two rows are the same for all organisms. The other numbers depend on the organisms' complexity
[b] Up to and including the level of the segregon, the information is arrangend sequentially. However, this does not mean that it must be read out in a single sweep
[c] Including punctuation codons
[d] Including intermediate regions

direction, it would not be possible to process information unambiguously. The only exception to this is the so-called "palindrome". Palindromes are sequences that can be read both forwards and backwards to give the same meaning. Examples are words such as LEVEL or ROTOR. The phrase "Madam, I'm Adam" can also be read in both directions.

Palindromes are relatively rare. Nevertheless, they also occur in genetic language. Here, palindromes are mainly used for coding particular recognition signals. With viruses, there are occasionally even overlapping encodings, so that one gene can encode more than one gene product (protein) because of shifts in the reading frame. For example, think of the words MADAM or NEARLY, which, shifted by one letter, also contain words: ADAM and EARLY.

It is a particular feature of language that the whole wealth of linguistic expressions can be generated with the help of only a few letters. However, the almost unlimited repertoire of linguistic expressions can only be exploited if the sequence of characters is mainly aperiodic. Only on this condition can the nearly endless variety of linguistic expressions be formed from a few different kinds of letter. In a mostly periodic sequence, however, in which the letters are strung together according to the same pattern, or similar patterns, it is not possible to encode any substantial information.

By the way, this is another reason why a crystal with its monotonous lattice structure is not suitable as a carrier of information and thus of possible life functions. The situation is different in the case of genetic information carriers. As expected, they fulfil the requirements of aperiodicity to the same extent that human language does, as this is the only way to encode enough information for the construction of living systems.

From a purely chemical point of view, the aperiodic structure of the hereditary molecules is by no means a matter of course. For example, plastic molecules, which are also chain-like molecules assembled from small-molecular building blocks, do not possess this property. They consist of a monotonous sequence of one kind of monomer only, which leads to homogeneous material properties excluding the possibility of carrying information.

There are also indications that genetic language has a recursive character. Recursiveness is a formal property of human language, which allows the creation of any number of sentences from a limited number of sentence elements and rules. A genetic analogy to recursive language generation is provided by the mechanisms of gene duplication and the recombination of genetic material, which make it possible to expand the genetic information repertoire to an unexpected extent. An overview of the parallels between genetic and human language can be found elsewhere [22, 24, 25].

All in all, the parallels go so far that it seems justified to describe the informational architecture of the genome as a form of language. Without the language-driven processes of instruction, information and communication, living beings could not exist at all. This view is, of course, in stark contrast to the philosophical doctrine that considers language to be the exclusive property of mankind. However, the findings of modern genetics seem to support and complete our previous arguments that language is a general principle of Nature, active at all organizational levels of living matter.

At the end of the nineteenth century Friedrich Miescher, the discoverer of the nucleic acids, had already suspected that the phenomenon of "language" must have far-reaching implications for an understanding of living matter. He developed this idea during his studies of stereoisomerism in chemistry. This phenomenon is caused by so-called "asymmetric carbon" in molecules (Fig. 1.8). The structures of such molecules, in which the central carbon atom has four different substituents, lead to their behaving like mirror-images of one another. Since biological macromolecules, as Miescher suspected, contain millions of such asymmetric carbon atoms, he thought that an equally large number of stereoisomers was conceivable, and that this could explain all the richness of life forms in one fell swoop. In a letter to his colleague Wilhelm His in December 1892 he wrote: "In the case of the enormous protein molecules [...] the many asymmetric carbon atoms allow such a colossal number of possible stereoisomers that all the richness and all the variety of the heredity factors could just as well find its expression in them as do the words and expressions of all languages in the 24–30 letters of the alphabet." [26, p. 117; author's transl.]

Today we know that the real cause of the enormous abundance of organic structures is genetic information and not their stereoisomeric properties. Nevertheless, Miescher had already intuitively recognized that the immense variety of possible genetic make-ups is based on a simple construction principle that in turn is based on a few essential elements. Miescher was downright visionary when he drew a parallel to the generative properties of human language. To put Miescher's thought in a nutshell: The organism does not function as a rigid clockwork, but rather as a "language-mechanism" [3]. We may add that this language mechanism consists of a set of rules that includes myriads of feedback loops through which the organism, in dialog with itself, organizes its life processes (Sect. 6.3).

Johann Wolfgang von Goethe, who observed Nature with the utmost meticulousness, also seems to have regarded this idea as definitely conceivable.

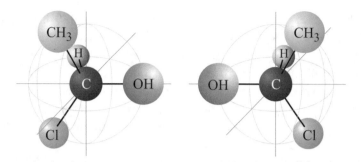

Fig. 1.8 Principle of stereoisomerism. Stereoisomeric molecules have the same chemical formula but differ in the three-dimensional orientation of their atoms (the figure shows the compound 1-chloroethanol). They are like mirror-images of one another. Stereoisomerism is caused by an asymmetric carbon atom (black sphere in the middle, a "stereocenter") to which four different types of atoms or groups of atoms are attached. Knowing the number of asymmetric carbon atoms of a compound, one can calculate the maximum number of its stereoisomers. According to a rule formulated in 1874 by Joseph Achille Le Bel and Jacobus Henricus van't Hoff, the number of stereoisomers of an organic compound is 2^n, where n represents the number of stereocenters (provided that there is no internal plane of symmetry). The fact that the number of stereoisomers increases exponentially with the number of stereocenters gave the physiologist Friedrich Miescher the idea that a similar principle might be the key to understanding the complexity and richness of living matter. [Image: Generalic, Eni. *Stereoizomer*. Croatian-English Chemistry Dictionary and Glossary]

In the foreword to his treatise *Zur Farbenlehre* ("Theory of Colors"), he writes enthusiastically about the language of Nature: "Thus, Nature speaks down toward other senses, to known, misjudged, unknown senses; thus, she speaks to herself and to us through thousands of appearances" [27, p. 315; author's transl.]. At the end he summarizes: "So multifarious, so complex and so incomplete this language may sometimes appear to us, its elements are always the same" [27, p. 316; author's transl.].

Goethe is not just referring to the language of Nature in some poetic way. He believed that natural phenomena genuinely reflect the elements of a universal language. This leads on to the question of whether one can really develop a consistent picture of language that on the one hand suffices to describe human language and on the other hand is so abstract that it can be applied to the "thousand appearances" of Nature.

In fact, it is a recognized finding of comparative behavioral research that primitive forms of communication exist in the animal kingdom [28]. Plants, too, can exchange information with one another by using a sophisticated communication system that employs signal molecules; in this way they can defend themselves against impending attack by pests (see [29]). We have long known that bacteria communicate by releasing chemical signals. Maybe they even use some kind of a linguistic code. However, most surprising is the fact that basic organizational structures of human language are already reflected in the molecular structure of the carriers of genetic information. As we have seen before, the analogy goes far beyond a mere vague correspondence—it comprises several almost identical properties that are, without exception, found in all forms of living beings.

If, in the light of this, we set out to uncover the "language of Nature", then we must first ask some questions about the "nature of language". This will, repeatedly and inevitably, lead us back to the language of humans, as this is paradigmatic for our understanding of language. However, we cannot simply extrapolate in an abstract way from human language. In trying to do so we would lose sight of the unique position of humans as beings that "understand" and thus also of some unique properties of human language, in which we *inter alia* pronounce judgments, formulate truths, express opinions, convictions and wishes. Moreover, at the level of interpersonal communication there are various different ways in which language appears: it may employ pictures, gestures, sounds or written symbols. Therefore, to develop a general concept of language, we have to construct it from the bottom up. We must try, on the basis of highly general considerations, to develop an abstract concept of language that on the one hand reflects human language and on the other hand is free from its complex and specific features.

1.6 Symbolic Forms

The assertion that language is an intrinsic property of living Nature does not call into question the unique position of human language. However, just as every product of natural evolution is unique in its own way and yet at the same time refers to a common origin, so does the language of man appear to be the particular and most sophisticated form of a universal principle of language that underlies all living matter. This statement is not intended to downplay the role of language, but instead emphasizes its true meaning. This understanding of language fits seamlessly into the image of a world that is recognizable to us because it has the structure of language.

Against this background, the study of signs, the so-called "semiotics", gains a particular significance. The roots of semiotics go back to antiquity. However, it was not until the end of the nineteenth century that it gained its real importance through the work of the philosopher, mathematician and natural scientist Charles Sanders Peirce [30]. He was the first to present a universal theory of signs, which describes the order of Nature as an evolutionary system of symbols that—as he believed—is continuing in the development of human sign systems.

Peirce's basic idea of understanding the development of Nature as a "semiotic" evolution appears highly plausible against the background of the findings of modern biology. The evolution of life is based essentially on the development of information and this, in turn, presupposes the emergence of meaningful signs. Peirce's theory of signs is also remarkable in that it is based on an understanding of signs according to which the meaning of a sign is determined by the context in which it is operating. Accordingly, every context in which a sign has an effect already possesses the function of an interpretant of the sign in question. This idea, which relates signs, their objects and their interpretants to each other, is referred to as the "semiotic triangle" (Fig. 1.9).

With his naturalistic view of signs and the model of semiotic evolution, Peirce established an essential epistemological starting point for modern biology, which is denoted as biosemiotics (see [31, 32]). That his ideas nonetheless play no significant role in the current debate of the fundamentals of biology may be due to the fact that he overloaded his epistemology with metaphysical interpretations that today appear outdated. However, the approach of tying the meaning of a sign in the broadest sense to a context of action, and thus of giving signs a pragmatic interpretation, captures an essential aspect of the modern understanding of signs (Sect. 6.8).

So far, we have considered signs more or less as abstract entities which are initially free of any meaning. However, strictly speaking, there are no signs without sense and meaning at all, since each sign already has an intrinsic meaning that is given to it by its mere property of "being a sign". At the next level of meaning, however, signs may have an instructing function, insofar as they may serve to mark out a fixed path, co-ordinates, or indices—just as traffic lights and signs control road traffic. In systems with communication, the rigid signs gain, depending on the level of complexity, more and more freedom to change their meaning. This aspect is immediately reflected in the context-dependence of the meaning of signs.

It is only in human language that signs can unfold all their wealth of meaning. In contrast to the sign language of Nature, in which signs acquired their meaning by natural evolution, man's symbolic language allows for an explicit agreement on the meaning of signs. At this highest level of sense and meaning, signs become symbols in which the mental aspects of our reality are manifested. Here, symbols refer not only to objects and processes; they also stand for actions, thoughts, emotions and the like. In their reciprocal

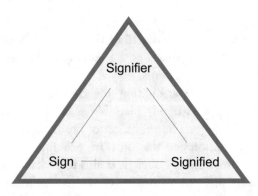

Fig. 1.9 Semiotic triangle. According to Peirce's semiotic theory, any action inevitably involves signs, their objects ("signified") and their interpretants ("signifier"). The interpretant is not necessarily a person who actively interprets the signs. Instead, any context in which any activities take place is regarded by Peirce as an interpretant. The term "semiotics" covers today all sign theories, not least those of modern linguistics

relationships symbols ultimately form a meaningful whole, just as they confront us in our cultural experience.

The genesis and function of symbols have been investigated above all by the cultural philosopher Ernst Cassirer. In his main work the *The Philosophy of Symbolic Forms* [33], he developed the idea that the human mind draws specific pictorial worlds of reality, which we experience as a variety of cultural manifestations. Beside science, these manifestations also include myth, religion, language and art. According to Cassirer, the different appearances of human culture are in each case nothing other than a specific symbolic representation of reality.

Cassirer took a simple example to explain how different mental images derive from the same thing. Following him, we set out from a particular perception experience: "from a drawing that we see in front of us and which we somehow grasp as an optical structure, as a coherent whole. Here we can first turn our attention to the purely sensual 'impression' of this drawing: we grasp it as a simple line. Now we change the point of view, and the spatial structure becomes an aesthetic structure: I grasp in it the character of a certain ornament, which, for me, is associated with a certain artistic sense and meaning [...]. And the type of consideration can change once again, if what was initially presented as pure ornament reveals itself to me as conveying a mythical or religious meaning." [34, p. 298 f.; author's transl.]

This example shows that a pure perception experience may already imply a variety of associations which, in turn, are an expression of the meaning variety of symbolic forms. According to Cassirer, it is precisely this variety that forms the basis for the diversity of cultural manifestations. Moreover, Cassirer believed that he could recognize in the symbolic forms the "veritable archetypes" of the human mind, i.e. forms of representation of human knowledge, which we imprint on reality and by which we shape reality.

Looking at science, Cassirer noticed: "As this insight develops and gains acceptance in science itself, the naïve *copy theory* of knowledge is discredited. The fundamental concepts of each science, the instruments with which it profounds its questions and formulates its solutions, are regarded no longer as passive images of something given but as *symbols* created by intellect itself." [33, p. 74 f.] Cassirer further explains: "It is one of the essential advantages of the sign—as Leibniz pointed out in his *Characteristica generalis*, that it serves not only to represent, but above all to *discover* certain logical relations—that it not only offers a symbolic abbreviation for what is already known, but open up new roads into the unknown." [33, p. 109]

In science, according to Cassirer, sensual perception recedes from a certain degree of abstraction to a position entirely subordinate to the symbolic activity of human language. "The abstract chemical 'formula', for example, which is used to designate a certain substance, contains nothing of what direct observation and sensory perception teach us about this substance; but, instead, it places the particular body in an extraordinarily rich and finely articulated complex of relations, of which perception as such knows nothing. It no longer designates the body according to its sensuous content, according to its immediate sensory data, but represents it as a sum of potential 'reactions', of possible chains of causality which are defined by general rules." [33, p. 108 f.]

It is the ability of language to abstract that makes possible the representation of those causal relations that are constitutive for the understanding of reality. From this perspective, the process of scientific cognition resembles a translation process in which the "language of Nature" is translated into the language of the human being. This process is by its very nature an open one in the sense that the meaning of the elements of a language always depends on their particular context. The context-dependence of language is also the reason why a word-to-word translation from one language to another is sometimes impossible.

While Cassirer emphasizes the autonomy and creative power of the human mind, here the language of Nature is emphasized, by which the symbolizing activity of the human mind is stimulated in the first place. Contrary to Cassirer, who highlights the constructivist character of his symbolism, we here refer to the natural background of this symbolism. Apart from the differences in perspective, Cassirer also seems to be close to the idea that there is a reason for unity in the variety of symbolic forms. He even saw it as an essential task to understand these forms as different manifestations of the same primary function of the human spirit. Consequently, Cassirer argued that a universal understanding of culture must encompass the endless series of its symbolic forms: their general and universal traits as well as their particular gradations and inherent differences.

Of particular importance is Cassirer's idea of a "grammar of symbolic functions as such, which would encompass and generally help to define its special terms and idioms, as we encounter them in language and art, in myth and religion" [33, p. 86]. In this context, culture is to a certain extent conceived of as an overarching language, i.e. as a form of collective understanding in and between social systems that transcends the individual languages of humankind. Insights into intercultural contexts would therefore only be possible from knowledge about the transformation of symbolic representations. This idea represents an extension of the concept of language to cultures, which in turn fits very well into the idea of an overarching principle of language as suggested here.

This approach is, for Cassirer, access to the unity of human knowledge. It is only in the understanding of all cultural differences that our world-view—in its rational and irrational, its consistent as well as contradictory, its familiar and unfamiliar features—fits into a unitary whole. Accordingly, Cassirer tried with the help of an operative understanding of symbols to combine the scientific interpretation of the world with the historical change of the world's images and to interpret the continual shift in world-views as changes in the use of symbols.

Cassirer's major achievement consists above all in his having made the creation and use of symbols the subject of a comprehensive philosophy of culture. After his ideas had been neglected for some time, contemporary philosophy has now again become aware of the importance of the use of symbols by man. Nelson Goodman [35], for example, has called for science and art to be considered from the symbolic perspective and to be interpreted as two equivalent forms of the cognition of reality. According to this understanding of culture, we generate in science the objects of research, using symbolic representations, just as in art we create aesthetic objects as symbolic products. Correspondingly, science and art do not differ from each other in their essential references to reality, but only in the different use of symbols. In other words, symbolic representation in art has a meaning for our understanding of reality comparable to the symbolic representation in science, expressed in theories, formulas and diagrams.

Cultural reality consists primarily of the creation, application, transformation and interpretation of symbols. This is the way, Goodman argues, in which we make worlds, or more precisely: world versions. In contrast to Cassirer, however, Goodman does not see the task of the philosophy of symbols as lying in a search for unity of world versions, but rather in the analytical exploration of types and functions of symbols and symbol systems, an exploration that should reveal the multiplicity of real worlds. With this controversial position, which Goodman himself termed "realism", we enter a philosophical discussion which—by its very nature—cannot bridge the gap between the one-world and the many-worlds version. A thoughtful analysis may at best contribute to the clarification of the logical and terminological status of our world description. Philosophy itself cannot solve any world-riddles. It can only fathom the breadth and depth of human thought by critically questioning our world-knowledge and by reflection on the world —albeit without ever reaching a final conclusion (Chap. 2).

1.7   Excursion into the Empire of Signs

The enormous significance of symbolic representations and actions in the cultural life of man is particularly evident in the Far Eastern cultures. Among these, Japanese culture occupies a particular position, since in Japan cultural structures are uniquely interwoven with social structures (see [36]). Even though this culture may at first seem strange and sometimes mysterious to Westerners, it is worthwhile to take a closer look at it. It quickly becomes apparent that Eastern cultures are based on an understanding of reality fundamentally different from that in Western cultures. At the same time, we see how much modern science is rooted in the tradition of Western thought and shaped by this tradition.

In the face of the profound differences between the Eastern and Western cultures, one may rightly ask whether mutual understanding is possible at all. The fact that the cultural and the social order in Japan are intimately intertwined, and are dominated to a large extent by symbols and symbolic actions, raises a fundamental problem: Since the meaning of symbols changes as soon as their cultural context changes, it will hardly be possible to grasp the meaning of such contextual symbols from the outside. In other words, every external view of Japanese culture is, as it were, a view through the spectacles of a foreign culture. For this reason, the symbolic language of Japanese culture will always be challenging for Westerners to understand.

In his book *Empire of Signs* [37], Roland Barthes tried to decipher the meaning of the signs that are used in Japanese culture. However, he finally gave up, because he had to realize that the signs defy any explanation as symbols. These symbols remain for the external observer more or less "empty" since he is by and large excluded from the overall structure of culture in which the symbols are embedded.

Given this restriction, the following consideration cannot be more than a wooden simplification of Japanese culture, which in fact is much more complex and multifaceted. Nevertheless, it has a primary feature that explains many of its other special features: the fact that the excessive self-centering of the personality, as is typical of Western societies, is foreign to Japanese culture (see [38]). This feature is deeply rooted in Zen Buddhism. According to the life philosophy of Zen, the contrast between the self and the world, subject and object, can be overcome by giving up the self-centering of the individual. Thus, Zen is a life practice directed at the experience of the unity of all being. The tool of Zen is meditation, in the course of which the individual puts itself into a state of total togetherness of body and mind. By this kind of enlightenment, so the promise of Zen, harmonization of the inner with the outer world is achieved. From this point of view, Zen is more a particular form of life practice than it is a philosophy according to Western tradition.

In contrast, subjectivity has an entirely different meaning in Western cultures. Here the concept of the subject is fundamentally shaped by Cartesian thought. Descartes, for example, interpreted matter as an extended, i.e., space-filling substance ("res extensa") while describing the mind as a thinking substance ("res cogitans"). According to Descartes, the mind as a thinking substance is not extended, and neither does it share any other property with material bodies.

The dualism of mind and matter, which according to the Cartesian doctrine cannot be bridged, is the basis for the so-called "Cartesian cut", i.e., the splitting of the world into subject and object, into the inner world and the outer world. Here we come across the irreconcilable difference between the Eastern and the Western understanding of science. While Cartesian dualism must appear strange to Eastern thinking, in Western science it is an absolute prerequisite for the claim of scientific objectivity. This is because objective knowledge presupposes a distance, and thus a separation, between the recognizing subject and the objects of cognition.

At the same time, the Cartesian cut outlines the path to the subject's self-assurance. According to Descartes, the subject becomes self-aware only in the sphere of the *res cogitans*, i.e., in the area of thinking, whereby systematic doubt functions as a regulative principle of thought: only that which withstands rigorous and continuing doubt can claim certainty. This hurdle is so high that it can only be overcome by the doubting subject itself. What, then, *is* the subject? Descartes gives the logical answer: The Ego, cleansed of all doubts, which experiences itself as being identical in doubt, affirmation, negation and the like, is nothing other than the pure thinking itself. This is the philosophical background of his famous formula "I think, therefore I am".

In contrast, Zen takes a fundamentally different position on the question of the subject. In this case, the subject's self-assurance is not achieved through the inwardly directed reflection of the thinking subject, but through outwardly directed experience, through the embedding of the subject in the life practice. For Zen, this view is in so far mandatory as only the externalization of the subject opens up the possibility of overcoming the contrast between subject and object, between inner and outer world, thinking and feeling. Consequently, the self-assurance of the subject focuses on the conscious levelling of the experienceable differences between the inner and the outer world.

While in the Cartesian view the certainty of the existence of the Ego is limited to the inwardly directed "I think, therefore I am", in Zen's life practice, the Ego gains the certainty of its existence through its externalization, through its expansion into the physical world. In contrast to the analytical subject, which Descartes extracted from the world context as "that which is without any doubt", the subject in Zen is holistically determined. Here, the subject is assumed to strive continually to overcome its oppositeness to the world of objects and thus to retrieve a balance between subject and object. In approaching this balance—this is the ideal of Zen—the Ego succeeds in tracing its world reference back to the essential.

For those who stand in the tradition of occidental thinking, it may be difficult to grasp that meditative access to the physical world should result in a "knowledge" that takes precedence over knowledge based on rationality and conceptual clarity. From the occidental perspective, one will tend to consider the meditative approach to knowledge rather as a mere bodily technique, the cognitive result of which is fundamentally different from analytically gained cognition. Meditative expertise cannot replace or complement analytical knowledge. Meditative experience has an entirely different character, which is often associated with the vague idea of "holistic knowledge". It is thus no wonder that in the Eastern world the sciences are also practiced according to the Western model.

Buddhism is a spiritual movement widespread in Far Eastern cultures. It elucidates the strange role which the concept of the subject plays in these cultures, but it does not explain the particular features Japanese society. Zen Buddhism was introduced from China to Japan in the twelfth century. There, Zen Buddhism encountered an already dominant religious movement, Shintoism, which focuses on experiencing Nature and soul, the worship of Nature, the ancestor cult and the adoration of a clan-deity. In the subsequent centuries, Shintoism and Zen Buddhism merged with elements of Taoism and Confucianism to form a particular religious movement, which in nineteenth century Japan finally became the state religion (see [39]). It is only against the background of these numerous influences that Japan's cultural and social features can be understood.

Shintoism is not a theological system, but a canon of values, thought and behavioral patterns that permeates all areas of life and society in Japan. As a kind of popular belief, Shintoism forms the cultural background that establishes the strict hierarchical order of Japanese society and at the same time gives priority to the community over the individual. Literally translated, Shinto is the "Path of the Gods", which leaves its traces in all things and phenomena of the world. Therefore, in principle, everything can be, or become, divine. The faith in the borderless transition between the divine and the secular, which has its roots in Shintoism, forms the basis for the prevailing intellectual current, according to which the transitions between the living and the non-living, between the natural and the artificial, between man and machine are fluid. This world-view is a direct consequence of the universal striving for harmony and balance which permeates all areas of Japanese life.

Only this attitude toward life makes possible the syncretism of tradition and modernism, Nature and culture, individual and society, man and technology, that is typical of life in Japan. Contrary to Western societies, the regulative principle of social behavior is not based on a transcendental morality, but on the harmonious relationship between the whole and its parts, the group and its individuals, the ruler and the ruled.

Harmony, consensus and loyalty are the fundamental values on which Japanese society rests. They motivate the Japanese to subordinate personality entirely to social ties. These may be the company, the employer or the family. For example, a Japanese who wants to signify his social position usually states the institution to which he belongs and not the type of his employment. The special hierarchy of his social ties also seems unusual. For example, the company ranks above the family in the hierarchy of social references: if a

Japanese uses the term "uchi" (my house), he is not referring to the home of his family, but rather to the company, university or other organization for which he works and to which he feels he belongs. Accordingly, the term "otaku" (your house) refers to the place of work of his neighbors.

Despite its strict hierarchical order, Japanese society is not characterized by a caste or class system. Indeed, this would contradict the spirit of Shintoism, which strives for the harmonization of differences, oppositions and contradictions. As the anthropologist Chie Nakane [40] points out, the social order in Japan is not a system of demarcation and exclusion, but rather a hierarchically ordered network characterized by a unique relationship between an individual (or group) and another individual (or group).

The Japanese attitude toward a life that is entirely geared to the social context is expressed in the philosophy of "basho". The basic idea of this philosophy, which was developed by Kitaro Nishida, is already seen from an analysis of the word "ba-sho". Even though the meaning of this word is complex in itself, its philosophical connotation can be explained in simplified form as follows: The word "sho" corresponds to the English word "place", the word "ba" to "relational field". Thus, "basho" refers to the place where a "ba", i.e., a relational field, exists or arises. This place can be, for example, an institution or the constellation of a particular event, which integrates some individuals to form a group. Above all, it is the place of pure self-experience of the Ego; however, not in the subject-centered, Western form "I am (myself)", but in a depersonalized, space-centered way "I am (here)" or "I am (it)".

Contrary to the group affiliation acquired through birth or performance, the affiliation to a "ba" is situational. Only the completely internalized awareness of belonging to a "ba", i.e. to a particular relational field, makes possible a world-view that overcomes subject-centered thinking in favor of collective consciousness, and which ultimately leads to a social order free of egocentricity. These are the consequences of a depersonalized mindset in which the subject is defined in the broadest sense by the "place" at which it stands.

Of course, the Japanese attitude to life, in which one self steps completely behind the intentions, desires and sensitivities of another self, is immediately reflected in the structure and use of the Japanese language. Even though Japanese allows sentence constructions centered on the subject, these are not represented from the position of the acting subject, but from the perspective of the context in which the subject acts (see [41]). Similarly, in Japanese, the designation of the grammatical subject is of subordinate importance in comparison with the indication of the interpersonal reference. Accordingly, the grammatical form of the sociative case—in which the differences in gender, status and rank—can be expressed, is at the center of the linguistic structure.

Beside, the use of language in Japan is tailored to articulate the diversity of polite expressions. In this way, communication supports those forms of reverence, restraint and modesty that are characteristic of Japanese society. On the other hand, however, subjecting a language to this kind of politeness leads to linguistic blurring, which—as one can observe in Japan time and again—can only be compensated for by cumbersome and extensive communication processes. Sometimes one even gets the impression that the use of language primarily serves the purpose of establishing or maintaining interpersonal relationships and not so much that of efficient communication.

In exaggerated form, one could say that a Japanese person does not use his language from the perspective of the independent individual. This fact, in turn, could be an interesting case study for a thesis drawn up by the philosopher Ernst Tugendhat [42], who claimed that only propositional language—Tugendhat understands by this the possibility of saying "I"—establishes the autonomy and freedom of the human being. However, considering the life philosophy practiced in Japan, the opposite interpretation seems more conclusive: resigning from one's Ego leads, in contrast to Tugendhat's thesis, to an increase in freedom. By abandoning its autonomy, the subject tries to free itself from the constraints and contradictions that follow directly from the Cartesian cut, namely, from saying "I".

While for Western thinking the concept of freedom is inextricably linked to the independence of the subject from the outside world, for the Japanese this freedom is defined by the balance and the associated harmonization between the inner and the outer world. It is precisely in the different interpretations of freedom that the distinction between Western and Eastern cultures becomes particularly apparent.

The psychologist Doi [43] interpreted the inclination of the individual to strive to confide and subordinate himself unconditionally to his social context as the desire for "freedom in security". Doi justified this in the frame of his "amae" theory as the right to dependence. The word "amae" is the nounal form of the verb "amaeru" which means roughly "letting oneself be indulged by adjusting". Doi uses this word to characterize the behavior of an individual that is aimed at "depending and presuming upon another's benevolence". However, this form of dependence is not, as one might think, a form of incapacitation, but rather a dependence in which the individual's characteristics are understood and acknowledged. It is a dependence in which the individual finds a protecting freedom to realize and live out its characteristics —just as a child, depending on its parents and at the same time under their protection, can develop its predispositions and allow a free run to its inclinations.

Doi's theory emphasizes an aspect of freedom that is often neglected in Western culture: freedom is not only freedom *from* something, but also freedom *to* (do) something. On the other hand, "freedom in security" also seems to have its darker side. Thus, the culturally-based distancing from the Ego is so highly rated in Japanese society that individuals are continually striving for self-authentication. They do this by entering into as many circles of relationships as possible, because the individual's characteristics become more pronounced, the richer its circles of relationships are. In fact, only through the multiplicity of its connections can the individual increasingly differentiate itself without breaking out of its social context. Or, to put it another way: only a complex field of reference leads to an increase in individuality. It should be mentioned in passing that this figure of thought is very similar to the idea used by Cassirer in an attempt to describe the genesis of the particular within the network of its general relations (Sect. 4.2).

Concerning the problem of individuality, one must also consider the role that Zen Buddhism plays in the Japanese life-world. For example, the main direction of Japanese Zen, the Buddhism of the Soto School, denies the substance of things and thus the existence of absolute being. Instead, this school of thought asserts that everything is related to everything else, and that everything arises from mutual connections. From the standpoint of such teaching, and from the fact that Japanese society only knows the relative Ego, but not the autonomous and absolute one, the individual and the particular are always seen as being dependent on the situation, which in turn can be narrowed down by (solely) employing various symbolic representations. Therefore, the world of the Japanese is a realm of signs and symbols, in which the relationship between the Ego and its world is depicted. The facial expression, the gestures, the tea ceremony, the arrangement of dishes and flowers, the design of gardens and much more represent a cosmos of signs and symbols, whose full meaning, as Barthes rightly pointed out, mostly remains concealed from Western thinking.

Let us finally highlight once more a particular feature of Eastern cultures that could (and I believe should) become a pattern in Western civilizations for dealing with scientific and technological progress. As mentioned before, Eastern cultures do not recognize sharp borderlines within reality, ones that separate the inanimate and the animate, the natural and the artificial, Nature and technology. Instead, such borders are thought of as fluid (Fig. 1.10).

This understanding of reality gives science and technology in Eastern— unlike in Western—societies greater latitude to develop and to exploit fully

Fig. 1.10 Japanese stone garden (Ryōan-ji garden, Kyoto). Stone gardens are minimalistic symbolic representations of the attitude to Nature that predominates in the Eastern world. Gravel raked into waves signifies water, a stream, a river or the sea. Stones embedded in gravel may, for example, symbolize either an island in the ocean or an animal in Nature. The ambivalent use of symbols is the consequence of a world-view that assumes smooth transitions between the inanimate and the animate. In Eastern societies, this assumption, which is in accordance with the findings of modern science, has a hugely positive influence on the social acceptance of technological innovations. [Photo: Bjørn Christian Tørrissen, Wikimedia Commons]

their potential for technological application. As a consequence, Eastern societies have a more liberal and affirmative attitude to scientific and technological progress than Western civilizations have. Moreover, the all-pervasive context-dependence of social behavior in Japan's life-world promotes thinking in relative categories. The latter, in turn, is a necessary condition for the development of a modern knowledge-based society that rests upon critical knowledge and is unrestrictedly open for scientific and technological innovation (Chap. 7).

1.8   Context-Dependence of Signs and Language

Signs and symbols, and thus language and knowledge, only gain their meaning in the context of a pre-existing field of meaning. This claim is the common starting point of all "contextualist" sign theories. However, it contains some philosophical explosive. If every sign ultimately refers to another sign, man's perpetually changing understanding of the world must comprise a continual re-evaluation of the meaning of all signs. This, in turn, seems to call into question the scope of human knowledge and its claim to truth. The perpetual re-evaluation of all signs would have the consequence that, ultimately, the ever-changing flow of signs does not denote anything at all. If one pursues this thought, then one is forced to draw the radical conclusion that in the codified form there can be no binding knowledge at all. From there it is only a short step to a borderless world-view such as was envisaged in Greek antiquity by Heraclitus in his famous phrase ("everything is flowing") and taken up again in modern times by Friedrich Nietzsche (Sect. 2.2).

Under the influence of Nietzsche, contemporary French philosophy, in particular, continued the idea of a relativism of signs. For Michel Foucault [44], for example, each sign that offers itself to interpretation is no longer the sign of an object, but is itself an interpretation of other signs. As a result, any interpretation would be forced to pick itself up, again and again, to interpret itself, so that no interpretation can ever be complete. Moreover, "the farther one goes in interpretation, the closer one comes at the same time to an absolutely dangerous region where interpretation not only will find its points of no return but where it will disappear as interpretation, perhaps involving the disappearance of the interpreter himself" [44, p. 274].

This reading of sign relativism seems to be an easy way of justifying the perspective and interpretative character of our world-understanding. It is, therefore, no wonder that sign relativism ultimately provided the basis for a philosophical movement which, under the label of "post-modernism", propagates the dissolution of all knowledge structures and thereby postulates the total arbitrariness of human knowledge. Here, by taking on the form of the absolute, sign relativism pushes itself to the point of absurdity.

In the mainstream of postmodern philosophy, an odd world-view evolved in the end, according to which the exact sciences are merely a modern myth that tells dark stories and fairy tales, just as the mythical poems did in times past. Among the critics and opponents of science and technology, such abstruse ideas naturally fall on fertile ground. Where the absolute, even if only in the perverted form of absolute relativism, is dogmatically brought into play against science and technology, ideologies will inevitably spread. Their common characteristic is intellectual narrow-mindedness, of which, unfortunately, there are many examples.

Let us return to the more general aspects of sign relativism. There is no doubt that knowledge exists only in the context of other knowledge and that it can only be interpreted and understood relative to this background or context. However, we are also confronted with the context-principle even below the level of linguistic forms of expression: for example, in the area of sensory perception. Cassirer's impressive case of the ambiguity of a simple perception experience verifies the perspective character of human perception and demonstrates that any interpretation of a sign depends on the particular background of the interpretation.

Frequently cited cases in the field of cognitive psychology are multistable figures (Fig. 1.11). If one considers, for example, a two-dimensional representation of a cube, then its contours suddenly seem to step out of the picture and flip back and forth between two possible variants. The context-principle can easily explain this phenomenon. The two-dimensional image that we initially perceive is evaluated against the background of our three-dimensional world of experience. In doing so, the brain tries to supplement the reduced (and incomplete) figure by falling back upon our previous perception experiences. However, since the brain cannot make any unambiguous assignment, the perceived shape of the cube moves back and forth between two possible modes of perception. Another impressive example is a bistable figure in which the face of a pretty young woman suddenly turns into the face of an old woman and vice versa. Last but not least, many phenomena are known in which some elements of a figure are perceived as belonging together (Fig. 1.12). These phenomena even seem to follow their own laws, which in Gestalt psychology are termed laws of perception (see also Sect. 3.9).

The most notable example of the context-dependence of signs is language, even though we are not aware of this in our everyday use of language. The strong contextuality only becomes apparent in ambiguous expressions, as for example in the phrase "dumb waiter". Such expressions can have entirely different meanings, depending on the underlying context. In the present case, it might mean a waiter who is incapable of speech, or it might mean a lift for conveying food from the kitchen to the dining-room. A decision as to which of the two meanings is applicable can only be made from the context in which the word or phrase in question is used.

The philosopher Günther Patzig [45] has taken examples from everyday language to illustrate that the act of speaking is always performed in the context of other actions. Thus, the strange word sequence "right, left, head raised, lowered, side, good, now rear left, right, brakes, fine" will only be understandable for someone who has brought his car have its lights tested. Without this context, garbled word sequences of that kind convey no meaning and are incomprehensible. Even fragments of speech such as "Two to London and back" or "Sausage, sir?" are examples of spoken expression,

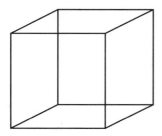 

Fig. 1.11 Multistable images. Left: The "Necker cube" is a reversible figure which goes back to the 18th-crystallographer Louis Albert Necker. It is a simple drawing of a cube that, for the viewer, seems to flip back and forth between two perspectives of perception. This cube is paradigmatic for the phenomenon of bistable perception. Right: What do you see in the picture? A young or an old woman? The figure is based on a drawing by the cartoonist William Ely Hill, "My Wife and My Mother-in-Law" (1915)

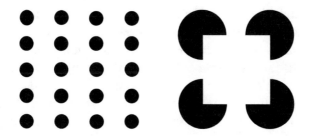

Fig. 1.12 Examples demonstrating different laws of perception. Human perception seems to follow certain "Gestalt laws" that make a selection among bistable perceptions. Left: The "law of proximity" affects our perception in such a way that neighboring elements of a figure are perceived as belonging together. In the present case, we view the grid points as vertical and not horizontal rows. Right: The "law of closure" leads us to perceive four circles covered by a white square and not four truncated circles. Gestalt psychology also knows laws that select among bistable perceptions those invoking impressions of similarity, symmetry, continuity and the like

the meaning of which is immediately apparent when the context of the action is already known—the one at a ticket window, the other in a restaurant.

The context-dependence of linguistic expressions is a typical characteristic of a language, which is already reflected in the language's inner structure. Thus sounds, words and the like stand in a highly sophisticated fabric of relationships one to another. This network is what makes linguistic expressions elements of a language in the first place.

The "inner" structure of human language was first studied systematically by Ferdinand de Saussure [46] at the beginning of the twentieth century.

From his research, a linguistic-philosophical school of thought emerged at various European centers—above all in Prague, Moscow and Paris—which later became known as "structuralism".

Structuralism presents the view that every language is a unique network of relationships between speech sounds, words and their meanings—one that cannot be reduced to its components because only the context of the overall structure defines the parts of a language. Thus, only within their network do the elements of the language coalesce into a linguistic system. This system provides the basis in which the language elements are demarcated from each other, and in which a set of rules assigns to each item its linguistic value.

Again, there are numerous examples which illustrate the dependence of language elements on language structure. Thus, in English, the sounds "L" and "R" differ because they have a function of differentiating the meaning within word-pairs such as "lack" and "rack" or "loot" and "root". In Japanese, for example, the situation is different: The "L" sound and the "R" sound do not differ because in the context of the Japanese language they have no function in differentiating between meanings. An opposite example is familiar to speakers of English. Here, there is an audible difference between the alveolar "R" and the rolled "R", but this is not associated with any difference in meaning. At most, it indicates which linguistic region the speaker hails from (for examples from German see [45, p. 101 ff.]).

What applies to speech sounds also applies to words. The internal reference of a word to the structure of the language in which it is used is also the reason why word-for-word translations are not possible. Thus, the English word "horse", the German word "Pferd", the Latin word *equus* and the Greek *hippos* all refer to the same biological species, but their linguistic meaning first crystallizes out in their demarcation over against alternative terms. It furthermore depends upon the corresponding linguistic structure. In English, for example, the meaning of the word "horse" becomes more precise in the context of alternative expressions such as "stallion", "mare", "gelding", "colt", "foal", "steed" or "nag".

Each of the many human languages has an individual structure (see [47]). However, this is demonstrably true only of the so-called "surface structure" of language. Studies by the linguist Noam Chomsky suggest that, in the deep structure of a language, a uniform grammar is present that lays down the rules for the constitution of all languages [48]. According to Chomsky, the universal grammar is an innate—i.e., genetically hard-wired—language structure that, among other things, enables man in his childhood to learn complex languages in a short time.

Chomsky's thesis is supported by investigations according to which there appear to be innate structures of language that allow the learning of individual languages. For example, it has been observed that deaf-and-dumb children who grow up in very different linguistic cultures still develop an almost identical sign language [49]. Such children are able spontaneously, i.e. without guidance, to combine specific gestures into meaningful linguistic messages according to uniform rules.

In fact, there is increasing evidence that the language ability of humans is genetically anchored [50]. However, there are also reservations, based on various arguments, about Chomsky's theory of language. For example, sophisticated computer simulations suggest that structurally identical forms of language can also originate by pure learning processes [11]. This could mean that the assumption of the existence of a deep uniform structure of language is no longer mandatory to explain the striking ability of children to learn languages. The presence of pre-linguistic sign systems and their indisputable relevance for human language acquisition appears to be a further objection to Chomsky's theory of innate linguistic structures. On the other hand, such critiques can be countered by the argument that Chomsky's theory likewise provides for an experiential learning game for the process of language acquisition—a learning game based on specific structures of pre-understanding.

Even though the idea of the existence of a universal grammar is still controversial, it points the way to how universal principles of language may possess a depth that reaches down even to the genetic level [51]. In fact, we will see that Chomsky's language theory is of the utmost importance for the understanding and justification of the language of living matter (see Introduction). In Chap. 6, we will go a step further, approaching language from molecular biology, and will demonstrate the structural universality of language beyond human language.

The idea of an overarching language of reality is the core of the structuralist understanding of the world. Structuralism regards every ordered relationship of structural elements as a linguistic structure. In this sense, the philosopher Gilles Deleuze explained: "In fact, language is the only thing that can properly be said to have structure, be it an esoteric or even non-verbal language. There is a structure of the unconscious only to the extent that the unconscious speaks and is language. There is a structure of bodies only to the extent that bodies are supposed to speak with a language which is one of the symptoms. Even things possess a structure only in so far as they maintain a silent discourse, which is the language of signs." [52, p. 170 f.] In his anthropological studies, Claude Lévi-Strauss [53] has consistently implemented this idea and interpreted the rules of human behavior, including the structures of social institutions, as structures of an overarching social language.

Out of a common interest in the abstract structures of reality, structuralism has developed into a powerful movement, primarily in France. Initiated and promoted by intellectuals such as Roman Jakobson, Roland Barthes, Gilles Deleuze, Claude Lévi-Strauss, Michel Foucault and Jacques Lacan, structuralist thinking has spread into linguistics, literature, sociology, ethnology, psychoanalysis and mathematics.

Regarding the depth and scope of structuralism, one can certainly have different opinions. This will depend not least on whether we regard structuralism as a comparatively precise tool of the humanities and social sciences, or whether we also see structuralism as the sole authority for our world-understanding. The latter case would stretch the "explanation" claim of structuralism to the utmost. Here one might justifiably ask whether an analysis of abstract structures can really grasp the entire richness of cultural phenomena, when the already living language of man seems to dissolve in the hands of the structuralists into a bloodless web of structural elements.

If structuralism recognizes the limits set for every science, it will not claim to capture reality in its totality. Instead, following its original intention, structuralism will examine the abstract structures of reality, i.e. irrespective of the forms in which these manifest themselves in reality. Accordingly, the descriptions and insight of structuralism can only lead to a "structural" knowledge. This statement puts into perspective the achievements of structuralism as regards the depth of the attainable understanding of the world, but not as regards its scope. Occasionally, the concept of structuralism has been abused, to lay claim—through an uncritical mathematical formalization—to the exactness of cognition. This has been rightly criticized by the physicists Alan Sokal and Jean Bricmont as *Fashionable Nonsense* [54]. Such undesirable outgrowths of structuralism, however, do not alter its fundamentally positive approach to developing a general representation of the issues of the humanities.

The importance of structuralism is also underlined by the development of a class of exact sciences that emerged in recent decades from the investigation of complex systems. These sciences, which may best be denoted as "structural" sciences, include *inter alia* such important disciplines as systems theory, cybernetics, game theory and information theory. The structural sciences developed mainly in close connection with biological questions. Since then, they have grown into powerful instruments for the investigation of complex phenomena in Nature as well as in society. In their methods of abstraction, they are very close to the idea of structuralism in the humanities, and they thus underline the vital role that structural analysis has for the understanding of complex phenomena (Sect. 4.9).

1.9 Contextual Aspects of Science

Let us now return to an aspect of context-dependence which we have just worked out in detail. This aspect refers to the fact that the elementary structures of language—speech acts (a concept explained in Sect. 2.9), phrases, words and even individual speech sounds—are indissolubly dependent on a hyperstructure, namely language itself. This is also relevant for science. In the broadest sense, the context in which scientific terms are used is the theoretical background of science. However, this background is by no means uniform, but it itself consists of a multitude of theories.

Each theory represents a specific canon of statements in which the terms of science acquire a precise meaning in the first place. Conversely, a theory is only sufficiently meaningful if it is based on clear terms. In general, this mutual dependence requires a long process of optimization, in the course of which the basic terms of a theory are refined by the continuous interplay between term formation and theory formation. However, with increasing sharpness of their terms, theories become increasingly precise instruments with which reality can be adequately described.

According to philosopher Stephen Toulmin [55], the context-principle is decisive not only for the development of the essential terms of science, but also for the development of entire systems of terms. He even compared this development with the process of natural evolution. According to Toulmin, the development of scientific terms does not occur in a targeted manner, but rather by the principle of trial and error. Comparable with the mechanism of competition and adaptation among living beings, Toulmin claimed, there is a competition among the systems of terms used in different epochs of science. This competition, he believed, leads ultimately to the gradual implementation of the most effective system. Decisive for this process would above all be the context in which the organization of knowledge takes place.

What here has been described here in a few words is in reality an extremely complex process that we only just begin to see through. Too many metatheoretical questions of science are intertwined here. These relate to the relationship between scientific language and observer language as well as to the logical structure of scientific theories and the problem of theory reduction. The latter issue refers to the interconnection between scientific theories that, in turn, leads directly to the question of the unity of scientific knowledge (see Chap. 4).

The close interplay of term formation and theory formation in science make it clear that the progress of knowledge is associated with a continual correction of language use. From the perspective of our working hypothesis, according to which we can only recognize reality if it has the structure of a language, this is also understandable. In this case, the cognition process must be seen as the continued attempt to translate the language of reality into the language of man. However, as explained before, the translation can never come to an end. Human language is not a rigid system of terms that directly depict the actual structures of reality. Rather, human language presents itself as a multiplicity of forms of expression, the significance of which is continually changing as knowledge progresses. It is precisely this adaptive process that becomes manifest in science in the formation of terms as well as in the formation of theories.

In scientific practice, on the other hand, one tries to bypass the problematic questions of the formation of terms as far as possible. Moreover, it would not be advisable for a scientist to engage in coining complicated term definitions in his daily work. It is entirely possible to carry out a good deal of basic research without making recourse to sophisticated terms. The term "life" is an example of this. Thus, it has been possible in the past to gain detailed insights into the structure and function of living matter without possessing a definite answer to the fundamental question "What is life?" The neurobiologists are in a similar situation. They gather more and more scientific insights into the essential structures and functions of the brain without having an explicit definition of mind and consciousness at all.

One could get the impression that there is a gap in basic research here. However, that alleged deficit is only the consequence of the methodical self-limitation of the exact sciences. Questions such as "What is life?", "What is time?" and the like cannot be answered exhaustively by science, because science is based on the methods of abstraction, simplification and idealization. For this reason, any definition must necessarily be incomplete.

Concerning the definition of time, for example, the physicist Richard Feynman stated laconically: "Perhaps we should say: 'Time is what happens when nothing else happens'." This definition, he continues, "also doesn't get us far. Maybe it is just as well if we face the fact that time is one of the things we probably cannot define (in the dictionary sense), and just say that it is what we already know it to be: it is how long we wait!", and then Feynman adds succinctly, "What really matters anyway is not how we *define* time, but how we measure it." [56, p. 5–1] These remarks express in an exemplary form the "positivistic" understanding of the exact sciences according to which the objects of scientific research are only the "positively" given, i.e., those that are accessible through observation, experiment and measurement. Consequently, all questions relating to the deepest nature of the objects under investigation are to be ignored from the positivistic point of view.

This methodological approach has been so extraordinarily successful in the past that there is no apparent reason to change it. However, science does not hover in the "void" space of a scientific method. It also serves our understanding of the world. Therefore, despite all methodical self-restrictions, science is also confronted with the question of the content and scope of its key terms and theories, which then inevitably also raises the issues of the nature of matter, space, time, life, mind and the like. Even if such questions go beyond the positivistic approach of the exact sciences, they are nonetheless indispensable, because they ultimately raise critical questions about the significance of scientific findings.

Having this in mind, let us examine the question "What is life?". How far can the phenomenon of "life" be specified within the framework of the modern life sciences? The first thing that comes into focus here is the science of the origin of life. One would expect precisely this science to make a clear distinction between non-living and living matter, on the basis of which the transition from the former state of matter to the latter should be explicable. A closer look at this concept reveals that the working definition used here is based on three characteristics of living matter: self-reproduction, metabolism and mutability. With these criteria, the main aspects of the transition from inanimate to animate matter can be fairly well described and, in its essential steps, understood. Since these properties are common to all living beings, the working definition also has the necessary degree of generality.

On the other hand, it is not hard to see that this definition does not yet fully characterize living matter. Other manifestations of living beings such as instinct, behavior, consciousness and the like are not considered at all. However, since these refer to particular life forms, there is no place for them in a description that is intended to be universally valid. In other words, the three characteristics stated above are indeed necessary, but not sufficient, criteria to capture the full breadth of life phenomena.

This consideration makes it clear that the explanatory power of the modern theories of the origin of life is narrowed down to a definition of life that comprises just three general criteria: self-reproduction, metabolism and mutability. These criteria are the decisive factors for the origin and evolution of life. Consequently, any theory based on these criteria can "only" explain under which conditions self-reproductive structures will form in inanimate systems, structures that can adapt to their specific environmental conditions —no more and no less. This restriction applies generally: Any theory that sets out to explain living matter must relate its explanatory power to its underlying definition of life (see Sect. 5.7).

We must take note of the fact that complex questions have no simple answers. Thus, a question like "What is life?" becomes a scientific question only by dividing it into manageable sub-questions, as was demanded by Descartes' methodology. This means nothing less than that we can have quite different perspectives on the same problem. As a result, there is no theory at all that can provide an exhaustive answer to the question "What is life?". Instead, the scope and explanatory power of any theory will depend on the degree of abstraction of its terms and the context of its questions. This restriction must always be considered if one is assessing the significance of life-science research.

The context-principle permeates all areas of human cognition and seems to result in a general epistemic relativism according to which any knowledge is bound to a specific context. From this point of view, the fundamental question of the reliability of scientific knowledge also arises. Are there principles beyond relativism that surpass all context-related cognition, from which we can order and evaluate scientific knowledge? Or, put more generally: is there an absolute—i.e., unrestrictedly valid—principle of cognition, which provides a consistent foundation for the claim of scientific knowledge to truth, or does all knowledge have only a relative character?

Such questions are not only the starting point of numerous philosophical controversies, but they also seem to divide the sciences themselves. One indication of this is seen in the different cognition values that are ascribed to the natural sciences and the humanities, and which are reflected in the self-image of these two major scientific currents of thought (Chap. 4). While the humanities aim to understand reality in its context of meaning, the natural sciences aim to explain it in terms of its causal relationships. The humanities claim that their understanding of reality transcends the limited world-view of the exact sciences. The natural sciences, in contrast, are aware of the fact that their findings always remain provisional, i.e. hypothetical.

Only the findings of mathematics seem to be incontestable, since its theorems, as soon as they are proven, are considered to be true for all time. For this reason, the sciences have always sought to formalize and mathematize their knowledge. However, while the natural sciences can fundamentally rely on mathematics, analytical procedures in the humanities are only applicable to a limited extent. Nevertheless, as structuralism shows, there have been encouraging attempts to develop formal and abstract theories for the humanities as well. Last but not least, Western philosophy has repeatedly used mathematical methods, since its beginnings in Greek antiquity, to secure the truth of philosophical insights. Such approaches were associated with the hope that an accurate picture of reality, which is free of deception, can be deduced through the logic of rational thinking alone.

Despite all fascination emanating from formal thinking, one must not overlook the fact that mathematics has also once experienced a deep crisis. When it turned out at the beginning of the twentieth century that provability and truth are by no means congruent terms, it became clear: Even in mathematics, there is no cognition without any preconditions. Even mathematics relies on basic statements, the truth of which is already assumed to be given (see Sect. 5.2).

In the light of this insight, the mathematical procedures themselves were thenceforth confronted with the question of truth, albeit in a sense different from that of empirical knowledge. Here, we encounter a first case of distinction within the truth concept. "Logical truth", to which mathematics refers, differs from "empirical truth" that serves as a regulator of our empirical knowledge. The one is independent of experience; the other is based on it. One rests upon the method of proof, the other on the practice of observation and experiment.

In the natural sciences, the two methods of cognition intertwine, so that it seems as if the findings of the natural sciences could come very close to the ideal of absolute certainty. This interpretation appears to be self-evident insofar as the ultimate goal of the natural sciences is to discover the universal, eternally valid laws of the world.

On the other hand, the natural sciences have had to revise and adapt their findings time and again in the course of their long history. This aspect of science may seem self-contradictory. However, the contradiction begins to dissolve when we distinguish between the "cognition" we have of an object and the object itself. Then it becomes understandable that knowledge can be incomplete and hypothetical, yet still gradually approach the truth. At the same time, this distinction would introduce a further concept of truth, which now refers to the object of cognition per se as the "true being" of the object. The philosophical literature describes this kind of truth as "ontological truth" (to Greek *on, ontos*: being).

Admittedly, this way of dealing with the problem of knowledge is only an auxiliary construction, which assumes that there exists a reality with objective characteristics independently of the recognizing subject. However, ultimately, such realism cannot be proved. We can never recognize the "true" nature of things because the methods of acquiring knowledge already configure the objects of knowledge (for a detailed analysis, see Chap. 2). The more recent results of quantum physics underline this (Sect. 3.6).

This limitation becomes evident not least at the methods of natural sciences, which rest, above all, upon the principle of objectification. For this purpose, however, the object of knowledge must be prepared methodologically in an appropriate way. Here, the natural sciences have several instruments at their disposal: abstraction, simplification, idealization and generalization. All these methods are characteristic of the analytical research method and aim at reducing the complexity of the objects under investigation. This research strategy only investigates particular, methodologically dissected aspects of reality.

What about the truth of the insights gained by this method? By their empirical nature, the natural sciences rely on observations, measurements and experiments. Thus, their claim to truth seems already to be somehow guaranteed, owing to their closeness to experience. Nevertheless, we encounter a severe problem here. The empirical sciences are by no means a pure and unadulterated distillate of observations and experiments. Rather, their observational and experimental techniques themselves depend again on other scientific insights, the validity of which is presupposed.

For example, if one wants to test a physical hypothesis, one usually uses specific measuring instruments. However, a measuring device can only be built on the basis of physical theories, the validity of which must be presupposed. In other words: Every measurement already rests upon theoretical background knowledge, to which all findings gained in the experiment are related. This is called the "theory-dependence of observation". Of course, this also shakes the positivistic self-conception of the natural sciences, insofar as they claim to be focused upon that which is immediately given.

Therefore, even the empirical sciences must take note of the fact that the validity of their conclusions can only be verified within the framework of tacitly accepted background knowledge. This in turn means that their findings are also contextual. Nevertheless, these findings are not arbitrary or non-binding. Instead, they are subject to permanent checking by experience, which repeatedly enforces the correction of prejudices, misconceptions and errors. As the writer Arthur Koestler rightly pointed out, the progress of science is not a flawless ascent along the path to truth, but rather is comparable to "an ancient desert trail, [...] strewn with the bleached skeletons of discarded theories which seemed once to possess eternal life" [57, p. 178].

1.10   Context Principle and Quantum Physics

There is no better evidence for the ever-changing course of progress in physics than the revolutionary findings of relativity theory and quantum theory. While relativity theory questions our traditional concepts of space and time, quantum theory forces us to revise radically our conventional notion of a physical object. It turned out that quantum objects may have properties that can no longer be understood within the rational concepts of our intuition.

Let us consider, for example, the properties of light. Since the seventeenth century, there have been two conflicting views about the nature of light. The one was developed by Newton, who assumed that light consists of a continuous stream of tiny particles. This idea forms the basis of the corpuscular theory of light. The other concept of light, going back to Christiaan Huygens, interprets light as a self-propagating wave. This idea is the basis of the so-called wave theory of light. However, only in the context of quantum physics has it become clear that these two views are not mutually exclusive, but that light has both properties at the same time (Fig. 1.13). Which feature appears in a given context or experiment depends only on the specific conditions under which light is investigated.

Niels Bohr, one of the founders of quantum physics, considered the wave-particle duality of light to be the essential nature of light and to be irreducible to either the one or the other aspect. The two aspects, he argued, are complementary to one another and thereby constitute the true nature of light. Accordingly, he argued, wave-particle duality must be taken as an

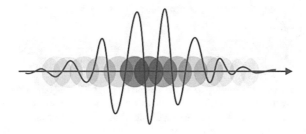

Fig. 1.13 Wave-particle duality of quantum objects. The illustration shows a wave packet (a wave pulse of varying wavelength) and the corresponding quantum particle as a superimposed set of spheres with varying brightness representing the particle's probability density. Which of the two manifestations is observed depends on the experimental arrangement. The wave-particle duality of light is beyond our everyday intuition. The question of how to reconcile this and other strange phenomena of quantum physics with our traditional view of reality has, still today, not been clarified conclusively

elementary fact of light—one that cannot be further resolved within the framework of classical physics. This thought is the core of Bohr's concept of complementarity. Bohr even generalized this idea by extending it to the physics of living matter. Later, however, in the course of the rapid development of molecular biology, Bohr found himself forced to revise his opinion regarding the physical interpretation of life (see Sect. 5.8).

It is strange that light can be described either as waves or as particles, and that it depends only on the specific context of the experimental arrangement which of the two aspects of light makes an appearance. Thus, there is no point in attributing exclusively the character of a wave or a particle to light. It is, rather, the observation apparatus that lets one or the other aspect come to the fore. Moreover, wave-particle duality is by no means limited to light. Instead, the duality turns out to be a general property of matter, which could be verified even for particles up to the size of molecules (Fig. 1.14).

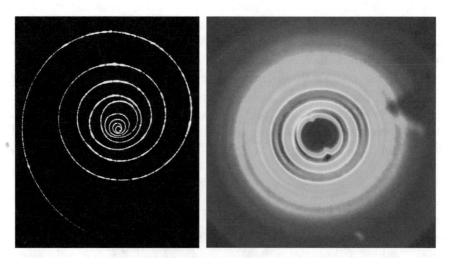

Fig. 1.14 Experiments revealing the wave-particle character of electrons. In the contexts of two different experimental arrangements electrons exhibit their dual nature. Left: Trajectory of an electron in a bubble-chamber experiment that can only be interpreted as the movement of a particle. The electron spirals in the magnetic field of the bubble chamber. As it loses energy in ionizing the liquid hydrogen in the bubble chamber, the electron becomes less and less resistant to the force of the magnetic field and curves inward in successively tighter curls (courtesy of Lawrence Berkeley National Laboratory). Right: Electron diffraction pattern demonstrating the wave character of electrons (courtesy of Brookhaven National Laboratory)

Wave-particle duality seems to be hard to understand outside its mathematical representation by quantum physics. Nevertheless, there are illustrative examples of a dual nature of structures that are similar to the wave-particle duality. These examples are the picture puzzles in which the eye "flips" between two possible images, which seem to bounce back and forth (see Sect. 1.8). We recollect that the same structure, a drawing for example, can invoke in us two perceptual images that are mutually exclusive. The classic example of such an image is the "Necker cube", shown in Fig. 1.11. Here too, we encounter a physical structure with two features that we cannot perceive at the same time. As with wave-particle duality, the structure's images always appear to us only in one form or the other.

On the smallest scale of matter, where quantum physics becomes relevant, one encounters the contextprinciple in its most elementary form. To gain deeper insight into the puzzling phenomena of quantum physics let us consider the level of atoms. The atomic model was developed in the early phase of quantum physics by Ernest Rutherford. Soon after, the model was improved by Bohr, who gave it the form generally accepted today. According to this model, the structure of an atom resembles that of a planetary system, with the atomic nucleus at its center surrounded by orbiting electrons (Fig. 1.15).

Following classical mechanics, calculation of an electron's orbit requires knowledge of the initial conditions (i.e., the electron's position and velocity at a given time), since only exact knowledge of these conditions will allow us to calculate the electron's motion. For example, on this basis the course of the planets can be predicted over millennia with impressive precision. Thus, if one is going to apply the laws of classical mechanics to an elementary particle, such as the electron in the atom, one has first to determine its exact position and velocity. However, is this at all possible for very small objects such as quantum particles?

The submicroscopic level of physics evades direct observation. Consequently, objects are only observable by employing technical aids such as electron microscopy, scanning tunneling microscopy, and the like. This in turn may lead to a problem of measurement. Since these objects are submicroscopically small, it cannot be excluded from the outset that the measurement exerts a certain influence on the object under observation. Thus, it could well be the case that the influence is the larger, the smaller the observed objects are. For particles such as atoms, electrons and photons, the disturbing power of the measurement could even become critical.

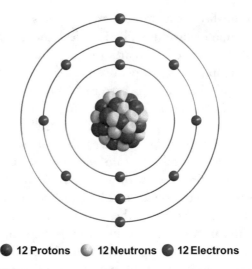

● 12 Protons   ◌ 12 Neutrons   ● 12 Electrons

Fig. 1.15 Model of the magnesium atom. Ernest Rutherford's atomic model forms the basis for the modern conception of the atom. According to this model, the atom has a positively charged nucleus that contains almost the entire mass of the atom. The nucleus is surrounded by negative charge carriers, the electrons, which together compensate for positive charge of the nucleus. In 1913, Rutherford's model was refined by Niels Bohr, who assigned to the electrons discrete orbits and thus different energy levels. The picture shows the structure of a magnesium atom, containing a nucleus of 12 (positively charged) protons and 12 (uncharged) neutrons, surrounded by 12 (negatively charged) electrons. The absorption or emission of light by an atom is explained by an abrupt jump of an electron between the discrete orbits. The model is reminiscent of the movement of the planets around the sun. According to classical physics, an exact statement of the position and velocity of the orbiting electrons should allow the motion of the electrons to be described in detail. However, the uncertainty relation, discovered by Werner Heisenberg, prohibits a classical description of the atom. Instead, only statements about the probability density for the position of an electron are possible. [Image: oorka, Can Stock Photo]

Werner Heisenberg investigated this issue by a simple thought experiment. In physics, thought experiments are not necessarily experiments that can actually be carried out. However, they often help to clarify theoretical problems. In the present case, Heisenberg argued, one has to specify the experimental conditions under which the position and velocity of a quantum object can be measured. To observe, for example, the position of an electron one has to locate it with an appropriate measuring device, let's say with a high-resolution microscope. For this purpose, one has to illuminate the electron with extreme short-wave light. This, however, creates another problem: owing to the so-called Compton scattering, light exerts a radiation pressure upon the electron that will significantly change its velocity (Fig. 1.16).

Heisenberg's thought experiment illustrates what will happen. The incident wave (the photon) will instantly cause a change of the electron's position, just as when two billiard balls collide. By a plausibility consideration, Heisenberg estimated that the product of the uncertainties of position and momentum cannot be substantially smaller than Planck's quantum of action [58]. Consequently, the more precisely the position of a quantum object is determined, the less precisely is its momentum known, and vice versa. Thus, it is not possible to determine simultaneously and with arbitrarily high precision the position and the momentum of a quantum object. Below this limit, as stated by the uncertainty relation, the words "position" and "velocity" lose all reasonable meaning (cf. [59]).

The uncertainty relation is the actual reason for the occurrence of statistical correlations in quantum physics. The fact that it does not play any role in common measurements, for example in a radar control (which measures simultaneously the position and the speed of a car), is because the influence by the observer is only noticeable with submicroscopic objects such as atoms, electrons and the like. The larger the objects observed are, the smaller are the inaccuracies of the measurement process. Even with dust particles the inaccuracies are negligible.

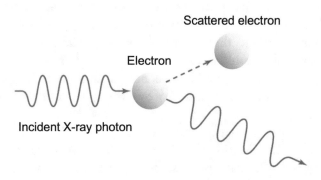

Scattered electron

Electron

Incident X-ray photon

Scattered X-ray photon

Fig. 1.16 Compton scattering. An incoming X-ray photon, hitting an electron, transfers some of its energy to the electron, thereby changing the electron's velocity. The scattered photon has lower energy, and therefore a longer wavelength, than the incoming photon. The unavoidable impact of Compton scattering on the measurement of submicroscopic particles caused Heisenberg to formulate the "uncertainty principle" of quantum physics. [Adapted from Encyclopedia Britannica]

It is a peculiarity of quantum physics that its phenomena elude our intuition. This is the main reason why the quantum world appears to us as full of riddles, the physical and epistemological meaning of which are still interpreted in different ways today. Above all, the so-called "Copenhagen interpretation" of quantum physics, which goes back to Bohr and Heisenberg, continues to be the main topic of discussion. According to their interpretation, the microscopic world of atoms cannot be described without including the measuring device in the description. It is only the entirety of the quantum object and the measuring device, as Bohr always stressed, that leads to the peculiarities of quantum phenomena.

The dependence of quantum objects upon the experimental arrangement can also be understood as a dependence of the observed object upon the observer—an interpretation that seems to shake the objectivity postulate of science according to which the objects of research as a part of reality are given independently of the humans researching into them. Two epistemological opposites seem to collide here. The realistic view, according to which all cognition starts from the object of cognition that is given at any instant, and the idealistic standpoint, according to which all cognition results from the cognitive subject—in this case the experimenting observer. We will discuss both philosophical currents in detail in Chap. 2.

There is still another problem. Heisenberg's uncertainty relation raises the question of whether the uncertainty in the determination of the initial conditions of a quantum object is due only to lack of knowledge, or whether it is of a fundamental nature. If it were the latter, then this would question physical determinism and thereby also the principle of causality. The consequences that might result from the violation of determinism were—for decades—a point of contention between Bohr and Einstein. In Sect. 3.6 we will enter more fully into the controversy over the correct interpretation of the quantum phenomena.

However, the context-dependence of the wave-particle duality of light already indicates the direction in which we may have to go for a deeper understanding of quantum physics. It appears as if the experimental arrangement used in any particular case might contain a piece of specific information that makes the one or the other aspect of light come to the fore. Evidently, the context information confers an unambiguous character upon an *a priori* ambivalent phenomenon. The context gives information its meaning. This was the result of our previous considerations. In order to understand the properties of quantum phenomena, one has to obtain physical access to the semantic aspect of information. We shall suggest a solution of this problem in Sect. 6.8. That will also shed some light on the meaning content of a quantum object.

In this connection, it is worth taking up once more the question of how we define objects conceptually. In Sect. 1.4, we argued that the essence of an object could be narrowed down conceptually by yes/no-decisions. To do this, one must decide a number of specific alternatives given, in turn, by the context of the question. In physics, this context is provided by the measuring apparatus. Thus, we ask questions by an experiment, so that Nature is forced to answer them. In quantum physics, however, the answer depends on the measuring device. This means that at the level of quantum objects, an unlimited and unambiguous sequence of decisions is no longer possible. Instead, physics here is approaching a final alternative, the wave-particle duality, which must be taken as a final fact.

This train of thought prompted Carl Friedrich von Weizsäcker to postulate the existence of "ur-alternatives", which he assumed to be the most elementary objects of quantum physics [20]. Ur-alternatives correspond to the so-called "observables" of quantum physics. They represent, as it were, one bit of *potential* information that becomes *actual* information by the quantum-physical measurement. Thus, ur-objects may be regarded as atoms of information, in which being and knowledge fuse together into an indissoluble unit. At this level, the physical world consists of ur-alternatives, which only by observation and measurement coagulate into the reality with which physics deals. Weizsäcker believed that the concept of ur-alternatives could be the starting point of an axiomatic reconstruction of quantum theory, which, by its very nature, would be a quantum theory of information. For further details, see Weizsäcker's *The Structure of Physics* [60]. Beside quantum theory, many other examples demonstrate the general importance of the concept of information for a comprehensive understanding of Nature [61]. This applies not least to modern biology. Without recourse to information theory, life's origin and evolution would not be understandable at all [22, 23, 62]. Therefore, it will be the main task of the following chapters to work out the deep structure of science under the guidance of the concept of information.

The context principle, extending from language to the phenomena of Nature, leads inevitably to a relativistic understanding of the world. In this connection, we will look with interest at the world-understanding of the humanities. The cognition method, which large parts of the humanities refer to, is "hermeneutics". Historically, hermeneutics was initially the art of interpreting and commenting on biblical and legal texts. In the twentieth century, hermeneutics finally developed into a general methodology of the humanities, given profound justification by philosophy.

Hermeneutically oriented humanities strive to understand reality in its meaning context. They argue in favor of a world-view that includes the comprehending subject itself in the overall context of understanding. In respect of this position, hermeneutics stands in contrast to the natural sciences with their pursuit of objectivity. For the natural sciences, the separation between the recognizing subject and its objects of cognition is a prerequisite for the cognition process. The threatening danger of a loss of objectivity, which the results of quantum physics seem to imply, is the reason for the lively discussion about the correct interpretation of quantum theory. In sharp contrast to the natural sciences, philosophical hermeneutics considers the unity of subject and object as a precondition for all real understanding (see Sect. 3.4). By placing the idea of unity at the center of scientific reflection, philosophical hermeneutics tries to fill an alleged gap in our scientific world-view. It claims to do what the analytical research method is not capable of doing: to consider reality as a continuous context of meaning and thus as an irreducible whole of understanding.

However, in the relevant treatises on philosophical hermeneutics, one will search in vain for methodological criteria that could secure hermeneutics' claim to truth. Its legitimization is based solely on the conviction that the whole *is* the truth. Thus, hermeneutics seems to satisfy a basic human need, which expresses itself in the search for a comprehensive and absolute world-understanding.

In contrast, the natural sciences have rigorously eliminated the idea of the absolute from their concepts and theories. Only this step has made possible the revolutionary advances in relativistic physics and quantum physics. Undoubtedly, a highlight of this development was Einstein's discovery that neither space nor time is absolute, but that the two are related to each other as coordinates in a four-dimensional world. Quantum physics, in turn, had to give up the idea of an absolute object which exists for and in itself and to replace it by a relative concept of the object, related to the observer. Also, in biology, there is a tendency to adapt its basic idea of genetic information. The context-dependence of all information provides further impetus to such adaptations [62].

Nevertheless, the idea of the absolute still seems to exert a particular attraction for some scientists. A prominent example was Max Planck, the founder of quantum physics. Throughout his life, Planck was never able to break away from the idea of the absolute. The search for the absolute seemed to him to be the noblest task of research. In an essay entitled *From the Relative to the Absolute* [63], Planck argued that with modern physics the absolute had not been completely done away with, but only deferred. The physicist, so

Planck, would need the belief in an independent absolute, because otherwise the relative would float in the air like a coat without a peg to hang it on.

Despite all willingness to acknowledge the relativity of scientific knowledge, Planck held fast his conviction that every system of knowledge needs an absolute fixed point, because in his view this would be the only possible point of reference in the fabric of relative relationships. This invariant frame represents itself as the general context of experience and its unchangeable laws. Here, Planck had not least the natural constants in mind, which together with the laws of Nature are considered to be absolute and immutable.

However, not even this can be regarded as certain, as physical theories are also conceivable in which the fundamental constants and the laws of Nature change over time. With this argument, we are approaching an epistemic position, according to which all knowledge of Nature is in principle open for adaptation. However, one must not interpret this position in the sense of absolute relativism, since the idea of relativism would cancel out in this case.

No matter how one looks at the arguments, the idea of the absolute cannot be justified in a logically flawless form (see Sect. 2.6). There is no Archimedean point beyond human thought from which the truth content of human knowledge can be justified. Therefore, every system of findings inevitably starts from a principle that determines the relationship between cognition and reality, but whose validity is no longer questioned within the system.

While this seems to us to be immediately apparent, we must be somewhat surprised by another fact: Our insights into reality seem to evolve truthfully, even though there is no definite proof for the truth of a particular cognition. It already appears to be sufficient for the progress of knowledge that ideas which we once thought to be correct can be refuted and thus classified as wrong. However, if the advancement of knowledge finds its confirmation only in the elimination of errors, what does progress then mean? Or does the continual correction of errors lead in the end to an inversion of the thought of progress? Is the progress of our knowledge restricted to knowing what is wrong?

Such doubts can only arise if we fail to distinguish carefully between quantitative and qualitative knowledge. Actual progress is not reflected in a quantitative increase of knowledge, but rather in the deepening of it. The refinement, in turn, is done by excluding false conclusions and ideas. By gaining everyday information about wrong ideas, the quality of our knowledge is continuously increasing. Regarding the fundamental questions of our world-understanding from this vantage point, we note that these have not changed significantly in the course of cultural history. The answers, however, have become more and more precise, detailed and sophisticated over time.

References

1. Aristotle (1933–35) Metaphysics. In: Tredennick H (ed) Aristotle in 23 volumes, vols 17, 18. Harvard University Press, Cambridge/Mass [Original: Tà metà tà physiká, 350 B.C.]
2. Bacon F (1875) The New Organon. In: Spedding J, Ellis RL, Heath DD (eds) The Works of Francis Bacon, vol IV. Longman, London [Original: Instauratio magna, 1620]
3. Blumenberg H (1986) Die Lesbarkeit der Welt. Suhrkamp, Frankfurt
4. Bacon F (1875) New Atlantis (1626). In: Spedding J, Ellis RL, Heath DD (eds) The Works of Francis Bacon, vol V. Longman, London
5. Descartes R (1983) Principles of Philosophy (transl: Miller VR, Miller RP). Reidel. Dordrecht [Original: Principia philosophiae, 1644]
6. Descartes R (1965) Discourse on Method, Optics, Geometry, and Meteorology (transl: Olscamp PJ). Bobbs-Merrill, Indianapolis [Original: Discours de la methode pour bien conduire sa raison, & chercher la verité dans les sciences: plus la dioptrique, les meteores, et la geometrie, qui sont des essais de cete methode, 1637]
7. Pennisi E (2000) Human genome. Finally, the book of life and instructions for navigating it. Science 288(5475):2304–2307
8. McKay LE (2000) Who Wrote the Book of Life? A History of the Genetic Code. Stanford University Press. Stanford
9. Gadamer H-G (2004) Truth and Method. Continuum, London/New York [Original: Wahrheit und Methode, 1960]
10. Galilei G (1960) The Assayer (transl: Drake S., O'Malley CD). In: The Controversy on the Comets of 1618. University of Pennsylvania Press, Philadelphia [Original: Il Saggiatore, 1623]
11. McClelland JL, Rumelhart DE (1985) Distributed memory and the representation of general and specific information. J Exp Psychol Gen 114(2):159–197
12. Nowak MA, Krakauer DC (1999) The evolution of language. Proc Natl Acad Sci USA 99(14):8028–8033
13. Fadiga L, Craighero L, Fabbri Destro M, Finos L, Cotillon-Williams N, Smith AT, Castiello U (2006) Language in shadow. Soc Neurosci 1(2):77–89
14. Fitch TW (2010) The Evolution of Language. Cambridge University Press, Cambridge
15. Gibson KR, Tallerman M (eds) (2013) The Oxford Handbook of Language Evolution. Oxford University Press, New York
16. Cantalupo C, Hopkins W (2001) Asymmetric Broca's area in great apes. Nature 414(6863):505
17. Tremblay P, Dick AD (2016) Broca and Wernicke are dead, or moving past the classic model of language neurobiology. Brain Lang 162:60–71

18. Shannon CE (1948) A mathematical theory of communication. Bell Syst Tech J 27(3):379–423 and 27(4):623–656
19. Shannon CE, Weaver W (1963) The Mathematical Theory of Communication. University of Illinois Press, Illinois
20. Weizsäcker CF von (1980) The Unity of Nature. Farrar, Straus & Giroux, New York [Original: Einheit der Natur, 1971]
21. Hartley RVL (1928) Transmission of information. Bell Syst Tech J 7(3):535–563
22. Küppers B-O (1990) Information and the Origin of Life (transl: Woolley P) MIT Press. Cambridge/Mass [Original: Der Ursprung biologischer Information, 1986]
23. Küppers B-O (2016) The Nucleation of Semantic Information in Prebiotic Matter. In: Domingo E, Schuster P (eds) Quasispecies: From Theory to Experimental Systems. Springer International, Cham, pp 67–85
24. Ratner VA (1974) The Genetic Language. In: Rosen R, Snell FM (eds) Progress in Theoritical Biology, vol 3. Academic Press, New York
25. Eigen M (1983) Sprache und Lernen auf molekularer Ebene. In: Peisl A, Mohler A (eds) Der Mensch und seine Sprache. Propyläen, Frankfurt, pp 181–218
26. Miescher F (1897) Letter to Wilhelm His, 17 Dec 1892. In: Histochemische und physiologische Arbeiten, Bd 1. Vogel, Leipzig
27. Goethe JW von (1981) Werke, Bd 13. Beck, München
28. Bradbury JW, Vehrencamp SL (2011) Principles of Animal Communication. Sinauer Associates, Sunderland
29. Kirsch R, Vogel H, Muck A, Reichwald K, Pasteels JM, Boland W (2011) Host plant shifts affect a major defense enzyme in *Chrysomela lapponica*. Proc Natl Acad Sci USA 108:4897–4901
30. Peirce CS (1998) The Essential Peirce: Selected Philosophical Writings, vol 2 (1893–1913). Indiana University Press. Bloomington
31. Barbieri M (2008) Biosemiotics: a new understanding of life. Naturwissenschaften 95:577–599
32. Emmeche C, Kull K (eds) (2011) Towards a Semiotic Biology. Life is the Action of Signs. Imperial College Press, London
33. Cassirer E (1965) The Philosophy of Symbolic Forms, vol 1. Yale University Press, New Haven & London [Original: Philosophie der symbolischen Formen, 1923–1929]
34. Cassirer E (1927) Das Symbolproblem und seine Stellung im System der Philosophie. Z Ästhetik Allg Kunstwissenschaft 21:295–322
35. Goodman N (1978) Ways of Worldmaking. Hackett, Indianapolis, Indiana
36. Coulmas F (2003) Die Kultur Japans. Beck, München
37. Barthes R (1983) Empire of Signs. Jonathan Cape, London [Original: L'empire des signes, 1970]

38. Kimura B (1995). Zwischen Mensch und Mensch: Strukturen japanischer Subjektivität (transl: Weinmayr E). Wissenschaftliche Buchgesellschaft. Darmstadt [Original: Hito to hito to no aida: seishin byōrigakuteki Nihon ron, 1972]

39. Lokowandt E (2001) Shinto. Eine Einführung. Ludicium. München

40. Nakane C (1972) Japanese Society. University of California Press, Berkeley

41. Okochi R (1995) Wie man wird, was man ist: Gedanken zu Nietzsche aus östlicher Sicht. Wissenschaftliche Buchgesellschaft, Darmstadt

42. Tugendhat E (2013) Egozentrik und Mystik: Eine anthropologische Studie. Beck, München

43. Doi T (1973) The Anatomy of Dependence. Kodansha International, Tokyo [Original: Amae no Kozo, 1971]

44. Foucault M (1998) Nietzsche, Freud, Marx. In: Rabinow P (ed) The Essential Works of Michel Foucault 1954–1984, vol 2. Penguin, Harmondsworth [Original: Nietzsche, Freud, Marx, 1964]

45. Patzig G (1970) Sprache und Logik. Vandenhoeck & Ruprecht, Göttingen

46. Saussure F de (1983) Course in General Linguistics. Duckworth, London [Original: Cours de linguistique générale, 1916]

47. Hjelmslev L (1961) Prolegomena to a Theory of Language. University of Wisconsin Press, Madison [Original: Omkring sprogteoriens grundlaeggelse, 1943]

48. Chomsky N (1957) Syntactic Structures. Mouton, The Hague

49. Goldin-Meadow S, Mylander C (1998) Spontaneous sign systems created by deaf children in two cultures. Nature 391(6664):279–281

50. Lai CSL, Fisher SE, Hurst JA, Varghe-Khadem F, Monaco AP (2001) A forkhead-domain gene is mutated in a severe speech and language disorder. Nature 413(6855):519–523

51. Searls DB (2002) The language of the genes. Nature 420(6912):211–217

52. Deleuze G (2004) How do we recognize structuralism? (transl: Lapoujade D). In: Taormina M (ed) Desert Islands and Other Texts 1953–1974. Semiotext(e)/ Foreign Agents. Los Angeles (distributed by MIT Press, Cambridge/Mass), pp 170–192 [Original: A quoi reconnaît-on le structuralisme?, 1972]

53. Lévi-Strauss C (1963) Structural Anthropology. Basic Books, New York [Original: Anthropologie structurale, 1958]

54. Sokal A, Bricmont J (1998) Fashionable Nonsense. Postmodern Intellectuals' Abuse of Science. Picador, New York

55. Toulmin SE (1972) Human Understanding: The Collective Use and Evolution of Concepts. Princeton University Press, Princeton

56. Feynman RP, Leighton RB, Sands M (1963) The Feynman Lectures on Physics, vol 1. Addison-Wesley, Reading/Mass

57. Koestler A (1967) The Ghost in the Machine. Hutchinson & Co, London

58. Heisenberg W (1927) Über den anschaulichen Inhalt der quantentheoretischen Kinematik und Mechanik. Z Phys 43(3):172–198

59. Heisenberg W (1931) Die Rolle der Unbestimmtheitsrelationen in der modernen Physik. Monatsh Math Phys 38:365–372
60. Weizsäcker CF von (2006) The Structure of Science. Springer, Dordrecht [Original: Aufbau der Physik, 1985]
61. Roederer JG (2005) Information and Its Role in Nature. Springer, Berlin/Heidelberg
62. Küppers B-O (1995) The Context-Dependence of Biological Information. In: Kornwachs K, Jacoby K (eds) Information: New Questions to a Multidisciplinary Concept. Akademie Verlag, Berlin, pp 135–145
63. Planck M (1929) Vom Relativen zum Absoluten. Naturwissenschaften 13:53–59

2

Truth: The Regulative Principle of Cognition

2.1 What is Truth?

The philosopher Friedrich Nietzsche once remarked contemptuously that truth is just an interrelated set of errors which our species cannot live without [1, pp. 844 and 915]. Are the humans in search of the true givenness of the world just pitiable, deluded people? Are the findings of science nothing more than a hodgepodge of errors? Is there truth in an objective sense, or does truth only exist in a subjective form, as the belief that something is true? Is there anything like truth at all? And if so, is there only one unalterable truth, or are there many forms of truth—that is, in reality, no truth?

It seems as though the philosophical problem of truth is a veritable maze in which one continually turns in circles, as all statements about the essence of truth must themselves be able to live up to their own claim to truth. On this premise, however, any justification of the concept of truth is already based on what exactly it is that needs to be clarified and justified, namely the claim to truth. Since no intellectual point of reference has proved to be accurate, from which the essence of truth could unequivocally have been determined, all attempts at justification of the concept of truth are inevitably circular. No theory of truth can evade this circularity.

An unsolvable problem, which leads to hopeless discussions, has since Greek antiquity been referred to as an *aporia*. Despite this unsolvability, we cannot avoid getting involved in the sensitive issue of truth, because somehow the whole thinking of humankind seems to revolve around the problem of truth. Nietzsche is undoubtedly right: Without the idea of truth we humans obviously cannot live. We must assume that "true" judgments about reality

© Springer Nature Switzerland AG 2022

B.-O. Küppers, *The Language of Living Matter*, The Frontiers Collection,

https://doi.org/10.1007/978-3-030-80319-3_2

are possible. Otherwise, reality would be nothing more than a world of illusions, in which no judgment was consistent with the facts and in which no judgment could reasonably be linked to another judgment. In a world without a claim to truth, a reasonable life would not be possible. Again, no theory of truth can evade this circularity.

Truth and the reference to the reality of our judgments are in fact inseparably linked. Only through accurate judgments can a reliable connection be established between our inner and outer world. This view is also expressed in a standard definition of the concept of truth, which can be traced back to the thirteenth century. Truth, as the Dominican theologian Thomas Aquinas formulated it at the time, is the concordance between the facts and the judging thought (*adaequatio rei et intellectus*). This conception of truth is called the "correspondence theory" of truth. The real originator of this theory, however, was not Aquinas, but Aristotle, whose writings were translated into Latin and commented on by Aquinas in the Middle Ages.

The correspondence theory of truth expresses an understanding of truth that is as simple as it is evident: thoughts and ideas, if they are true, represent reality as it is. However, correspondence theory only makes sense on the assumption that there is an objective reality independent of the judging thinking. The correspondence theory of truth justifies an epistemic position that is denoted as "realism."

Against the backdrop of its two thousand years of history, the correspondence theory of truth seems so attractive because—above all—it comes very close to our intuitive understanding of truth, since we believe that judgments about reality are right when they are consistent with the facts. Last but not least, the empirical sciences draw upon this understanding of truth. They also assume that their theories represent reality the more truthfully, the better they agree with the facts.

On closer examination, however, the correspondence theory of truth proves to be by no means unproblematic. This is because our judgments of thinking that underlie our understanding of reality and with which we grasp and explain the facts of this world can only be expressed in the sentences of a language. Therefore, in the opinion of the logician Alfred Tarski [2], the truth of such judgments can, strictly speaking, only exist in the correspondence between sentence and fact. However, sentences and facts, the mutual agreement of which is required by the correspondence theory, are fundamentally different from each other. They cannot easily be compared or even equated with one another. This is like trying to compare apples with oranges.

In what sense can we then still speak of the correspondence between sentence and fact? Aristotle does not give a clear answer to this. In chapter 9 of his *De interpretatione* he says that "statements are true according to how the actual things are" [3, 19a 35]. However, this wording allows different interpretations [4]. One the one hand, Aristotle could have meant that true sentences *are* the way things are. On the other, he might alternatively have thought that true sentences *say* the way things are. In the first case, sentences would have an imaging function, and in the second case, a symbolic representation function. However, which interpretation did Aristotle prefer? And that's not all: The thesis of the correspondence between sentence and fact is not only ambiguous, but it also raises severe philosophical questions for which there are no simple solutions (see for example [5]).

A possible way out of this dilemma could be the assumption that reality itself already has the structure of a language, as proposed in Chap. 1. In this case, the correspondence between sentence and fact would merge into the correspondence of two linguistic structures, which express "linguistic" facts. Now, one could argue that this approach is not a real solution at all, as the incongruence of sentence and fact is only avoided because the fact-content of reality is interpreted as linguistic structure from the outset. However, the force of this objection is moderated if one considers that sentences are also facts, more precisely: linguistic facts. The moment we become aware that the relationship between a true proposition and the fact it expresses can be understood as the structural equivalence of two propositions, the sharp juxtaposition of proposition and fact, as implied by the correspondence theory of truth, vanishes.

Since every language structure is context-dependent, the truth criteria of a sentence will shift. Truth will no longer appear as the selective correspondence of sentence and fact, but as the truth-content of an ensemble of statements that refer to particular events related to each other. In other words: the question of truth no longer concerns isolated facts, but rather a fabric of facts in which some aspects of reality are expressed. Consequently, the truth-content of an ensemble of statements will be judged by its coherence, i.e., whether the statements about reality are free from contradictions. This view on the problem of truth is ultimately a direct consequence of the structuralist understanding of language according to which each element of a language structure acquires meaning only in its overall linguistic context (Sect. 1.9).

We see here in outline how difficult it is to find a consistent definition of "truth". One way of circumventing this difficulty might be merely to look for an operational definition. This is because the question of whether or not judgments about reality are right is not only a theoretical question—one of the relationship between sentence and fact—but also a practical problem, insofar as a judgment's claim to truth is decided not least by the success and failure of human actions.

The roots of this approach, which judges the truth of beliefs, ideas and insights by the extent to which they prove themselves in human actions, lie in the philosophy of "pragmatism". The Greek word *pragmatiké* means the art of acting correctly. However, one can only act correctly on the basis of true judgments. Otherwise one would open the door to chance. The concept of truth understood in this way leads ultimately to "pragmatic" truth.

Modern pragmatism—founded by Charles Sanders Peirce, William James and John Dewey (Sect. 2.9)—has had a significant influence on the contemporary philosophy of science. This is not least because pragmatism is of direct importance for the natural sciences, which derive their claim to truth *inter alia* from their practical relevance in the solution of technical problems.

In the form developed by Peirce, pragmatism assumes that in the process of cognition, final—i.e., absolutely correct—knowledge can never be achieved, but only a gradual approach to truth can be made. As for René Descartes (Sect. 1.2), the method of doubt is a central element of Peirce's theory of knowledge. In contrast to Descartes, however, who only wanted to accept that which withstands radical doubt, Peirce assigned a vital cognitive function to the convictions based on inward assumptions and beliefs that every human being carries. According to Peirce, the human being must hold fast to certain beliefs, which are initially excluded from doubt, so that in a world in which in principle everything can be questioned there is a fixed point of reference.

This argument makes sense, as a reasonable co-existence of people is only possible if they share common convictions. Nevertheless, Peirce was aware of the fact that all convictions are based on assumptions that may one day prove unsustainable or at least need to be corrected. That is why, according to Peirce, dubious beliefs are replaced time and again by less questionable ones. However, this would require a supervisory body that can communicate something about the quality of our beliefs. For pragmatism, this supervisory body is the practice of human action: we can only act successfully on the basis of well-founded rational convictions. Wrong convictions, on the other hand, lead to failure.

It is evident that within the framework of pragmatism, the idea of absolute truth is abandoned in favor of hypothetical truth. If the claim to truth of our convictions can only be seen in the success of our actions, and if our beliefs have to be revised time and time again, we can no longer speak in the strict sense of "truth", but only of the "temporary validity" of our convictions. According to pragmatism, once our existing beliefs are refuted by experience, we will correct them by replacing them with other beliefs that are better suited to experience. In this way, inadequate or false beliefs are eliminated step by step. Because all beliefs can be improved in principle, an idea's claim to truth can only have a hypothetical character. The method of error correction described by Peirce is also called "fallibilism". It is central not only to the theory of knowledge in pragmatism, but also for our present-day understanding of scientific rationality.

According to the world-understanding of pragmatism, the inner world of our subjective convictions is linked to the outside world of objective things by our actions. At the same time, this conjunction guarantees and specifies the semantic content of our convictions. Nevertheless, even the pragmatic approach must be in principle be expressible linguistically, or (more generally) by signs. Otherwise, the existence of beliefs would solely be manifested in the execution of actions, and we would not be able to speak of our ideas in any meaningful way. Thus, ultimately, the pragmatic concept of truth also includes the linguistic level or the level of signs.

Of course, an astute thinker like Peirce recognized this. Therefore, in his theory of signs, he was particularly interested in the question of how signs gain their significance [6]. His conclusion was: signs that refer to an object only acquire their meaning in the context of "conceivable" actions that we can associate with the signs in question.

Peirce's methodology is immediately apparent for terms like "tensile" or "fragile". Such terms presuppose a causal knowledge, which in turn finds its expression in exemplary actions: "If a rubber band is pulled at both ends, then it becomes longer", or "If a glass is thrown on the floor, then it breaks." Accordingly, our terms do not refer to any absolutely understood objects of the outside world, but only to further terms which in turn refer to contexts of action with similar effects. In other words, there is no direct correspondence between sign and object. Instead, there is a connection between the two only in the context of conceivable actions that adopt the role of the interpretant or "signifier" (for details see Sect. 1.6).

According to Peirce, the interpretation of a sign, i.e., the assignment of its meaning, always presupposes a context of possible action. This context is thus the actual interpretant of a sign. This idea led Peirce to develop the model of the "semiotic triangle", which consists of three interrelated elements: the "sign", the "object" which is represented by the sign and the "interpretant" which establishes the sign/object relationship. In Peirce's semiotics, these three elements form the basic scheme of cognition.

If, however, as Peirce assumes, signs and objects are always assigned by an interpretant, then the correspondence theory of truth loses its foundation.

Thus the certainty of cognition is not based on the direct accordance of sign and object, but rather upon their pragmatic assignment by the interpretant.

Thus, within the framework of pragmatism, the truth-problem now becomes a question of practical relevance directed at the interpretant of signs. In the idealized cognition model of pragmatism, the final instance which decides about this is the (in principle infinitely large) community of all rationally acting subjects. This community chooses whether, and to what extent, their convictions are to be considered true in solving real-world problems because they lead to successful actions. The model of truth derived from this is referred to as the "consensus theory" of truth.

Peirce himself regarded pragmatism primarily as a methodology by which the meaning of ambiguous words and abstract terms can be clarified. Others finally developed from Peirce's pragmatism a popular truth theory, which reduces the truth value of human knowledge exclusively to its usefulness and practical success (Sect. 2.9). However, a concept of truth that focusses so one-sidedly on human action can at best prove the value of practical truths. However, it will hardly be able to advance to the roots of our world-understanding.

2.2 The Search for True Being

It can be assumed that the question of truth arose for the first time when humans detached themselves from traditional mythical ideas and began to search for natural explanations of reality. For example, we know that early Greek philosophy searched for the truth behind physical reality—after, beyond or behind all being, becoming and passing (Greek: *tà metà tà physikà*).

However, one has to be aware that of the early phase of Greek philosophy only text fragments, indirect quotations, summaries by philosophers in antiquity and the like are available. The earliest largely authentic texts known today are those of Plato and Aristotle from the fourth century BC, even though these texts are also just copies made—at best—more than a thousand years later. Nevertheless, despite these limitations, some general features of the philosophical thinking of that time can be reliably reconstructed from the fragmentary sources.

Metaphysics reached its first peak between the sixth and fourth centuries BC. This period of philosophy is generally referred to as "pre-Socratic". Against the background of the distinction between adequate and inadequate knowledge, the pre-Socratic thinkers were already searching for criteria by which true and enduring knowledge can be distinguished from the apparent knowledge that tangible reality seems to present to us.

First and foremost, Parmenides of Elea and his pupils developed a doctrine of true being, which influenced the direction of thinking in antiquity. In his famous didactic poem *On Nature*, Parmenides describes his encounter with the goddess Dike, who allegedly revealed to him two paths of cognition. The first path led through the world of appearances. However, the cognition gained here would always remain hypothetical and illusory. It would be nothing better than the cognition of the ordinary mortal, which consists of mere assumptions and arbitrary opinions. Behind the appearing world—so the divine revelation—there was yet another world, to which only the second path of cognition could lead. This world was the world of true being, free of any deceptions. Only cognition of this world could claim to be valid. Since Parmenides pretended to invoke disclosure by the goddess Dike, his doctrine was obviously meant to appear to his public as objective truth and not merely as Parmenides' own subjective opinion.

Parmenides' didactic poem deals with two kinds of knowledge, which we nowadays denote as "empirical" and "metaphysical" knowledge. Empirical knowledge arises from experience and has, owing to the continual change in experience, always a provisional character. Metaphysical knowledge, on the other hand, is independent of any experience. It refers, as Parmenides assumed, to something that is behind appearances and that exists in itself. The metaphysical world is the world of true being. Any knowledge of this world is unchangeable and absolute.

At the same time, Parmenides pursued the idea that metaphysical knowledge can only be based on logically true propositions such as "being is" or "non-being is not". In the philosophical tradition, such sentences are referred to as "analytical" sentences or judgments. They have the characteristic of being independent of experience and of being certain in themselves. They share this characteristic with other tautological sentences like "a rainy day is a wet day". Tautological sentences do not broaden our knowledge, because their truth or falsehood is already determined by the mere inspection of the concepts referred to within them.

According to Parmenides, metaphysical judgments about reality are always right, because statements such as "being is", cannot be negated without contradiction. However, the situation is different for empirical judgments, i.e. statements that refer to individual cases of tangible reality. These are never absolutely true or false; instead, they can, at any time, be negated in a meaningful way. For example, the statement "it is raining", as well as its negation "it is not raining", can be right on a case-by-case basis in the one form or the other.

Statements that are certain in themselves or which proceed from logical inference are termed statements *a priori* ("from that which is before"). They have a character completely different from that of statements *a posteriori* ("from that which is after"). which are based on experience and observation. According to Parmenides, only judgments which are based on logical inference are unrestrictedly valid and immune to all conceivable experience-based refutations or augmentations. In contrast, empirical judgments about reality always remain doubtful and hypothetical.

A doctrine of knowledge that is cut off from any experience must inevitably come into conflict with reality. And indeed, one finds numerous examples of this in the philosophy of the Eleatics (see below). Thus, the analytical determination of "being" as something that "is" does not allow any other temporal determinations of being such as "being was" or "being will be". Consequently, Parmenides saw himself compelled to deny any change in reality and to regard it as a mere illusion of human perception, while the "true" world rests upon eternal and unchangeable being.

The principles of Parmenidean ontology are only correct because of their tautological wording. Since they have no factual content, they apply to all possible worlds. Nevertheless, from their ontological doctrine, the Eleatics drew far-reaching conclusions about the real character of the world. They believed, for example, that they could deduce from the phrase "non-being is not" the claim that there is no empty space. However, if there is no empty space ("that which is not") that can intersperse being ("that which is"), then there can only be a single, coherent beingness that has neither inner nor outer boundaries. From this, in turn, the Eleatics concluded that being has not arisen and is not transient; that it is unchangeable and indivisible.

To us today, such an interpretation of reality may seem abstruse. However, the peculiar world-view of Parmenides is the consequence of the fact that the world of sensory experience did not play any part in the ontology of the Eleatics. On the contrary: all changes of reality were considered by the Eleatics to be mere illusions and deceptions. Instead, the Eleatics believed that they could build an adequate knowledge of reality solely from logically correct sentences—a knowledge that, in turn, led inevitably to the doctrine of the immutability of true being.

The ontology of the Eleatics finally culminated in the famous brain-teasers of Zenon of Elea, who, with the help of sophisticated paradoxes, tried to persuade his contemporaries of the existence of an unchangeable world. Zenon was able to develop his amazing proofs because he played a tricky game with the concept of infinity (Sect. 5.1). In fact, in the ontological system of Parmenides, everything both possible and impossible could be explained, since his world model was based exclusively on tautological statements, which by their very nature are completely decoupled from experience. As the philosopher Karl R. Popper pointed out, all philosophical doctrines that begin with a tautological premise about being are "void", because nothing of interest can be deduced from them [7].

The doctrine of immutable being was presumably a counter-reaction to a philosophical thought attributed to Heraclitus, who is said to have claimed that all things are subject to constant flux. If, however, everything that exists is in a state of flow and changes all the time, how can one even speak of the existence of continuous things at all? With his flux doctrine, Heraclitus had questioned not only the consistency of being in the unstoppable change of its properties but also the very identity of things.

The philosophical questions which result from the flux doctrine were addressed by Plato in his famous dialog *Cratylus*. At the center of this dialog is the Greek philosopher of the same name, who was not only Plato's teacher but presumably also a disciple of Heraclitus. In Plato's dialog, Cratylus reflects, among other things, on the question of how one can talk meaningfully about the persistence of an object if one takes Heraclitus' flux doctrine seriously. For, as Cratylus argues, it is impossible to descend into the same river twice. One can speak of the same river in so far as it retains its shape within certain periods—provided it does not alter its course substantially. On the other hand, however, the positions of all water particles change every moment, so that the river never remains the same. Heraclitus himself is said to have claimed that "Upon those who step into the same rivers, different and ever different waters flow down" [8, fragment 22b12].

The philosophy of Heraclitus is one of the essential sources of the relativistic doctrine of cognition. If the attributes or properties of things are continually changing and we can no longer meaningfully speak of the identity of things, then the diversity of appearances contains no invariant reference point that could provide the basis for a consistent understanding of reality. Instead, any knowledge is relative to some other knowledge, whereby the chain of relativeness would continue indefinitely (Sect. 1.8). In the final consequence, nothing is positively determinable—a thought that already comes close to a nihilistic world-view. Although Heraclitus himself may not have drawn this radical conclusion, it is nevertheless a potential interpretation of the flux doctrine.

In antiquity, on the other hand, it was still assumed that there must be something objectively persistent in the flow of time. Aristotle called this something the substance of things. However, the alternative notion—that the identity of things is perhaps only a property that the thinking subject assigns to them—was still far from the mind of the philosophers of antiquity. In modern philosophy, Nietzsche finally pushed the flux doctrine to its limits when he made the paradoxical assertion that it is precisely in the flow of things, and thus in the non-identity of things, that their identity exists. For Nietzsche, therefore, the idea of "true being" was nothing more than an empty fiction: "The 'apparent world'", he stated, "is the only one; the 'true world' is just a *lie* that has been added." [9, p. 75; author's transl.]

From the opposed teachings of Heraclitus and Parmenides, the great questions of ancient philosophy emerged: What is the essence of being? How is the identity of a thing guaranteed in the light of its perpetual changing? Is there a timeless knowledge of the world in the changing circumstances of reality? What are the ways of recognizing the world, and what leads to valid cognition? Does the world disintegrate into an ideal world of absolute truth and a world of provisional knowledge, or must we seek the ideal world in the apparent world itself?

The various philosophical currents of the pre-Socratic era demonstrate how, time and again, fresh attempts have been made to combine the advantages of the Parmenidean ontology with those of the Heraclitan flux doctrine, to overcome the apparent weaknesses of each tradition. This era of philosophical thinking shows clearly that the emergence of new ideas follows a mechanism that is similar to the biological mechanism of genetic recombination. In the same way, as genes are recombined to yield new genetic entities, novel ideas emerge by recombination of a set of basic ideas that appear as the *a priori* of creative and progressive thinking (Sect. 3.2). We will show in the following paragraphs that the main upheavals in the history of philosophy can be traced back to such a mechanism of recombination. In a figurative sense, we could even speak of "a periodic system of ideas" from which the whole richness of philosophical thinking has emerged in a manner similar to that in which chemical substances originate by the combination of chemical elements.

An example of how the combination of ideas that, taken separately, are mutually incompatible can lead to new and fruitful ideas, is the development of the concept of elements and atoms in antiquity. As outlined above, two opposed concepts dominated thinking in the early phase of philosophy: the idea of unchangeable being and the idea of steadily changing things, expressed by the different ontological viewpoints of Parmenides and Heraclitus. These contradictory positions posed the question of how, in an unchangeable world of eternal being, the becoming and decaying of things could be understood at all. Parmenides circumvented this question because he denied *a priori* any change in reality and attributed the idea of change to the mere fleeting appearances of a deceptive world. However, such an interpretation, which results in total timelessness of all being, loses all reference to reality. Conversely, Heraclitus asked the tricky question of how something unchangeable can be conceived of when everything is changing at every moment.

Succeeding generations of philosophers recognized the deficiency of Parmenidean ontology. Nevertheless, they did not want to give up the idea of unchangeable being. Instead, they tried to harmonize the different views of Parmenides and Heraclitus by including in the Parmenidean ontology the apparent changes of reality as an objective characteristic of all things.

Empedocles, for example, developed a philosophical doctrine in which he tried to combine the actual changes of reality with the assumption of an unchangeable being. On the one hand, Empedocles adhered to the idea of Parmenides, according to which there is no emergence and no decay in an absolute sense. On the other, he considered the changes of observable things by interpreting these changes as the association and segregation of ultimate elements of a primaeval matter—elements that themselves can neither originate nor decay. Empedocles denoted these elements as "roots".

Moreover, Empedocles assumed all things to be aggregates of the four ultimate elements, which he associated with "Earth", "Air", "Fire" and "Water". These essential ingredients, he believed, enter into things in various proportions and thereby determine the things' observable properties (Fig. 2.1). Empedocles himself did not discuss the question of whether these are ultimate, indivisible parts of matter. It was only later on that this question found an answer in the doctrine of atoms developed by Leucippus and Democritus.

Thus, already in early antiquity it had been recognized that in the ontology of Parmenides the existence of empty space could only be denied meaningfully if one tacitly associated with the notion of "being" the trait of corporality. However, if one distances oneself from this understanding of being and admits that something non-corporal may also exist, then the existence of the empty space can well be substantiated without invoking a contradiction to the proposition "non-being is not".

In that way, the Greek atomists developed the idea of atoms as the indivisible units of everything that exists. The atoms are separated from one another by empty space. They can neither come into being nor pass away, but their movements are the basis of explanation for all changes in reality. Thus, the intrinsic progress of human knowledge about the nature of things

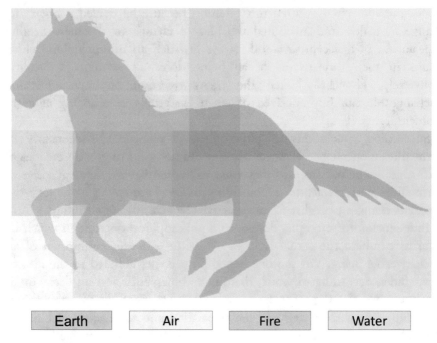

| Earth | Air | Fire | Water |

Fig. 2.1 The antique doctrine of elements. In the fifth century B.C., Empedocles developed his doctrine of ultimate elements by combining two conflicting ideas concerning the essence of being. Starting from the theoretical viewpoint that the emergence of being in an absolute sense is impossible, Empedocles understood the change of observable things as a relative emergence (and passing away). Thus, he assumed that things originate by the recombination of four ultimate entities which, for their part, neither originate nor decay. The term "ultimate element" is a theoretical one. In contrast to the things of the everyday world, the ultimate elements cannot be observed. Nevertheless, Empedocles associated these elements with earth, air, fire and water. He furthermore believed that the elements enter into things—into a horse, for example—in different mixing ratios and thereby lend substance to the observable differences in the qualities of the things

consisted in the combination of two initially irreconcilable world-views, namely, the metaphysical dogma of the immutability of true being and the irrefutable experience of a permanently changing reality.

The idea that all matter is built up of unchangeable atoms has been taken up again in modern times. Moreover, the development of the concept of atoms over its two thousand years of history shows how scientific concepts become refined, step by step, through the interplay of term and theory formation. For example, the older atomists, such as Democritus and Leucippus, did not yet have an abstract concept of the atom. Instead, they resorted to visible images by describing atomic particles as "spherical," "angular" and the like. Only modern physics has succeeded in developing a satisfactory picture of the atom. However, this is so hard to visualize that it can only be adequately represented in the abstract language of mathematics.

Surprisingly, even in Greek antiquity, there was already an approach to describe atoms mathematically. In his dialog *Timaeus*, which is devoted to the philosophy of Nature, Plato had postulated, that the four elements earth, air, fire and water are composed of regular bodies: cube, octahedron, tetrahedron and icosahedron (Fig. 2.2). He furthermore suspected that these bodies consist of geometric archetypes that had the shape of triangles. Moreover, Plato developed the idea that the four elements can be transmuted from one to another, with each of the elementary triangles capable of rearranging to give another element, so that air, for example, could be converted into water.

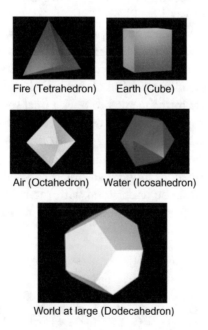

Fig. 2.2 The platonic solids. Plato postulated that the four elements earth, fire, water and air and the world at large are composed of regular bodies that possess the geometric form of the tetrahedron (four faces), the cube (six faces), the octahedron (eight faces) and the icosahedron (twenty faces). The world at large, the celestial aether, Plato associated with the dodecahedron (twelve faces) believing that each side corresponds to one of the twelve stellar constellations. Furthermore, Plato suggested that these bodies are composed of archetypes which have the shape of triangles

With his geometric interpretation of the elements, Plato had already given the older models of Empedocles, Leucippus and Democritus a formal description. One can also assume that Plato's mathematical ideas were influenced by Pythagoras and his school, even though Plato and his successors reinterpreted most of the Pythagorean ideas.

It cannot be overlooked that Plato's natural philosophy combined traditional philosophical ideas and new thoughts. Thus, for example, the Pythagoreans had already come up with three of the five regular bodies that Plato later placed at the center of his natural philosophy: the tetrahedron, the cube and the dodecahedron. Moreover, Pythagorean philosophers had developed a geometric theory of (natural) numbers, which they associated with symbolic and philosophical issues (Fig. 2.3). It is possible that the Pythagoreans understood numbers as the primal principle of the world. According to Aristotle, the Pythagoreans even ascribed a material property to numbers. Be that as it may, the "number-atomism" of the Pythagoreans already points in the direction of a formal understanding of Nature—a view which was also appropriated by Plato.

For Plato the formal understanding of Nature must have been attractive from two points of view: On the one hand, it seemed to prove the presence of reason in the world and thereby to confirm his philosophical doctrine of ideas. On the other, the interpretation of material changes as a rearrangement of primordial geometric forms seemed to offer a final solution of the problem of becoming and decaying that had been discussed for such a long time among the pre-Socratic philosophers.

From today's perspective, the atomism of ancient times seems to have anticipated a central aspect of modern atomic theory, according to which the essence of matter can only be understood with the help of formal structures. Werner Heisenberg, for example, went so far as to see in modern atomic

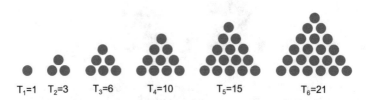

$T_1 = 1 \quad T_2 = 3 \quad T_3 = 6 \quad T_4 = 10 \quad T_5 = 15 \quad T_6 = 21$

Fig. 2.3 Pythagorean number theory. Within the framework of the Pythagorean theory of numbers, the (natural) numbers are represented graphically by dots. The triangular number T_n, for example, is the number of dots in a triangular arrangement with n dots on a side. Shown here are the first six triangular numbers. Other forms of representation are squares, rectangles and pentagons

physics the completion of the path outlined by Plato when he stated that the ultimate root of phenomena is not matter, but rather mathematical law, symmetry and mathematical form [10, p. 22].

The fact that eminent physicists such as Max Planck and Werner Heisenberg felt so strongly attracted to Plato's natural philosophy must be surprising at first glance, for Plato and Parmenides both took the idea of an absolute and timeless reality underlying appearances as the starting point of their philosophy. However, it is precisely quantum physics that teaches us that the observer cannot be taken out of the physical description of the world and that knowledge in physics must therefore always be considered in relation to the observing subject (Sect. 1.10). Yet, the belief that behind reality an ideal world is hidden, recognizable only its relative relations, seems to exert a powerful influence on many physicists. One can only speculate about the reasons for this, unless—as was the case with Planck—religious motives are recognizable in the background. Maybe Plato's philosophy falls on fertile ground among mathematicians and physicists because it claims that mathematical objects and structures have their own existence in an ideal world, an existence that is independent of human thinking.

2.3 One or Two Worlds?

Like the pre-Socratic philosophers, Plato also dealt with the fundamental question of how to speak in a meaningful way of a true, unchangeable being in a world of continual change. However, Plato did not try to resolve this contradiction; instead, he assumed the very existence of two separate worlds from the outset: the world of appearances, which is accessible to experience, and the world of immutable being, which lies beyond time and space. The first world is the world of becoming and decaying. It embraces things, in their particular form, and their changing properties, as we perceive them by our senses. The second world, in contrast, is the world of immutable being, the world of "forms" or "ideas", and it is, according to Plato, accessible exclusively to reasoned thinking.

Accordingly, true cognition manifests itself in knowledge of the eternal, changeless ideas. However, how do we acquire such knowledge? Plato's answer to this question is that the human being has the unique ability to recognize ideas intuitively and that, in this way, he can attain knowledge of the truth. Furthermore, Plato believed that things in the world of appearances participate in ideas, in that they "imitate" the ideas. However, as mutable objects, things can never perfectly represent the ideal. This thought can be illustrated by the case of geometric figures. The circle, for example, is an idea to which a circle we draw may approximate very closely, but which can never be represented in its ideal form.

Plato's doctrine of ideas is a further example that shows how new philosophical concepts are created by combining existing ones. In this case, too, the world-views of Parmenides and Heraclitus were combined, but without melding them together. This approach inevitably leads to a two-world doctrine in which the ideal world of immutable being, truth and normativity on the one hand, and the physical world of the appearances, change and relativity, on the other hand, are opposed to each other. Last but not least, the concept of ideas shows that Plato again takes up the thought of the Pythagorean school according to which the essence of things is expressed in geometric and arithmetic form.

The epistemological position, founded by Plato's philosophy, is that of "ontological idealism". It should explain why the things of the experiential world coincide with those of the ideal world. If one follows Plato, the consistency is based on the fact that the things of the experiential world participate in the ideas and imitate the patterns given by ideas. Already in antiquity, it became clear that Plato's two-world doctrine endangers the unity of knowledge. This is shown by, among other things, the discussion on the relationship between the general and the particular, as initiated by Plato's student Aristotle. For Plato, true being was the general principle that determines the essence of the things experienced in their specific manifestations. However, since the general (as Plato believed) belongs to the realm of ideas and thus to a higher level of reality, the general and the particular seemed to be the expression of two separate spheres of existence. Aristotle, on the other hand, did not sympathize with this thought. He believed, in contrast to Plato, that the general does not transcend the realm of experiential things, but rather that it must be sought in the things themselves.

In the Renaissance, when efforts were made to revive the cultural achievements of Greek and Roman antiquity, the different views of Plato and Aristotle were impressively staged by Raphael (Fig. 2.4). His fresco *The School of Athens* shows Plato and Aristotle in discussion with their students. Plato points upwards with his right hand toward the heaven of ideas as a symbol of the absolute and the general. Aristotle, in contrast, points downwards and thereby refers symbolically to the earthly world of appearances. The realist Aristotle here faces the idealist Plato.

However, the differences between Aristotle and Plato should not be exaggerated. It would be wrong to present Aristotle as a pure empiricist,

Fig. 2.4 In the fresco *The School of Athens*, painted at the beginning of the sixteenth century, Raphael impressively staged the factional dispute between the Platonic and the Aristotelian views of the world. Plato and Aristotle, surrounded by their students, demonstrate by gestures their different positions. Plato points upwards to the world of ideas, which he took to be autonomous entities that exist beyond tangible reality. He considered ideas to be the true essence of all being. Aristotle, in contrast, points downwards to the world of things that are perceivable by our senses. He sought the essence of things in the things themselves. [Details from the fresco at the Raphael Rooms, Apostolic Palace, Vatican City]

devoted solely to the experiential world. Although Aristotle described Plato's references to archetypes as "empty words" and "poetic metaphors", he nonetheless adhered to certain idealistic concepts of his teacher Plato. This included, for example, the teleological world-view, which is oriented toward thinking for purpose. Aristotle was also a Platonist inasfar as, for him, true cognition consisted in the comprehension of general facts. Unlike Plato, however, Aristotle sought the aspects of the general in the things themselves.

According to Aristotle, general cognitions result from abstraction and idealization of concrete things. For this purpose, he introduced specific categories such as substance, quantity, quality, relation and the like. With the thesis that there could also be an authentic knowledge attainable by experience, Aristotle adopted a position fundamentally different from that of Plato. For Aristotle, Plato's ideas were nothing more than general concepts under which things can be subsumed. Here, the methods of modern science, based as they are on experience, observation and generalization, are already heralded.

Nevertheless, the ideal of gaining knowledge by pure reason has survived until modern times. This is due not least to the Neoplatonism of late antiquity, which played a significant role in shaping the foundations of the Christian Middle Ages. More remarkable is the fact that the ideal of pure and perfect knowledge was already challenged in the pre-Socratic epoch with good arguments. At that time, criticism came above all from the philosophical school of the "sophists" in the fifth and fourth centuries BC. The sophists were regarded as particularly eloquent, and they used the instrument of rhetoric skillfully to disseminate their skeptical positions.

In addition to Protagoras and Gorgias, the two leading sophists, Cratylus too must be numbered among the sophists. In late antiquity, this critical philosophy developed into the so-called "skepticism", which was not only critical of religious and transcendent truths but also went so far as to deny the possibility of finding the truth at all.

The sophists had already sharply attacked the doctrine of Parmenides, which they thought to be untenable. Like Gorgias, they criticized not only the logic according to which the Eleates developed their ideas of the true being, but they also doubted that there was anything such as being at all. Protagoras, on the other hand, criticized the opinion of the Eleatics according to which everyday experience does not allow any insight into true being. He argued that the ordinary mind could very well open up to man the possibility of adequate knowledge, since the question of being could not never be answered independently of the issue of how we humans experience being. Consequently, Protagoras dissociated himself from the teaching of true being with the famous words: "Of all things the measure is Man, of the things that are, that they are, and of the things that are not, that they are not" [8, Fragment 80b1]; this is usually shortened to the pithy "Man is the measure of all things."

This phrase has entered the history of philosophy as the so-called *homo-mensura* sentence or dictum. Since Plato, this sentence has been interpreted as the expression of a relativistic position of knowledge, claiming that there is no universal truth. Instead, things are to be taken to be true as they appear to the judging person, as each case arises. Following the sceptic Sextus Empiricus, the *homo-mensura* dictum could even be interpreted in the sense of a bare subjectivism. In this case, every phenomenon could be considered by each human as being true. Which of the two possible interpretations Protagoras may have favored can no longer be reconstructed reliably. However, in any case, the *homo-mensura* dictum emphasizes the fundamental dependence of knowledge upon man and his ways of acting.

The sceptics of antiquity did not only deny the existence of an absolute truth independent of man; they also criticized radically the traditional values of man, his religious and political views. They thus laid the foundations for a practical and epistemological relativism without which human knowledge would hardly have been able to develop effectively. However, all the objections of the sceptics, no matter how well-founded they may have been, were unable to shake the belief in the possibility of absolute knowledge. The vision of penetrating the foundations of reality and recognizing in it its eternally constant, ideal structures seems to be a central need of human thought that no criticism of knowledge can eradicate.

2.4 Empiricism Against Rationalism

According to Parmenides, true knowledge is only possible on the basis of unequivocal propositions. This idea was taken up again in modern times by Descartes. The latter also pursued the goal of using pure thought, trusting reason alone, to find an approach to reality that is free from all kinds of illusions and deceptions. As a core model for the rational reconstruction of reality, Descartes used a procedure that was oriented toward the method of geometry as developed in antiquity by the Greek mathematician Euclid. In his major work *Elements*, which consists of thirteen books, Euclid had summarized the entire mathematical knowledge of his time and presented it systematically. In doing so, he was able to derive mathematics from a few, directly plausible principles, the so-called axioms, according to strict rules.

The axiomatic-deductive method of Euclidean geometry became very popular in the seventeenth century. It seemed to be the royal road leading to safe, complete, and universally valid knowledge. Descartes shared this opinion. He described his expectations of the "geometric" method in downright euphoric words: "The long chains of reasonings, every one simple and easy, which geometers habitually employ to reach their most difficult proofs had given me cause to suppose that all those things which fall within the domain of human understanding follow on from each other in the same way, and that as long as one stops oneself taking anything to be true that is not true and sticks to the right order so as to deduce one thing from another, there can be nothing so remote that one cannot eventually reach it, nor so hidden that one cannot discover it." [11, p. 17 f.]

In the seventeenth century, the idea of reconstructing reality by the axiomatic-deductive method was the source of the great rationalistic systems of knowledge. Their proponents were all convinced that one could derive systematically and in a logically consistent way, from only a few self-evident principles, a comprehensive understanding of reality. The rationalist method of gaining knowledge by type of geometry (*more geometrico*) seemed able to claim the same evidence and certainty that its mathematical paragon could. However, the basic principles had to be evident, i.e., beyond any doubt. Only that what is uncontested, Descartes claimed, could be the basis of well-founded knowledge.

Systematic doubt thus became the outstanding instrument of the Descartes' method of cognition. Accordingly, Descartes considered all judgments to be fundamentally dubious—with one exception. This was the assertion: I think, therefore I exist (*cogito ergo sum*). Obviously, the fact that I exist while I am doubting cannot be questioned, as I could not doubt if I did not exist—or could I? Only the ability to doubt seems to give us that certainty of existence which reveals us to be beings gifted with reason. However, uncompromising doubt must have appeared dubious already to Descartes himself, as he finally took refuge in the idea that God would not deceive us when we look at His work of creation. This meant nothing more or less than that Descartes tried to explain the reliability of our perceptions and ideas by appeal to the truthfulness of God.

In fact, a life without certain convictions that are, at least initially, excluded from doubt, would be largely unoriented. It is evident that human life must be borne along by a fundamental existential certainty if it is not to become intolerable, or, as Carl Friedrich von Weizsäcker put it: "We, who cannot live together without a certain minimum of confidence in the world and in each other, do not live in absolute skepticism. I daresay that anyone who is alive does not doubt absolutely. Absolute doubt is absolute despair." [12, p. 176]. This may well be the reason why many people hold fast to the conviction that there must be an absolute reference point of human knowledge, one that must not be questioned.

Baruch de Spinoza, who, like Descartes, was one of the outstanding thinkers of the seventeenth century, believed that existential certainty consisted in the consubstantiality of mind and matter, which he considered respectively as "thinking" and "extensive" substance. In contrast to Descartes, who interpreted the *res cogitans* and the *res extensa* as different substances, Spinoza considered both to be attributes of one and the same substance, namely of the Divine or of Nature. For Spinoza, God and Nature were necessarily the same, since everything that exists follows inevitably from the same indivisible and infinite substance. Since nothing could exist in logical independence of this one substance, there was, according to Spinoza, neither any free will in the mental world nor anything random in Nature.

According to the doctrine of Spinoza, no phenomenon can exist on its own. Instead, everything is determined by the logic of the one substance, which is the logic of the Divine. Therefore, Spinoza assumed knowledge of the world to be synonymous with reconstruction of the divine logic of being. As a result, his epistemology inevitably focused on the ideal of the geometric method as a paradigm for logical conclusions. Only in this way, Spinoza believed, could ontological truth be reconstructed with strict consistency. In the Latin title of his major work *Ethica ordine, geometrico demonstrata*, published posthumously in 1677, Spinoza even expressly points out that his representation of epistemology follows the geometric method.

The ideal of antiquity, according to which pure reasoning without recourse to experience makes possible an absolute knowledge about reality, dominated the philosophical doctrine of cognition until well into the nineteenth century. However, all attempts to gain real understanding of the world by pure thought have an inherent weakness, from which the philosophy of Parmenides already suffered: no knowledge of the experiential world can be gained from logically correct sentences alone. With mere definitions of terms and logical operations, no references to reality can be established. Without taking experience into account, every cognition inevitably remains empty.

In the seventeenth and eighteenth centuries, a critical countermovement to rationalism evolved, which shifted the focus of its doctrine of cognition to experience and common sense. This philosophy of knowledge, known as "empiricism," took up critical arguments that had already been prepared in antiquity by the Sophists and the Skeptics. The issue of empiricism was not some speculative reality hidden behind appearances, but the appearance of reality itself and the sensory experience of it. Accordingly, the empiricists did not even search for laws of reason or metaphysical reasons for being. Instead, they placed the cognitive capacities of the human mind at the center of their philosophy. The empiricists pursued, among other things, the goal of redefining the relationship between man and his existential values such as truth, freedom, moral action and the like.

The decisive impetus for the development of empiricism was given by John Locke in his treatise *An Essay Concerning Human Understanding* published in 1690. It is no coincidence that this publication came at the time of English Liberalism, in which the Glorious Revolution and the Bill of Rights of 1688/89 accompanied a turning away from the absolutism and Catholicism of James II. With his political and philosophical writings, Locke influenced liberalization to a great extent in his time. His idea of popular sovereignty formed not only the basis of English democracy but also exerted—spread by

Voltaire and Montesquieu on the European continent—a significant influence on European theories of state.

Locke's liberalism is also reflected in his doctrine of cognition, which must be seen as the prototype of an enlightened theory of knowledge, endeavouring to free itself from the constraints of the rationalist philosophy of knowledge. Because of his liberal attitude, Locke resisted any form of dogmatism. This mainly affected the traditional principles of metaphysics. He was only prepared to acknowledge three principles as certainties: the existence of God, his own existence, and the truth of mathematics. However, he rejected the doctrine of innate ideas, as propounded by the English Platonists of his time.

According to Locke, human consciousness at birth is like a blank sheet of paper, which only begins to fill up as the individual acquires outer and inner experiences. Outer experience, as Locke saw it, is based on our sense perceptions ("sensations"), and inner experience on our self-perception ("reflections"). "We have nothing in our minds," Locke claimed "that didn't come in one of these two ways." [13, II, chapter 1, §5] Thus, in contrast to the Platonist concept of ideas, Locke considered that ideas arise already through the act of perception. He furthermore believed that only the activity of the mind—remembering, comparing, distinguishing, connecting, abstracting, generalizing and the like—finally assembles the simple ideas into complex ones. Therefore, in Locke's strict empiricism, the mind plays only a subordinate part in the actual act of cognition, since its activity already presupposes the existence of simple ideas. The only thing left for the mind is the task of combining simple ideas into complex ones.

By tracing human understanding entirely to pure, direct perception, Locke's empiricism took an extreme counter-position to the rationalism which dominated the seventeenth century. However, his model of cognition also provided cause for criticism. For example, the perception of the directly given—as presupposed by empiricism—seems to be an idea that is just as unsustainable as the idea of an observation that is free of any theory. Instead, the question arises as to whether sensory perception does not also make use of rational tools by employing, for example, the capacity of the human mind for abstraction. Undoubtedly, this is a weak point in Locke's philosophy, and Gottfried Wilhelm Leibniz also drew attention to it when he countered the empiricist claim that "nothing is in the intellect that was not first in the senses" with the words "except the intellect itself" [14, Book II, p. 36]. This objection by Leibniz suggests that the directly given cannot be justified without contradiction: any theory of cognition that ignores this problem risks ending in naïve sensualism.

What about empiricism's claim to truth? The rationalists had no problem with this question, because all their findings were based on judgments that are either self-evident or depend on other evident judgments. Their insights seemed to be necessarily true because they could rely on logical truths. The starting point for empiricism was different. According to the empiric doctrine, all judgments result from experience. As experience-based judgments, they contain factual knowledge. Such statements, however, can never be self-evident or be derived from evident judgments, for facts are not necessarily as they seem, and, therefore, no experience-based statement can be right in the sense of "compellingly" true. If empirical judgments were already true from the outset, the notion of experience would lose its meaning. In the empirical model of cognition, the question of truth is thus reduced to the validity of "empirical" truth, for which, however, there is no timeless certainty.

At the same time, empirical epistemology abandoned the ideal of the axiomatic-deductive method. It was recognized that new empirical facts cannot be deduced from judgments that are not themselves self-evident. From empirical judgments which may change from time to time, anything arbitrary can be deduced. According to Locke, only mathematics fulfils the axiomatic ideal, since in mathematics, as he said, it is only the relation between mathematical statements that counts. In that case, it is irrelevant whether the content of a statement is true or not.

Locke assumed that our ideas are real and that they depict the outside world as it is. This conclusion necessarily suggests itself if all ideas, as Locke thought, are composed of simple elements, which in turn come from pure and direct perception. Moreover, since Locke started from the premise that the order of things was a steadfast order created by God, he concluded that the same things would always generate the same ideas in us. Thus, like Descartes, Locke trusted that the Creator would not deceive us when we observe his work. In this belief, ironically, a remnant of the dogmatic legacy emerges that Locke had actually intended to eradicate.

With his doctrine, Locke had wanted to stress the autonomy of the role of common sense for human understanding. His philosophy, however, was not as stringent as would have been desirable from the empiricist point of view. In particular, George Berkeley and David Hume felt compelled to reformulate and accentuate the significance of perception in the process of cognition. Berkeley, for example, argued from the fundamental principle "being is something which is perceived" that there are material things only insofar as they appear to us as a regulated and lawlike sequence of sensory complexes or ideas. He denied, however, that there exists a material body world outside the human mind. He even believed that one could verify his assertion in a logically stringent manner.

Hume, in turn, tried to justify with astute arguments the assertion that—except for direct observation, logic and mathematics—all human cognition is fundamentally uncertain. He shared Locke's view that complex ideas are composed of simple ones, and that these ultimately arise from impressions gained through perception. However, in contrast to Locke, who had considered that the relationships which human mind establishes between events perceived have an objective basis in the outside world, Hume believed that our ideas are merely the expression of certain habits of thought, which have an exclusively psychological basis.

Let us consider an example: The observation that the sun rises every morning is so familiar to us that we are convinced that the sun will rise tomorrow as well. Hume, however, challenged this belief. He asserted that there was no objective reason for this in the outside world at all. The only argument supporting our conjecture would be the fact that, until now, the sun has always risen every morning. According to Hume, the human mind tends to associate perceptions of the same kind and combine them into a purportedly regular phenomenon. This, in turn, invokes in us the impression of an objectively existing context of experience, although in fact it can only be explained psychologically.

According to Hume, causality is not an objective law of the outer world, but rather a regularity of the inner world. Mere association evokes the impression of causality, just as other forms of association evoke in us ideas of similarity, proportionality, equality and so forth. The certainty that we associate with our ideas, therefore, is nothing more than the power of habit—that is to say, sheer belief in the continuation of what we have repeatedly experienced in the past. In other words, the truth that pure experience seems to convey to us is nothing other than an imagined form of truth which our psyche presents to us. Thus, ultimately, Hume's empiricism could justify the truth content of knowledge just as little as the rationalists could explain the empirical content of their logically derived truths.

2.5 The Copernican Revolution of Thought

Given the seemingly irreconcilable contrasts between rationalism and empiricism, it was an outstanding philosophical achievement of Immanuel Kant to combine the two positions into a fruitful epistemology. Kant also examined critically the two forms of knowledge: *experience-independent* and *experience-based* knowledge. In contrast to his predecessors, however, he did not play off the two kinds of knowledge against each other, but instead tried to relate them to one another. Again, the philosophical progress that Kant achieved was only possible by creating a combination of two fundamental ideas that seem mutually exclusive. To this end, Kant first addressed the general question of whether and in what form a metaphysics that meets the claims of enlightened and critical thinking is even possible.

Since its appearance in Greek antiquity, metaphysics has seen itself as a doctrine of the fundamental principles that stand behind the world of phenomena and which thus elude sensory experience. Aristotle even regarded metaphysics as the "prime philosophy", i.e., as philosophy *par excellence*. However, in the course of its long history, metaphysics had led to crude and absurd speculations. The development of empiricism was ultimately the counter-reaction to a metaphysics that had eluded all critical control, but which at the same time asserted a dogmatic claim to truth. Locke, for example, who was clearly annoyed by some metaphysical speculations of Leibniz, wrote to a friend: "You and I have had enough of this kind of fiddling" (after [15, p. 555]). Consequently, Locke opposed the rationalists' view with a theory of cognition that was completely oriented toward the perception of the experiential world. However, in his one-sided fixation on sensory perception, empiricism seemed to be no less dogmatic than the metaphysics it rejected.

That was the background against which Kant began his investigation of the problem of knowledge. Kant likewise criticized the orthodox doctrine of metaphysics. However, in contrast to the empiricists, who outrightly rejected metaphysics, Kant held onto metaphysics, but he tried to put it onto a sound basis. Of course, Kant was aware that this required a radical reorientation of metaphysical thinking. Traditional metaphysics, he criticized, had always assumed that our knowledge depends on objects of cognition. Accordingly, metaphysics would have considered it as its primary task to reconstruct the knowledge of these objects as a knowledge *a priori*. Kant, however, argued that it could also be possible that our knowledge does not depend on the objects of cognition at all, but, conversely, that these could hinge upon our cognitive faculties, that "we can cognize of things *a priori* only what we ourselves put into them" [16, Bxviii].

This thought seems to cast an entirely new light on the problem of knowledge. In contrast to the rationalists, who were seeking *a priori* knowledge in the objects of cognition, Kant claimed that this knowledge can only be found in the properties of our cognitive apparatus. Kant was aware that this was a revolutionary approach. In the preface to the second edition of the

Critique of Pure Reason of 1787, he confidently compared his approach with that of Copernicus: "This would be just like the first thoughts of Copernicus, who, when he did not make good progress in the explanation of the celestial motions if he assumed that the entire celestial host revolves around the observer, tried to see if he might not have greater success if he made the observer revolve and left the stars at rest." [16, B/XVI]

Starting from this idea, Kant investigated the question of the extent to which metaphysical knowledge is possible. An essential preliminary stage of his exploration was to clarify the status of metaphysical statements or "judgments." To this end, Kant studied all conceivable forms of judgment in detail and classified them according to logical criteria. This analysis led him at first to distinguish between "analytical" and "synthetic" judgments. The difference between these forms of judgment can be illustrated with a simple example. Consider the statements: "A rainy day is a wet day" and "this rainy day is a cold day." The first one is a mere restatement, and is thus void: the characterization of a "rainy" day as a "wet" day is already contained in the understanding of "rainy". Kant denoted such tautological statements, which are self-referential and which do not carry any substantial information, as "analytical". Incidentally, these are the same judgments on which Parmenides had based his metaphysical theory of knowledge (Sect. 2.2).

For the second judgment ("this rainy day is a cold day"), the situation is different. In this case, the designation of a "rainy" day as a "cold" day provides additional information, which is by no means already included in the concept of a "rainy day": a "rainy" day can just as well be a "warm" day. Judgments of this kind, which widen our cognition, Kant denoted as "synthetic".

However, this classification does not yet cover all forms of judgment. It is further necessary to distinguish between judgments that are independent of experience (judgments *a priori*) and such that are dependent on experience (judgments *a posteriori*). If one takes these kinds of judgment into account, one can differentiate between four basic classes of judgments: "synthetic judgments *a priori*", "synthetic judgments *a posteriori*", "analytical judgments *a priori*", and "analytical judgments *a posteriori*" (Fig. 2.5). However, it is easy to see that analytical judgments *a posteriori* are impossible. By definition, analytical judgments are judgments that only explain the meaning of the terms they are built from. Such judgments do not depend on any experience or sensory impressions. They are, by nature, always judgments *a priori*.

From the above classification, it becomes evident that metaphysical judgments cannot belong to the class of analytical judgments. This is because analytical judgments do not convey any real knowledge, but merely a sham knowledge, as indeed is already evident in the teaching of Parmenides.

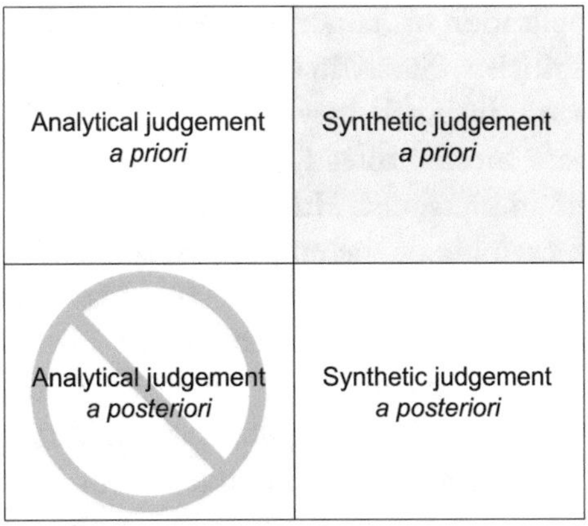

Analytical judgement *a priori*	Synthetic judgement *a priori*
Analytical judgement *a posteriori*	Synthetic judgement *a posteriori*

Fig. 2.5 Kant's classification of judgments. Investigating the foundations of human cognition, Kant came up with four basic types of judgments. Statements that are independent of any experience he denoted "judgments *a priori*". Statements based on experience he termed "judgments *a posteriori*". Moreover, Kant differentiated between statements which extend our knowledge ("synthetic judgments") and those which are only self-referential ("analytical judgments"). Actually, however, "analytical judgments *a posteriori*" cannot exist, because analytical judgments are by definition independent of any experience or sensory impressions. Kant concluded that genuine knowledge, one that is independent of any experience, could only consist of "synthetic judgments *a priori*". This consideration finally led him to the core question of his epistemology: How are synthetic judgments *a priori* possible? In contrast to the rationalistic philosophy, which sought a solution to this problem in the *a priori* structures of the world, Kant concluded that the roots of our knowledge are the *a priori* structures of our cognitive apparatus itself

Genuine knowledge, on the other hand, is only imparted to us in synthetic judgments. Therefore, Kant had inferred that meaningful statements of metaphysics, if they are to be valid over and above every experience, can only fall into the class of synthetic judgments *a priori*. This raised the question: How are such judgments possible? This problem, Kant replied, can only be clarified by a critical exploration of our faculty of judgment. Thus, ultimately, we are faced with the question: How do we arrive at judgments at all?

In this connection the Empiricists had referred to sensory perception. In their opinion, our judgments are due to the immediate perception of the world that appears to us. Kant, on the other hand, believed that we reach judgments by applying concepts and words or terms denoting them. The words describe phenomena. These, in turn, are subject of our intuition. Therefore, it looks as if concepts are formed by abstraction from our intuition. At the same time, this would mean that concepts are always given *a posteriori* and that our judgments and thus our general ideas ultimately emerge from the application of "concepts *a posteriori*". Precisely this had been the view of the empiricists. Since, in their opinion, all knowledge is derived from direct perception, they had inevitably to assume that concepts are also constructions that are formed directly from perception and with which we merely describe our impressions. Hume's interpretation of the concept of causality is a prime example of the empiric view of this issue.

However, Kant disagreed entirely on this crucial issue. He believed that the concept of causality can be applied to experience, but cannot be abstracted from intuition. For Kant, the concept of causality is instead a prerequisite for our ability to articulate empirical judgments at all. The idea of causality, he concluded, is not a concept *a posteriori*, but a concept *a priori*. One must necessarily presuppose causality for the cognitive process since otherwise, we would not be able to establish in the abundance of perceptions any context of experience.

On the one hand, *a priori* concepts are independent of experience. On the other hand, however, they make possible experiential knowledge. How does this fit together? What, then, do the concepts *a priori* refer to at all? Kant gives the surprising answer that these concepts describe appearances that are also part of our experience, but which we cannot perceive with our senses. Such *a priori* appearances are space, time and causality. Kant believed that he had a striking proof to hand for his thesis. He argued that we can very well imagine an empty space. On the other hand, we cannot imagine that there is no space at all. If space were a manifestation *a posteriori*, which we only abstract in retrospect from the set of our experiences, we must be able to imagine these experiences also without space. However, that is impossible. To illustrate this: We can readily imagine grey and non-grey elephants before our eyes. However, it is impossible to imagine elephants that are not spatial.

According to Kant, space, time and causality are *a priori* forms of intuition. They are necessary conditions for the possibility of experiential knowledge at all. The traditional view that "all knowledge starts from the object of knowledge" was thus restricted by Kant to the extent that the subject also contributes to the structures of the recognized object through *a priori* forms of intuition. Accordingly, Kant distinguished between two sources of knowledge: sensory perception and reason. Without sensory perception, he stated, no object would be given to us, and without reason no object could be thought of—or, to use Kant's own words: "Thoughts without content are empty, intuitions without concepts are blind." [16, B75] It is the power of judgment, Kant continues, that mediates between the two faculties of cognition. In all three components of the cognitive act Kant believed that he could discover experience-free elements, which he regarded as the *a priori* conditions of the possibility of cognition.

To sum up: According to Kant, objects of knowledge reach our consciousness only in the pre-designed form of our cognitive faculty. Things that lie outside our imagination, "things-in-themselves," are as such not recognizable. They affect our imagination, but they only appear in the form in which our cognitive faculty has processed them. Therefore, we see reality only through the glasses of our *a priori* faculty of cognition, without ever being able to discard these glasses.

Kant himself did not take a clear position on the question of whether things-in-themselves are to be regarded as real, objective things or merely as a logical necessity, as an explanatory reason for appearances. Sometimes he tended to a more realistic, then again to a more idealistic interpretation of the thing-in-itself.

We cannot have direct knowledge of things-in-themselves, because they always encounter us in the pre-structured form of the human mind. Consequently, all cognition is bound to the *a priori* conditions of our cognitive faculties and must, therefore, be related to these conditions. Or, to put it positively: Human knowledge can rely on aprioristic forms of cognition and finds in them timeless and unchanging access to a perception of reality that is free of deceptions.

2.6 Speculative Rank Growth

Kant had sent the cognitive problem off in an entirely new direction: Our cognitive faculty is not guided by the objects of cognition, but, conversely, the objects of cognition are determined by our cognitive faculty. Only the assumption that the subject itself contributes to the object of cognition, in Kant's view, allows us to speak meaningfully of *a priori* knowledge. However, despite the great appreciation and admiration that was bestowed upon Kant for his philosophy, there were also philosophers who accused him of having embarked on the right path but not having gone down it consistently. This criticism came mainly from the philosophers Johann Gottlieb Fichte, Friedrich Wilhelm Joseph Schelling and Georg Wilhelm Friedrich Hegel. They found fault above all with Kant's thought that our knowledge is affected by things-in-themselves and thus depends ultimately on the objects of cognition—a thought that seemed to be directed against the spirit of his own philosophy.

Fichte and his younger contemporaries Schelling and Hegel formed the famous philosophical triumvirate at the university of Jena where, at the beginning of the nineteenth century, they founded the influential philosophy of German Idealism. It is characteristic of these philosophers that they applied in every respect "absolute" standards to philosophical thinking. Not only did they lay claim to completing Kant's philosophy, but they also tried to substantiate systematically all knowledge that can be attained by man. Philosophy itself was regarded as a doctrine of absolute knowledge and thus as science *par excellence*.

An example of the idealistic understanding of philosophy is Fichte's *Wissenschaftslehre* ("Doctrine of Science") from 1794. The philosophical problems dealt with here, first by Fichte and later taken up and continued by Schelling and Hegel, all revolve around the central question of how philosophy can arrive at an absolute knowledge. Accordingly, the philosophical discourse of the day was dominated by the idea of the "absolute", on which all conceptual and theoretical issues were based.

The most important instrument of the philosophy of German Idealism is the sense-analysis of concepts. For example, to gain an idea of the absolute, all terminological determinations running counter to the idea of the absolute are excluded by a sense-analysis. The typical argumentation pattern runs as follows: Knowledge which is to have "absolute" validity must be valid for all time. There must be nothing that can question its claim to truth. In particular, such knowledge cannot be based on experience, because otherwise it could change in the context of progressive experience. Consequently, absolute knowledge must be an "unconditional" knowledge per se, i.e. knowledge which is conditioned by nothing else than itself. The unconditioned thus became the key concept of the philosophy of the absolute.

In fact, the demand for unconditionality seems to be indispensable, as all conditional knowledge inevitably changes when the conditions change under which this knowledge was gained. For this reason, Fichte rejected Kant's thought that our ideas are affected by things outside the subject, by the things-in-themselves. For if the external things and our knowledge of these things actually stood in a cause-and-effect relationship to one another, then this knowledge would only be relative, and not unconditional.

From this, Fichte believed the conclusion to be inevitable that all true knowledge emanates from the subject. Only a subject whose cognitive faculty is in this sense absolute, he argued, can be the source of true knowledge that in no way depends on any things-in-themselves. With this idea, the dogmatic claim of traditional metaphysics to truth was reintroduced into epistemology through the back door. This prompted Kant to make the comment that he considered Fichte's doctrine of science to be a completely unsustainable system and that he felt no desire to follow the metaphysics of Fichte [16, p. 370].

In fact, Fichte's philosophy must have seemed like a relapse into the old speculative metaphysics. Fichte was not only in search of absolute knowledge, but he also based his arguments on the traditional method of deductive cognition. Like Parmenides two thousand years earlier (Sect. 2.2), Fichte believed that he could deduce philosophical knowledge entirely from a single logical truth; this he considered to be the supreme principle of his doctrine of science.

To work out this principle, Fichte made use of the instrument of sense-analysis of terms or statements. At first, he looked for the simplest statement that is true in itself. He believed to have found it in the formal-logical identity "$A = A$". Since the proposition "$A = A$" is a logical truth, it cannot be doubted. Fichte, however, demanded even more from the supreme principle of science, namely that it is "absolutely" true. This means, in Fichte's understanding, that the supreme principle must be completely unconditioned. However, is this requirement really fulfilled by the identity relationship?

This question seems justified, in that statements are always statements by subjects. Thus, the statement "$A = A$" obviously presupposes a subject, which places A in an identity relationship to itself (i.e., to A). In this way, however, the formal-logical identity becomes an item which, contrary to the demand of Fichte's doctrine of science, no longer has the property of being unconditional—with one exception: the only content of A that gives the formal-logical identity the character of the unconditioned is the Ego or the I itself: "$A = I$". It is the statement of the absolute I which places itself in an identity relationship to itself: "I am I" or in brief "I am".

Fichte believed that he could derive from the abstract proposition "$A = A$" by a mere sense-analysis an absolute content, namely the statement "$I = I$". This seemed to confirm a central idea of his doctrine of science, according to which form and content of the supreme principle are mutually dependent. The next question is how the absolute I relates to the world of its cognitive objects. Fichte's answer can be summarized as follows: The objects of cognition arise by the self-limitation ("negation") of the autonomous I in that the (divisible) I again and again places a (divisible) non-I in a relationship over against itself. In this way, an increasingly fine network of borderlines arises between the I and the non-I that constitutes the detailed world of cognitive objects.

The loss of reality associated with this kind of philosophy seems to be unavoidable, as one encounters here the same problems that arose in the ontology of Parmenides. Here again, it becomes apparent that no empirical content can be derived from a logical truth alone. Above all, this philosophy does not answer the critical question of the truth of our ideas. On the contrary, the problem of reality is even trivialized. Since the world of the non-I, as Fichte assumed, emerges exclusively by the productive intuition of the I, the outside world must necessarily coincide with the inside world.

No wonder that Fichte's subjective idealism soon became the target of vicious criticism. Schelling, for example, who was initially an enthusiastic follower of Fichte, remarked ironically that the divine works of Plato, Sophocles and other great minds were actually his own works, as they—if one interprets subjective idealism consistently—were engendered by him through productive intuition [17, p. 639]. On the one hand, Schelling recognized Fichte's achievements in having restored the subject-object identity as a central issue of philosophy. On the other, however, he criticized him for considering the identity of subject and object as a mere subjective identity emerging from our consciousness. After extraction of all substance from the speculation, he argued, in the end nothing more is left of the identity principle than "empty chaff" [18, p. 396]".

In his *Ideen zu einer Philosophie der Natur* ("Ideas for a Philosophy of Nature"), published in 1797, Schelling attempted to correct this deficiency by first objectifying the subject-object identity and not, as Fichte had done, by regarding it as an identity proceeding exclusively from the subject. Moreover, according to Schelling the subject-object identity must be considered as absolute. This means that subject and object are not two separate entities that stand in an identity relationship one to another, but rather that the entire subjective *is* at the same time the entire objective, and the entire objective *is* at the same time the entire subjective. Only when subject and object unambiguously and reciprocally depict one another, one to one, was there—as Schelling believed—no inconsistency between the inner and the outer world.

Unlike Fichte, Schelling regarded the real world as more than just an epiphenomenon of the ideal world. Rather, he saw conceptual and material appearances as two manifestations of one and the same entity, and understood this as an absolute subject-object identity. At the same time, he realised that he had to pass beyond the concept of Fichte's "Doctrine of Science" and to regard the I as an all-embracing world concept, one that encompassed both the entire subjective and the entire objective. Schelling admittedly retained Fichte's idea that the absolute I engenders the world of its own objects in an act of free will ("freie Tathandlung"); however, at the same time Schelling interpreted the absolute I as being the highest level of existence of a self-creating Nature, which in the human mind is conscious of itself.

In this interpretation, Fichte's central statement "I = I" becomes the self-expression of an autonomous and unconditioned Nature: "I = Nature". Consequently, the acts of free will on the part of the absolute I are interpreted by Schelling as objective acts of creation of an all-encompassing activity of Nature, which are raised into the realm of human consciousness. Schelling summarises: "Nature is the visible mind, mind is invisible Nature." [18, p. 380; author's transl.] This is to be taken as meaning that the subject can regard itself in Nature as in a mirror. Nature is visible mind. Conversely, mind is invisible Nature, insofar as mind mirrors Nature at the highest level of its being. Thus, mind in Nature and Nature in mind can contemplate one another. According to this view of Nature, the actual task of natural philosophy is the reconstruction of Nature's self-construction.

The philosophy of Nature, which Schelling developed in critical demarcation from Fichte's philosophy of the subject, is based on the absolute identity of mind and Nature. Thus, if one knows the principles of reason, one also knows the corresponding principles of Nature and vice versa. For Schelling, these principles are the superordinate authority that decides how to order and interpret experience. Accordingly, empirical knowledge will only become reliable knowledge if its logical place in the system of natural philosophy is recognized. In this sense, he claims that natural philosophy is a "higher knowledge of Nature", a "new organ of contemplation and understanding Nature" [18, p. 394; author's transl.]. Furthermore, the principles of natural philosophy cannot be refuted by experience, but can only be confirmed by it. If Nature's principles could be refuted by experience, the principles of reason would be refuted as well. This, in turn, would reduce the possibility of cognition as such to absurdity.

In Schelling's philosophy of Nature, two basic thoughts of ancient metaphysics re-emerge. On the one hand, his philosophy is based on a supreme, unconditioned principle, from which he attempts to derive the entire system of knowledge of Nature. Here, he continues a tradition that leads from Parmenides to Descartes, Spinoza and Leibniz through to Fichte (Table 2.1). On the other hand, one also encounters in his philosophy the idea that the world of appearances, at which the perception and observation of Nature are directed, is merely the surface of an underlying "true" Nature. Schelling interprets this "true" Nature as an acting subject, which he also denotes as *natura naturans* (literally: growing Nature). This phrase is a metaphor taken

Table 2.1 Cognitive systems of rationalism. The rationalistic approach to the world, which goes back to the philosophy of Parmenides, experienced a boom in the seventeenth century, culminating in the great rationalist philosophical systems of Descartes, Spinoza and Leibniz. Rationalism sets out from a prime principle of human reason, considered to be the ultimate truth. Influenced in particular by Spinoza's monism, Fichte and Schelling developed a philosophy of identity; Fichte interpreted this as a subjective identity, whereas Schelling attempted to objectify the identity

Philosopher	Primary principle	Philosophical feature
Parmenides	Being is	Ontological truth
Descartes	*Cogito ergo sum*	Certainty of existence
Spinoza	Singular substance	Consubstantiality of mind and matter
Leibniz	Monad	Ultimate unit of existence in the universe
Fichte	I = I	Subjective subject-object identity
Schelling	I = Nature	Objective subject-object identity

from medieval philosophy, the so-called scholasticism. It describes Nature, in the sense of the Aristotelian notion of *poiesis* (Sect. 1.2), as being creative, as a nascent Nature, characterized by its unlimited productivity. According to Schelling, the task of natural philosophy is the discovery of the driving forces behind Nature's productivity. In contrast, empirical research, according to Schelling, is directed exclusively at the products of Nature. These characterize Nature as Object, as *natura naturata* (literally: grown Nature).

However, to maintain subject-object identity as the supreme principle of his philosophy, Schelling must avoid the disintegration of Nature into a Nature as Subject and a Nature as Object. For this purpose, he made use of an artifice. He assumed that Nature's productivity has not extinguished itself in its products, i.e. in Nature as Object, but only infinitely delayed. As in the philosophy of the Eleatics, the idea of infinity must serve to save the consistency of the cognitive model.

Let us single out, from the numerous ideas that Schelling developed in his natural philosophy, one thought that still plays a major part in the discussion about science and technology today. It is the idea of "wholeness", to which Schelling gave an idiosyncratic but effective historical meaning. Schelling developed his concept of wholeness from the principle of identity, which he understood as the "absolute" identity of Mind and Nature. In its absolute identity, the absolute, as Schelling put it, is "cloaked in itself". However, to become accessible to the act of cognition, the absolute must "expand" into the spheres of the ideal and the real. Only through this act of "divisiveness", Schelling believed, do Mind and Nature step out of their absolute identity and thereby recognize themselves as being an essential unity. By its externalization, however, the absolute must not lead out of the absolute. According to the determination of its essence, the absolute must remain identical with itself in all its absoluteness.

From this abstract figure of thought, Schelling believed he could derive the self-similar character of the world, or of Nature. "Every part of matter", he claimed, "must carry in itself an imprint of the whole universe." [19, p. 413; author's transl.] Following Schelling, this aspect is expressed in the self-similarity of Nature and the living organism. The organism epitomizes the essence of Nature, Nature the essence of the organism. Consequently, the All-Organism "Nature" and the living organism must be organized according to the same principles. This, in turn, means that the organismic understanding of Nature becomes the explanatory reason for natural phenomena, and in particular all physical and chemical processes. However, this idea turns the usual direction of explanation in natural sciences, which leads from simple to complex phenomena, more or less on its head.

Schelling's philosophy of Nature has not remained the only attempt to prescribe science's path to the knowledge of Nature. However, none of the subsequent designs for this even approached having the effect of Schelling's natural philosophy. With his speculative understanding of Nature, Schelling had initiated a countermovement to empirical research, which became particularly popular among the opponents of the mechanistic world-view. This was because Schelling's natural philosophy made the mechanistic understanding of Nature appear only as a perspective narrowing of an image of Nature, which, in reality, is organic. Thus, in the first half of the nineteenth century, a romantic understanding of Nature as an organic unity of man and Nature emerged, and this still plays a major part in the debate about our dealings with Nature today.

However, the organismic conception of Nature led straight into the mist of a romantic transfiguration of Nature. How vague the organismic way of thinking really was can be illustrated by its concept of wholeness. According to Schelling, the phenomena of Nature form an organic and thus irreducible whole, as Nature is the epitome of the absolute. However, the idea of wholeness understood in this way is anything but transparent. The term "wholeness" cannot even be explained in a meaningful way, since, in the end, wholeness can only be determined tautologically—namely, as that which is in its essence a whole.

Here, an inherent and inevitable defect in the philosophy of the absolute comes to light. If Nature, as Schelling believed, is the expression of the absolute, then the essence of the absolute is also reflected in Nature, for example, in the essence of its wholeness. However, the absolute stubbornly resists analysis. As the unconditioned per se, the absolute eludes any relationship of conditionality, not least that of reflection. There is no Archimedean point outside the absolute, one from which the absolute might be determined. The absolute as such cannot be determined. For precisely this reason, one repeatedly encounters in Schelling's philosophy empty wording according to which one is asked to think of the absolute, for the time being, as "pure absoluteness" [18, p. 386; author's transl.].

In the passages where Schelling still endeavours to define the absolute, his thoughts finally dissolve into poetry. The introduction to his *Ideas of a Philosophy of Nature*, in which he repeatedly attempts to express the inexpressible, is rich in samples of poetic word-creations and pictorial comparisons. However, they get lost (in the most real sense of the phrase) in nebulousness. One reads, for example, that the absolute "is enclosed and cloaked into itself", that it "is born out from the night of its being into the day". There, Schelling speaks of the "aether of absolute ideality" and the "enigma of Nature" [18, p. 387 ff.; author's transl.].

The poetic language that Schelling makes extensive use of is the inevitable accompaniment of a philosophy in which human thinking perpetually seeks to transcend itself. Only thinking about the absolute can be reflected upon, but not the absolute itself. Nevertheless, Schelling tried to circumvent this *aporia* by the concept of "intellectual intuition". This concept, already used by Fichte, should allow one to understand how one can gain access to the absolute: namely, by the act of introspective self-intuition. In contrast to Fichte, however, Schelling withdraws the subject from the contemplation, and thereby only considers the purely objective part of this act. In the end, intellectual intuition appears like an inwardly inverted Archimedean point, from out of which the absolute was supposed to be made comprehensible. In this way, a quasi-meditative, almost occult element crept into Schelling's natural philosophy and remained permanently and annoyingly stuck to it.

Nevertheless, despite its romantic superelevation by the organismic interpretation of Nature, Schelling's philosophy is, deep down, even more mechanistic than the mechanistic sciences that he often criticised so violently. Although Schelling attempted to replace the linear causality of the mechanistic view of Nature by a cyclic causality, all events of Nature were subjected by him to the almost mechanically operating logic of deductive philosophy. The fact that this philosophy marginalized the experience and discovery of Nature infuriated (for example) Goethe and turned him against the romantic philosophy of Nature.

Goethe had initially been sympathetic to the aims of these philosophers and supported them actively, but he now turned away, "shaking his head" about their "dark," "ambiguous," and "hollow" talk "in the manner of prophets" [20, p. 483 f.; author's transl.]. He even went so far as to see in their speculative philosophy a "grotesque appearance", something "highly strange and at once dangerous", because here "the formulae of pure and applied mathematics, of astronomy, of cosmology, of geology, of physics, of chemistry, of natural history, of morals, religion and mysticism [...] are all kneaded together into a mass of metaphysical speech", with the consequence "that they are putting the symbol, which indicates an approximation, in the place of the thing; that they turn an implied external relationship into an internal one. In this way, they do not depict things, but instead they lose themselves in allegories." [20, p. 484; author's transl.]

Even the closest comrades-in-arms of the Jena Circle of Romanticists criticised the notion that pure speculation could provide the basis for any profound knowledge about the world. "Schelling's philosophy of Nature," Friedrich Schlegel reasoned, "will inevitably give rise to contradiction by the crass empiricism whose destruction it was intended to bring about." [21, p. 50; author's transl.] And physicist Johann Wilhelm Ritter insisted that "pure experience ... is ... the only permitted artifice to gain pure theory" [22, p. 122; author's transl.] We shall "approach true theory imperceptibly, without searching for it. We shall find it by observing what actually happens. What more do we desire of a theory than that it tell us what is actually happening?" [22, p. 121; author's transl.]

Ritter rejected the claim of Schelling's philosophy to deliver an *a priori* foundation of the phenomena of Nature. Nevertheless, at the same time, he moved toward Schelling's philosophy because he made its speculative theses the guidelines of his experimental studies. Other scientists, such as Hans Christian Ørsted, Lorenz Oken and Carl Gustav Carus, moved in a similar direction. From this trend there finally emerged a research field, which its proponents regarded as the "romantic study of Nature". Ultimately, they were searching for the mysterious force that, as was assumed at the time, permeates and connects all organic and inorganic matter and which Schelling had designated by the ancient term "world-soul" [19].

Toward the end of the eighteenth century, the field of choice for experiments in romantic research into Nature was above all electrochemistry. At the centre of Ritter's studies, for example, were the observations made by Luigi Galvani that a suitably dissected frog's leg could be made to move by applying electric currents (see Sect. 3.7). Everything seemed to point to galvanic electricity as the key to understanding living matter. Moreover, the newly discovered electrical and magnetic phenomena appeared to confirm Goethe's conception of polarity and enhancement ("Steigerung") as the "two great driving wheels of all Nature", which was highly popular at that time [23, p. 48; author's transl.].

For the experimenter Ritter, the phenomena of polarity and enhancement became the central guiding principle of his research. His urge to discover rose to the degree of an obsession, as he began to conduct electrochemical experiments on his own body. The fact that Ritter's self-experimentation ultimately brought about his death can be regarded as a macabre climax of a dogmatic understanding of Nature in which the borders between self-recognition and the cognition of Nature, between the human body and the body of Nature, were denied with blind fervour [24].

Schelling's natural philosophy promised to allow systematic access to natural phenomena, free of any possible refutation by experience. It was this promise that made his philosophy so attractive to the romantic naturalists. However, apart from a few exceptions, their contributions to the progress of science were modest. Even though some of them claimed that they owed their discoveries to Schelling's philosophy, this must seem more than questionable from today's point of view. Rather, they seem to confirm the well-known fact that chance, and even false ideas, can lead to ground-breaking discoveries. Thus, the successes of Ritter and Ørsted—one discovered ultraviolet radiation, the other the relationship between electricity and magnetism—were definitely not based on a necessary linkage between philosophical insight and experimental findings. Apart from a superficial use of analogies, natural philosophy yielded no theoretical background knowledge that could have guided empirical research in any meaningful way.

Apparently, the romantic naturalists achieved their results in the same way as a blind man using a lamp to look for his key and, quite by chance, finding it. The physical chemist Wilhelm Ostwald, likewise, could not explain the curious successes of the romantic naturalists in any other way than by the "freedom from every impulse to shy away from the absurd, of which the philosophers of Nature made such a liberal use." Only in that way, he wrote, were the romantic naturalists in a position to "find analogies that in fact were present, but which had escaped the attention of the naturalists' contemporaries because of their unusual nature. This is what their discoveries were based on." [25, p. 8; author's transl.]

Already in Schelling's time, the speculative philosophy of Nature was harshly criticised by leading mathematicians and naturalists. For Friedrich Wöhler, Justus von Liebig, Friedrich Wilhelm Herschel, Rudolf Virchow, Hermann von Helmholtz and many others—as Carl Friedrich Gauß put it representatively—their "hair stood on end at the thought of Schelling, Hegel, Nees von Esenbeck and their consorts" [26, p. 337; author's transl.].

This withering criticism is very easily understood if one considers, for example, Hegel's philosophy of Nature. Not only did Hegel dismiss, without further discussion, Newton's theory of light as "nonsense" but he also challenged the most elementary insights of the science of his time. Thus, for example, he denied that lightning is an electric discharge or that water is a compound of hydrogen and oxygen. "The healthy person," Hegel claimed, "does not believe in explanations of that kind." [27, p. 146; author's transl.] Instead, Hegel constructed a world in which solar eclipses influence chemical processes, in which magnetic phenomena are dependent upon the presence of daylight, and in which the solar system is an organism—to mention but a few of his obscure inspirations.

With Hegel's "derivation" of the order of the planets—which, scarcely promulgated, was immediately refuted by observation—the speculative philosophy of Nature finally began to verge upon the ridiculous. Hegel's flop demonstrates the pitfalls of a dogmatic philosophy according to which nothing can be the case that philosophy does not admit to be the case. The polemics with which Schelling had once castigated the allegedly "blind and unimaginative kind of natural science that has taken hold since the deterioration of philosophy and physics at the hand of Bacon, Boyle and Newton" [17, p. 394; author's transl.], now returned like a boomerang and hit the philosophy of Nature squarely. For Liebig, this kind of philosophy was a "dead skeleton, stuffed with straw and painted with make-up", "the pestilence, the Black Death of our century" [28, pp. 23, 29; author's transl.]. Wöhler saw in its promulgators "fools and charlatans", who did not even themselves believe the "hot-headed stuff" that they preached [29, p. 39; author's transl.]. Matthias Schleiden lost his patience over "the crazy ideas of these caricatures of philosophers", whom he brushed off as philosophasters [30, p. 36; author's transl.].

There remained just one single, initially almost impregnable bastion of natural philosophy: romantic medicine. The human organism, with its unimaginable material complexity, organised as it is in a seemingly infinite number of cause-and-effect loops, appeared completely resistant to any attempt at a mechanistic analysis. Consequently, the organismic research paradigm seemed to offer a valid way of regarding the organism, one that would not so quickly be caught up with by mechanistically oriented research. Romantic medicine thus succeeded in establishing itself alongside academic medicine as an independent paradigm for medical research. Even today, holistic medicine and its more esoteric fringe doctrines carry significant weight in the shape of the so-called complementary and alternative medicine.

This is not the place to trace the entire history of the impact and public perception of the romantic philosophy of Nature with all its ramifications—not least because this philosophy had no lasting effect upon the natural sciences, and because its impact and public perception had already exhausted themselves in the polemic debates between this kind of philosophy and exact science. "That shallow twaddle", wrote Schleiden, the founder of cell biology, "had no influence whatever upon astronomy and mathematical physics, but for a while it confused the life sciences; however, once they had grown out of these penurious ideas, they rightly branded Schelling as a 'mystagogue', and he disappeared in the smoke of his own mythological philosophy" [30, p. 35; author's transl.].

For the modern, empirical sciences, which base their claims to truth and validity upon the corrective power of experience, the dream of absolute knowledge must seem more like a nightmare. The idea that one might be able to derive from a single fundamental principle the rich, living content of reality —without any reference to observation or experience—comes over as a complete delusion. The mere analysis of the meaning of terms like that of the "absolute", from which the romantic philosophy of Nature drew its sole justification,   can   never   replace   the   knowledge-guiding   and knowledge-founding function of experience.

The assumption that pure thought already accords with the outside world in the absence of all experience turned out to be a grave error. It led to the momentous fallacy that experience could be systematically subordinated to thinking. "This error," Ostwald criticised, "we must avoid at any price." The romantic naturalists "tried to derive experience from thought; we shall do the opposite and let our thought be governed everywhere by experience" [25, p. 7; author's transl.]. Thus, the natural sciences followed the empiricist path consequently, while the romantic philosophy of Nature, snarled up in its aprioristic constraints, was pushed aside by the scientific development that took place in the nineteenth and twentieth centuries.

2.7 Knowledge and Pseudo-Knowledge

Kant's critical philosophy had been undermined by the German idealists. In the form of a constant stream of new ideas about the absolute, they created an all-encompassing world-view, which culminated in Hegel's concept of the "absolute world-spirit". Toward the end of the nineteenth century, a group of

German philosophers responded to this development by returning to the authentic form of Kant's philosophical program and making it the basis of modern scientific thinking. In Germany, two regional focal points of Neo-Kantianism emerged, the "Marburg School" around Hermann Cohen and Paul Natorp, and the "Southwest German School" around Wilhelm Windelband and Heinrich Rickert.

Closer examination, however, shows that these two schools pursued different goals. The Marburg School, for example, was concerned with accessing and exploring the human cognitive faculty through the rules of formally correct thinking. Following the exact sciences, they attempted to reconstruct the *a priori* structures of our cognitive faculties from logical and mathematical standpoints. The Southwest German School, on the other hand, argued that our cognitive faculties are dependent not only on mathematical and logical structures, but also on the cultural factors which guide the recognizing subject. Consequently, the conditions for the possibility of cognition, which Kant had addressed in his philosophy, would also have to include the context of values and norms into which the recognizing subject is embedded. With the inclusion of the problem of values as a constitutive element for the substantiation of human cognition, the framework of pure cognition logic was extended by adding in individual and psychological aspects.

The Neo-Kantians had hoped to complete Kant's program by gaining deeper insights into the *a priori* structures of our knowledge. However, their aims were not generally accepted. Instead, some philosophers saw Neo-Kantianism as a continuation of a philosophy the time of which had long since passed. The psychologist Wilhelm Wundt dismissed Neo-Kantianism as a "philosophy of authority", and the philosopher Eduard von Hartmann dismissed it as "Kantomania"; the critics still regarded *a priori* forms of knowledge as a speculative element that had nothing to do with modern epistemology.

Given the tremendous upturn in the natural sciences in the nineteenth century, the path toward scientifically founded epistemology seemed to lead alongside the positivistic ideal of cognition, which is oriented to observation, measurement, experiment and the like (Sect. 1.9). Thus, shortly after the First World War, prominent philosophers, physicists, mathematicians and logicians joined forces in Vienna under the guiding principle of positivism to form a discussion group which, known as the "Vienna Circle", unleashed a powerful current of contemporary philosophy that kept up its momentum in the subsequent decades.

The founders of the Vienna Circle included Moritz Schlick, Rudolf Carnap, Otto Neurath, Richard von Mises, Philipp Frank, Hans Hahn,

Herbert Feigl and Viktor Kraft. All its members shared the positivistic cognition ideal, even though their philosophical positions differed to some degree. They were all convinced that real knowledge could only be gained by recourse to experience, i.e. by verifiable observations and controlled experiments. On the other hand, findings based on mere speculation had no claim to validity. Such speculative findings were rejected as pure pseudo-knowledge.

One of the focal points of the philosophical discussion were synthetic *a priori* judgments, of which Kant had believed that they would provide factual knowledge, independently of experience (Sect. 2.5). As expected, the members of the Vienna Circle rejected this idea. From a positivistic point of view, all factual knowledge is always knowledge *a posteriori*. The positivists did not even want to accept *a priori* statements for mathematics, as the notion that the basic mathematical concepts are imprinted once and for all in the human mind contradicted the positivistic understanding of science from the ground up. This would mean "that a specific chapter of science will never change" [31, p. 129]. If there was any knowledge *a priori*, the positivists argued, then this could at best be contained in analytical sentences. However, even analytical sentences do not convey any factual knowledge. For example, the sentence "It is raining or it is not raining" could give the impression of being an empirical statement. In reality, however, it is an empty statement. To verify its truth, no observation is necessary: the statement is true by its linguistic construction alone.

This is precisely the problem that the philosophers of German idealism did not want to take note of. It is not possible to deduce all the richness of reality from analytical sentences that are empty by their nature. No wonder that the idealists' plans were doomed to failure. In fact, instead of empirically based theories, they designed untenable metaphysical systems; instead of real problems, they merely solved pseudo-problems. In effect, the idealists did not practice any real science, but only pursued some kind of pseudo-science. Consequently, the philosophers of the Vienna Circle understood it as their foremost task to distance themselves from metaphysical speculations of any kind.

Initially, it was believed that the demarcation of genuine science from pseudo-science could be achieved by carefully distinguishing between meaningful and meaningless empirical statements. For example, it was assumed that empirical statements are meaningful when at the same time the conditions can be stated under which they can be verified. In contrast, statements referring to the "absolute", the "nothing", the "divine" and the like were regarded as pointless, because such statements cannot be tested empirically. The occurrence of pseudo-problems was accordingly explained by asserting that, in the cases in question, the sense criterion was not applied strictly enough.

However, the method of verification led to a fundamental difficulty in its application to statements concerning natural laws. From a logical standpoint, natural laws are universal statements in the sense that they refer to an in principle infinite number of events. Thus, the general verifiability of such statements is excluded from the outset, even if one presupposes the verifiability of individual cases. In order to avoid the classification of statements about the laws of Nature as meaningless, however, various repair proposals were made for the verification model, but none of them led to any fundamental improvement.

Even if the idea of distinguishing between meaningful and meaningless statements turned out to be impractical in the end, it nevertheless sharpened consciousness for speech-analytical thinking. In that way it ultimately contributed to the so-called "linguistic turn" of philosophy. From then on, the human language, and with it the fact that the world is opened up to us through language and that our world-understanding is expressed by language, moved to the center of philosophical thought (Chap. 1).

Ludwig Wittgenstein described the paradigm shift with the words: "All Philosophy is a 'critique of language'." [32, 4.0031] This is almost the same wording as had been used by the physicist and philosopher Georg Christoph Lichtenberg in the Age of Enlightenment to describe the task of philosophy. It should be mentioned here that the logician Alfred Tarski had also carried out his logical-semantic investigations into the truth-problem against this background—investigations with which he wanted to develop an adequate and formally consistent concept of truth, free from any metaphysical dependence (Sect. 2.1).

In addition to the linguistic aspects, the members of the Vienna Circle were particularly interested in the logic of the cognition process. Therefore, their philosophy is often referred to as "logical positivism". The philosopher and logician Rudolf Carnap played a leading role here. Carnap not only spoke up as a harsh critic of metaphysics, but he also developed a modern, logically based variant of empiricism. Carnap's logical empiricism assumes that the overarching contexts of experience in science are gained through "induction," i.e. through the generalization of individual experiences. However, in contrast to the classical empiricism of Hume, who considered the inductive method to be merely an expression of a habitual activity of the human mind, Carnap attempted to justify the inductive method by scientifically employing an "inductive logic" [33].

However, the inductive theory of knowledge also seemed to stand on feet of clay, because the inductive method of gaining general statements is always based on probabilities. Thus, such a logic of cognition is not able to prove statements to be unrestrictedly valid. For example, the observation of countless white swans does not allow the conclusion that all swans are white. Indeed, conversely, it is enough to observe one swan that is *not* white to refute the general hypothesis "All swans are white."

The philosopher Karl Popper, who, like Wittgenstein, was close to the Vienna circle, drew upon the preceding example to question the inductive method generally. Popper rejected the idea that knowledge is gained inductively and that the truth of such inductively acquired knowledge, as Carnap believed, could be ensured by a "confirmation-function". According to Popper, knowledge is gained in a deductive way by starting from specific hypotheses and then checking them by experience [34]. If, however, a hypothesis comes into conflict with experience, we would usually abandon the hypothesis and replace it with new and better one. In other words: we gain knowledge through the creation of hypotheses, the truth of which can never be confirmed, but only refuted. Therefore, our knowledge remains at any time hypothetical. Here again, we come across with the method of error correction (Sect. 2.1); Popper now uses it as an instrument to distinguish scientific from pseudo-scientific findings. Following Popper, genuine scientific findings are characterized by their ability also to specify the conditions for their possible refutation. Such conditions would primarily include precise predictions that can be checked experimentally.

Popper's "fallibilism" takes over the idea of Peirce according to which there can never be absolute certainty regarding our beliefs, presumptions, hypotheses, theories and so forth. Fallibilism claims that our scientific knowledge is always provisional. Even if our knowledge has proven itself excellently in experience, it still remains uncertain. The application of fallibilism to science by Popper was undoubtedly an essential step toward the solution of the so-called demarcation problem, which demands a criterion to distinguish genuine science from pseudoscience. According to Popper, a "real" scientific hypothesis, theory or the like must also specify the conditions for its potential refutation. In contrast, a hypothesis or a theory that is basically immune to any refutation (because, for example, it makes no predictions that can be subjected to observation or tested by experiment) is classified as pseudoscience.

Today we must regard Popper's considerations as a crucial contribution to the clarification of the scientific character of empirical theories. However, there are indications that the scope of the demarcation criterion is limited. Thus, for example, Darwin's theory of evolution, which is definitely recognized in science, allows testable predictions only to a limited extent. Yet Popper went so far as to deny evolutionary theory's scientific character. For him, Darwinism was merely a "metaphysical research program—a possible framework for testable scientific theories" [35, p. 244].

The phrase "metaphysical research program" is undoubtedly an exaggeration, because Darwin's understanding of natural evolution had and has nothing to do with metaphysics in the traditional sense. However, Popper is right when he describes Darwin's concept as a "research program", which implies that it still has to be filled with strict, testable theories. If one looks for such theories, one would probably first think of the concepts of population genetics and, above all, the theory of molecular evolution, which provides a mathematical foundation for selection and evolution, including making testable predictions [36]. It must also be mentioned that—in the light of the progress made by molecular biology—Popper later partly revised his opinion.

Popper's research logic is revealing concerning, not least, the problem of truth. Its main points are: Human insights cannot claim absolute certainty. At any time, they have only a hypothetical character. The progress of knowledge does not consist in confirming our hypotheses, but in recognizing false assumptions as such and eliminating them. In this respect, the cognition process is in principle open. Consequently, all the knowledge that we have acquired through the empirical sciences is, and can only be, provisional. The only certainty that we can associate with scientific findings is that they may turn out one day to be mistaken.

The cognition method, described by Popper, resembles the deductive method of rationalism. However, unlike traditional rationalism, which set out from judgments that are certain in themselves, Popper's model takes as its starting point hypotheses that can in principle be wrong. For this reason, their validity must be perpetually checked by experience. Thus, in contrast to traditional rationalism, Popper fully recognizes the knowledge-guiding function of experience. In his eyes, error correction by experience is in fact the real driving force of the cognition progress. Therefore, it makes complete sense to denote Popper's epistemological approach as "critical rationalism".

According to Popper, knowledge progresses by formulating hypotheses, by testing them in experience, and—if they prove to be false—by replacing them with new hypotheses. In this way, we gradually adapt our cognitions to actual reality. This process resembles the evolution of life forms on Earth: the hypotheses (life forms) on trial are new variations of hypotheses already existing. Just as for organisms, new variants originate by random mutation and are subsequently tested by selection for their degree of adaptation. Those variants that do not pass the test are eliminated. In this way, the existing living forms (organisms or hypotheses) are steadily replaced by variants that are better adapted to the ambient conditions than their predecessors were.

Popper himself drew a parallel between the evolution of knowledge and the evolution of life. Making recourse to biological terms, he spoke of the "natural selection of hypotheses", the "fitness" of which has been manifested in the fact that they have survived perpetual attempts to disprove them [37]. The fact that Popper refers to the mechanism of biological evolution is, considering his reservations concerning Darwinism, quite remarkable. However, the parallels cannot be overlooked. Already at the end of the nineteenth century, the philosopher Ernst Mach and the physicist Ludwig Boltzmann had pointed out the analogies that exist between the evolution of knowledge and the evolution of life.

However, Popper's fallibilism leaves some questions unanswered. For example, one would want to know under what conditions the attempt to falsify a hypothesis can be regarded as successful. This question is not so easy to answer, because all observations and experiments already contain, on their own part, hypotheses or even theories. Thus, for example, measuring instruments can only be built on the basis of physical theories whose validity is not questioned in the experiment for which the instruments are used. In other words, hypotheses and theories can only be tested within the framework of other hypotheses and theories that are implicitly assumed to be correct.

There are further problems. Suppose that a contradiction between theory and experience arises. How do we deal with this? Can one identify parts or even single statements of the theory that are responsible for the contradiction, or does one have to call the whole theory into question? This also raises a general question: Is it possible at all to draw a sharp line between scientific theories, when everything, in reality, is interrelated with everything else? Questions of this kind prompted the logician and philosopher Willard Van Orman Quine to set up a holistic theory of knowledge according to which all statements of a theory are logically related, so that the experimental examination could only refer to the theory in total, and ultimately only to knowledge as a whole [38].

It has also been argued against Popper that the development of our knowledge in fact follows a course entirely different from the one prescribed to it by his falsification logic. Scientists by no means always try to refute their own hypotheses. Rather, they adhere to their hypotheses for as long as possible and are only prepared to abandon them when they definitely turn out to be wrong. Some critics of Popper even claimed that scientists try to immunize their hypotheses and theories against possible falsification attempts instead of exposing them to refutation.

From this perspective, which is oriented toward actual research practice, the historian Kuhn [39] developed his widely acclaimed model of the development of science by paradigm shifts. According to this model, the evolution of knowledge does not take place steadily in small steps, but abruptly in much greater leaps of development. Nevertheless, the fact that the actual progress of scientific knowledge is not as continuous as Popper's research logic would suggest in no way calls his cognition model into question. Popper was less interested in the actual course of scientific development than in the logic according to which knowledge progresses. For this reason, his main work on the philosophy of science [34] appropriately bears the title (translated literally from the original German) *Logic of Research*.

Modern analytical philosophy continues to pursue the goal of the Vienna Circle, that is, to clarify philosophical problems employing formal logic in conjunction with linguistic analysis. This approach has not only influenced decisively the philosophy of science of the twentieth century, but it can be found nowadays in almost all areas of philosophy.

2.8 Approaching Truth by Evolution?

All efforts to justify the truth content of human thought are confronted with the problem that thinking can only analyze itself in the mirror of thinking. Given this, a theory of knowledge has raised high hopes in recent years. It attempts to take a look "behind the mirror". The basic idea was described some years ago by the behavioral biologist Konrad Lorenz [40]. It is as simple as it is fascinating. If the human brain is the product of biological evolution, then its cognitive structures must also have been adapted to reality. They must—as Lorenz explains—fit in with the real world just as the fin of a fish fits the water or a horse's hoof fits the ground of the steppe. This in turn would mean that our findings must be (more or less) true, as only realistic representation of the outside world by the human cognitive apparatus could ensure the survival of man in his long history of development.

In the context of such considerations, the question arises whether also Kant's apriorism can possibly be interpreted as the result of an evolutionary adaptation to reality. Does an evolutionary theory of knowledge ultimately lead to the solution of the age-old cognition problem? This question already arose at the end of the nineteenth century, when for example the zoologist Ernst Haeckel wrote with admirable clarity: "The curious predisposition to *a priori* knowledge is really the effect of the inheritance of certain structures of the brain, which have been formed in man's vertebrate ancestors slowly and gradually, by adaptation to an association of experiences, and therefore of *a posteriori* knowledge. Even the absolutely certain truths of mathematics and physics, which Kant described as synthetic judgments *a priori*, were originally attained by the phyletic development of the judgment, and may be reduced to constantly repeated experiences and *a priori* conclusions derived therefrom. The 'necessity' which Kant considered to be a special feature of these *a priori* propositions would be found in all other judgments if we were fully acquainted with the phenomena and their conditions." [41, p. 11]

Previously, the physicist Ludwig Boltzmann, an early advocate of Darwin's theory of evolution, had developed similar ideas. Nevertheless, it took until the middle of the twentieth century for these thoughts to be taken up again by Lorenz and developed by modern philosophy of science into a naturalistic epistemology that strives for clarity and stringency of argument. Evolutionary epistemology is based on the central thesis that our cognitive apparatus images the world (more or less) realistically because it has developed by evolutionary adaptation to the real world. Accordingly, evolutionary epistemology sees it as its task to expose the phylogenetic roots of our faculties of cognition and to trace the claim to truth of our cognitive structures back to their evolutionary adaptation to reality. Here, adaption means the correspondence of our ideas of the outside world with the actual circumstances of the outside world—in a sense comparable to that envisaged by the correspondence theory of truth (Sect. 2.1). In the present case, however, the claim to truth has only a preliminary character, because the process of adaptation is not complete as long as evolution is proceeding. Therefore, the structures of the human cognition apparatus, and thus also the truth content of human cognitions, may change over time. As expected, evolutionary epistemology rests on the same "hypothetical realism" that underlies Popper's fallibilism (Sect. 2.7).

The program of evolutionary epistemology has met with a lively response in philosophy. For the first time, the possibility seems to be emerging of a naturalistic basis for the justification of the truth content of human knowledge. However, evolutionary epistemology is repeatedly accused of being based on circular reasoning. It employs, according to its critics, findings of evolutionary biology and thereby already presupposes the validity of that which is to be justified. Of course, evolutionary epistemology cannot break the logical barrier that surrounds the truth problem. Even an epistemology that is scientifically well-founded cannot provide any final justification for the truth content of human knowledge. Nevertheless, the evolutionary theory of knowledge seems to be much more attractive than all the artfully elaborated theories of knowledge that ignore the natural conditions of human cognition.

However, in addition to its advantages, the close connection of evolutionary epistemology to Darwin's theory of evolution also has disadvantages. These arise because the problems of evolutionary explanation are transferred automatically to its philosophical offshoot. Strangely enough, this has so far received little attention in the debate about the possibilities and limits of evolutionary epistemology. The perhaps most challenging problem is mentioned here only briefly. To illustrate it, let us consider an idea from evolutionary biology that is termed the "fitness landscape". This idea describes in illustrative form the basic features of an evolution based on self-reproduction, variation and natural selection.

The fitness landscape is a relatively simple idea. Imagine, in a thought experiment, that every organism can be assigned a numerical value that indicates the quality of its adaptation to environmental conditions. This value is equivalent to the selection value of a living being, because it determines its chance of survival in the selection competition. Well-adapted organisms have high, poorly-adapted organisms have low selection values. If one plots these numerical values as altitude above a plane whose coordinates are assigned to the corresponding organisms, then a net-like surface, the so-called fitness landscape unfolds above this plane. It is comparable to a natural mountain landscape with its summits, valleys and ridges (Fig. 2.6).

The model in the sketch illustrates, despite its simplicity, the essential features of evolutionary development. For example, evolution can only follow paths in the fitness landscape that lead upwards. This feature is firmly established in the optimization mechanism. In contrast, the exact route is not fixed. Instead, evolution resembles a hike through a mountain landscape, which has no concrete goal, but which is under the general guideline that the walker's path must always lead uphill. As with a real mountain hike, there are also resting periods in the evolution process. Technically speaking, the system then finds itself in selection equilibrium. However, these equilibria are metastable, i.e. temporary. As soon as new advantageous variants come into play as a consequence of mutation processes, the balance collapses. Then, figuratively speaking, the walk through the mountain range will be continued to the next (local) peak until the highest peak has been reached.

It is evident that the dynamics of evolution depend significantly on the structure of the mountain range. This structure is fixed if all selection values remain constant. However, this is an ideal case that certainly does not occur in Nature. It is much more reasonable to assume that the selection values change continually over time. There are various reasons for this. First, the environmental conditions may alter because the physical parameters, like temperature and so forth, have changed. This may in turn change the functional requirements imposed on living beings and thereby also change their selection values. Moreover, ecological couplings may arise between different species, so that

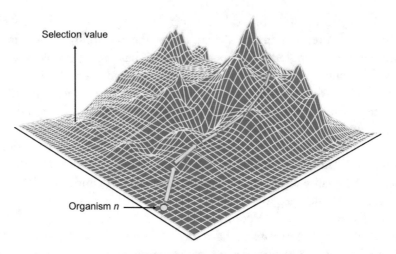

Fig. 2.6 Fitness landscape describing the evolutionary optimization of a population of organisms. The model is based on a two-dimensional grid whose nodes are each assigned to a specific organism. If the selection value of each organism is plotted above the corresponding point of the grid, then a surface is obtained that maps the differences in the degree of adaptation of the organisms. On the basis of this model, the evolution of life can be described as an optimization process which—always ascending—leads from one local maximum to the next local maximum, until finally the optimum—i.e. the highest possible—peak of the mountain chain is reached (yellow arrows). This deterministic model, however, does not yet consider the influence of random processes. If such processes are included, the details become more complicated and multi-faceted. To present the fitness landscape more precisely, one would also have to extend the picture to a hugh number of dimensions, which is only possible in a mathematical analysis. Details can be found elsewhere [42]

their selection values also depend on those of other species. However, every selective change in the population number of a particular species will entail a deformation of the whole "fitness landscape".

Consequently, the fitness landscape is not rigid, but possesses a certain plasticity, which depends on the evolutionary process itself. Thus, a peak in the fitness landscape that the evolution process is heading toward may dissolve after some evolutionary steps, only to re-emerge at some other place. As soon as the selection values depend on the evolutionary process itself, goal and targeting are inextricably linked. In Fig. 2.6 the goals, i.e. the peaks of the fitness landscape, are defined at any time. However, they change with each evolutionary step, which in turn influences the route of the adaptation process.

The fact that the fitness landscape changes in the course of evolution also has implications for evolutionary epistemology (Fig. 2.7). On the basis of the idea of (hypothetical) realism, evolutionary epistemology presupposes the

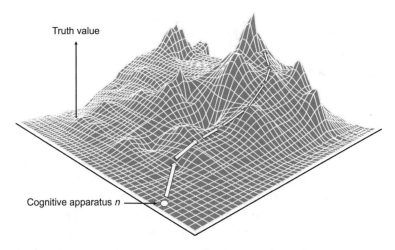

Truth value

Cognitive apparatus n

Fig. 2.7 Fitness landscape illustrating the idea of evolutionary epistemology. The model assumes—in analogy to Fig. 2.6—that each step in the evolution of the cognitive apparatus can be ascribed a truth value which reflects the degree of adaptation of the cognitive structures to reality. Moreover, evolutionary epistemology necessarily assumes that the cognitive structures have a quasi-continuous range of truth values, i.e., the cognitive structures represent reality more or less correctly. Consequently, evolutionary epistemology is based on a "hypothetical" realism. Nevertheless, this model of truth also raises critical questions. What are the features of cognitive structures? Do they refer to single cognitions, to an ensemble of cognitions, or do they relate to basic judgments like, for example, Kant's judgments *a priori*? Apart from this, the cognitive apparatus may feature cognitive structures that have different truth values. How they are weighted in the process of natural selection? Finally, the question arises: How can one speak at all of approaching truth by evolution, when the truth values of the cognitive structures continuously change because the fitness landscape changes in the course of evolution?

existence of a sufficiently stable reality to which the cognitive structures have adapted in the course of evolution. This reality, however, is to a great extent the biotic environment of man, which itself arose through evolution. In other words, the co-evolution of man and his environment entails the risk of tautological explanations, because the evaluation scale for adaptation is highly volatile.

To avoid the risk of tautology, one should focus on those adaptation processes that are primarily determined by the abiotic environment. Presumably, one has to go back to a point before the development of the central nervous system and the formation of a musculoskeletal system, which together have enabled organisms to move in a controlled manner and thereby become independent of the local conditions of their habitat. Indeed, there is some evidence that the gain of spatial freedom was a vital driving force in early evolution of higher organisms. This view is not only supported by neurobiological studies on the close connection between cognitive and motor function, but it also coincides with the analysis of the emergence of functionality, described in Sect. 1.5.

Even if evolutionary biology itself still has pressing problems to solve, the foundation of epistemology in evolutionary terms certainly presents an auspicious research program. As far as the problem of truth is concerned, evolutionary epistemology is based on correspondence theory. Here, "correspondence" means the agreement of the (subjective) cognitive structures with the (objective) structures of reality. However, what is meant by agreement or by "fit" remains unclear. For example, when the evolutionary biologist Lorenz says that the cognitive structures fit the real world just as the fin of a fish fits the water or a horse's hoof the steppe, then this can only mean the agreement of structure and function. However, is it possible at all to establish a correlation between structure and function in the sense of a fit? Is this not basically the same problem as we discussed in Sect. 2.1 concerning the correspondence between sentence and fact?

Strictly speaking, evolution theory does not even substantiate any correspondence between the evolving structures and the environment. Instead, it only makes the weaker assertion that the evolving structures must not be in contradiction to the environment. Accordingly, within the framework of evolution theory, we cannot specify what is right, but only what is wrong. In this sense, all the structures that were eliminated in the course of natural selection became displaced because they turned out to be false. However, not even evolutionary error has ultimate meaning, as the yardsticks for evaluation in evolution can change at any time. From this, we see that even an evolutionary epistemology cannot eliminate the fundamental difficulties that the concept of truth entails. Therefore, for good reason, evolutionary epistemology is based on a "hypothetical" realism—a realism that avoids any reference to absolute truth.

2.9 Practical Truth

If we review the many attempts to grasp the concept of truth or even to prove the existence of eternal truths, Nietzsche's mockery of "truth-seeking man" does not appear to be completely unfounded. It has become all too clear how poor our notions of truth actually are. We do not possess an Archimedean point, outside of thinking, from which the truth content of human thought could be justified beyond doubt. More to the point, the concept of truth proves to be a borderline term in the continuous process of human cognition, one which our thinking can align itself to, but which can never be caught up with by thought.

This is the reason why, in all philosophical approaches, the notion of truth itself ultimately remains empty. On the bright side, only the essential indeterminacy of the idea of truth has made possible all the richness of approaches that constitute our sophisticated and nuanced understanding of reality. It is the repeated failure of attempts to find final truth and certainty that have determined the paths of human thought at all times. Accordingly, the history of philosophy is essentially the story of the everlasting search for truth by man.

If, however, there is no theoretical assurance of what we believe to be true, then only the practice of everyday life remains as the final authority for assessing our findings' claims to truth. This brings us back to the starting point of our discussion, the pragmatic concept of truth. We remember that from the viewpoint of pragmatism, truth is not something timeless and absolute, but rather an operational term that must always be understood in the context of human actions. Thus, in pragmatism, the eternal truths of reason are supplanted by the time-bound practical truths defined by the particular circumstances of human action.

It is in the nature of the pragmatic concept of truth that it can be instrumentalized for the individual goals and desires of man. For example, the psychologist and philosopher William James considered true cognitions to be merely useful means of satisfying human needs. In his view it was not even relevant to ask whether cognition *de facto* enables us to solve practical problems. For example, he considered an idea—such as the hypothesis of the existence of God—to be true for the sole reason that it is useful or satisfying for our lives to believe in it. In this somewhat unusual interpretation of the truth concept by James, truth is not an independent category, but merely "one species of good" that can make people happy, alongside health for example; he also speaks of "truth's pragmatic cash-value"—an attitude that reduces the concept of truth to a utilitarian criterion that is only directed toward benefit [43, p. 97].

Later, the pedagogue and philosopher John Dewey developed pragmatism further into the so-called "instrumentalism". Dewey shared the utilitarian interpretation of pragmatism according to which knowledge is primarily an instrument, providing the means for solving and coping with practical questions of life. Proceeding from this maxim, he developed a pragmatic logic of knowledge which describes the cognition gain as a situation-related, problem-solving process, oriented toward actual life practice. For many decades, Dewey tried to put into practice the core idea of pragmatism, which views cognition and action as an indissoluble unity, in his pedagogical and political activities. This gave instrumentalism an immense significance, above all in the U.S.A., its country of origin.

Even though the concept of truth in pragmatism has lost much of its original splendor, it looks as if it has regained a certain down-to-earth quality —one that it had quite obviously lost in the glass-bead game of the truth theoreticians. This impression, however, is deceptive. By the change from the theoretical to the practical level, the problem of truth is not solved, but merely deferred, for instead of the concept of truth we are now dealing with the no less problematical concept of action. This is due not only to the multitude of possible forms of activity, but also to the fact that they often interact in a complicated way.

Even the briefest look at the numerous forms of human activity makes it clear that the concept of action does not help us to solve the problem of truth. Beside "poietic" action, which Aristotle had rated so highly for the foundation of our knowledge (Sect. 1.1), the concept of action encompasses all forms of activity that determine the life-world behavior of man: social, moral, strategic, symbolic or communicative actions, to mention but a few. The fact that the pragmatic concept of truth reflects all the dimensions of freedom of human activity makes practical truth situation-dependent. In this case, truth inevitably disintegrates into a colorful mixture of specific truths each of which is solely defined by the success of their respective action.

Moreover, when the success of an action is only subjectively determined, the concept of truth dissolves into arbitrariness. Truth atrophies here, as James' pragmatism illustrates, to a mere subjective truth. Even though James asserts, in the sense of the correspondence theory of truth, that only those actions can be successful which are appropriate to the facts, or at least do not contradict them, this form of objectivation is far from abolishing the subjectivity of his truth concept.

The pragmatic concept of truth can only claim intersubjective validity if an overarching and universally valid yardstick can be stated for the "success" of an action. However, this presupposes that a reasonable consensus can be established between the subjects, who are acting, and judging, out of different interests. For this purpose, two conditions must be fulfilled. On the one hand, there must be a fundamental objective truth at all, which can be approached through the process of reasonable understanding. On the other hand, the community of acting individuals must share basic values and abilities, such as truthfulness, openness, linguistic competence, professional competence and the like. Only a society that continually strives to examine critically its arguments under these conditions, and that tries consistently to exclude all errors and sources of error, can come closer to the truth. Needless to say, this is at first only the ideal image of a society striving for truth and truthfulness; realizing it may be quite another matter.

Like correspondence theory, the consensus theory of truth has a tradition that goes back to Greek antiquity. Its archetype is Socratic dialog. Socrates had argued that everything the individual can see and recognize has initially the character of appearances. Consequently, the perceived object is at first merely reflected, in inadequate knowledge, in mere opinion (Greek: *doxa*). Actual reality, Socrates believed, is brought to light only by a critical examination, in dialog with an interlocutor, about the perceived. According to this idea, the truth finally emerges from the unconcealment (Greek: *aletheia*) of being (see also [44]). Accordingly, the task of Socratic dialog is to uncover the truth, which is inherently present but still hidden, through speech and counter-speech.

A modern variant of this idea, which likewise accentuates the role of dialog for ascertaining truth, is *The Theory of Communicative Action* developed by the philosopher Jürgen Habermas [45]. His theory emphasizes, above all, social discourse as an instrument for finding the truth. This follows, in essence, the basic idea of Peirce's consensus theory, which describes the process of truth-finding as a consensual communication within a reasonably acting society (Sect. 2.1). Beyond that, however, Habermas' theory demands that the participants in the discourse must be committed to specific rules of verbal communication, so that the validity of arguments can be judged reasonably. According to Habermas, the ideal "speech situation" is given if all participants of discourse have the same chance to utilize specific types of "speech acts".

The concept of speech acts, developed by John Langshaw Austin [46], John Searle [47] and others, has its roots in "ordinary language" philosophy. Speech acts designate the basic units of verbal communication. The question of which factors, rules and the like are decisive for successful speech acts is the subject of the so-called "speech act theory". Relying on the concept of speech acts, Habermas believes that he has to make four demands on the ideal speech situation. (1) All participants in the discourse must be able to open up discussions at any given time and to continue them through speech and counter-speech, question and answer ("communicative speech acts"). (2) They must be able to establish interpretations, assertions, recommendations, explanations and justifications as well as to expound problems, justify them or refute their claim to validity ("constative speech acts"). (3) They must be able to express attitudes, feelings and desires, and thus to disclose their inner nature, thus ensuring their credentials as discourse partners and persuading others of their own truthfulness ("representative speech acts"). (4) They must be able to make and accept promises, to give and demand an account of actions taken, to allow and to forbid ("regulative speech acts").

The pith of these four speech acts lies in the fact that Habermas connects with them in each case a particular aspect of reasonable speaking. These aspects are truth, correctness, veracity and comprehensibility. In his interpretation, speech acts become pragmatic universals that enable the discourse community to develop valid and correct arguments. According to Habermas, a true consensus arises when, after the exchange of all arguments, a particular argument cannot be refuted any more and, moreover, is clear to all participants. Here, the ideal speech situation is of central importance, as it has to guarantee that the discourse is free of domination. This means that there is no distortion of speech and counter-speech. It furthermore means that the discourse is free from all constraints upon action and grants the participants in the discourse every conceivable freedom of experience.

2.10   Truths Without Truth?

With the "discourse theory of truth", Habermas adopted a fundamental idea of hermeneutics, according to which the relation of man to the world is established through language. By connecting this thought directly to the practice of human action, Habermas draws up the image of a society that bases the norms of its actions on a reasonable understanding within an ideal communication community. Viewed critically, however, the theory of communicative action proves to presuppose an unrealizable utopia. The assumption that the ideal speech situation is a sufficient precondition for finding true arguments is unconvincing. This is because the discursive search for truth makes truth appear as the result of a democratically organized process of weighing, selecting and bundling arguments. If, however, the truth of arguments can be traced back to the observance of certain norms of communicative understanding, then precisely these norms determinate ultimately which argument is to be regarded as true. However, who decides on the validity of such norms?

The idea of discourse free of domination also seems to be problematic. This idea implies that all participants have the same language competence, as only in this case can they debate with each other on an equal footing. Otherwise, the discourse will inevitably be dominated by the most articulate participants, merely because they can speak and argue more precisely and persuasively. This argument also applies if the participants of the discourse have to abide by certain norms of speaking. Equal articulateness, however, can never be achieved—not even if the participants belong to the same language community and have the same level of language proficiency. Spoken language is part of an activity that is influenced decisively by the life-world of the speaker. Within discourse theory, this would mean that the truth of statements depends sensitively on the cultural context of the participants in the discourse.

The fact that the life-world background of human language is reflected in the meaning of linguistic expressions was emphasized primarily by Wittgenstein in his theory of "language games". According to that, speech acts cannot be understood in isolation, but only within the context of the life-world. Or, in Wittgenstein's words: "What determines our judgment, our concepts and reactions is not what *one* man is doing *now*, an individual action, but the whole hurly-burly of human actions, the background against which we see any action." [48, §567] This means, inverting the argument, that "words are also deeds" [49, §546].

If one agrees with this view, then human language can no longer be regarded as a rigid, monolithic reference point of our world-knowledge. Instead, the essence of language must be interpreted dynamically, that is, as an activity. This understanding of language has to keep in consideration the fact that the meaning of linguistic expressions actually depends on the fluctuating and inextricably intermeshed web of human actions, in which every individual human activity is embedded. The fact that the content of linguistic expressions is fundamentally shaped by the use of language highlights once more the pragmatic dimension of information, language and communication, discussed in Chap. 1.

However, once one has set out down this path, the idea of expressing eternal truths in human language has to be abandoned altogether. If language, as outlined in Sect. 1.9, is context-bound and thus relative by its very nature, then this applies also to the truths that are conveyed by language. This objection has been raised by the philosopher Richard Rorty [50]. He pointed out that "truth is a property of linguistic entities, of sentences." The existence of sentences, however, depends on vocabularies and these are in turn created by human beings. This would mean that man's truths depend on man himself. They would be contingent, i.e. depend on a specific cultural context, in the same way as we saw to be the case for human language and human actions.

For this reason, Rorty pleads that man should free himself from the ideal of truth and the platonic ideal of absolute knowledge. Given the contingency of our reality of life, Rorty argues, it is pointless to want to fathom absolute truth and the essence of things. The goal of enlightened thinking, however, must be to abolish all oppositeness between true and false, being and appearance, absolute and relative, subject and object and the like.

The philosophy of "cognitive relativism," conceived by Rorty and others, strictly rejects all attempts at an ultimate justification of truth. Rorty cannot see any sense in the demand that one should strive for truth for its own sake. For him, it is only essential that we obtain knowledge that enables us to deal practically with our world and improve our living conditions. Like James, Rorty reduces the ideal of truth to its pure utility function. For him, too, truth only appears to be that which is better for us to believe in [51].

The loss of truth that is becoming apparent here is the inescapable consequence of a world-understanding that seeks to reverse any opposition between true and false in favor of a complete relativization of the concept of truth. However, any thinking that abandons the idea of truth so rigorously must itself lose any claim to truth. Such thinking remains aimless and arbitrary.

Attempts to banish the idea of truth from our thinking can only be taken seriously if they make a claim to truth and thus implicitly presuppose what they reject, namely the existence of truth. This is *one* aspect of an *aporia* of truth that no critical truth theory can resolve. Conversely, no truth theory can establish the existence of absolute truth, since the problem of truth can only be illuminated from the inward perspective of human thought. That is the *other* aspect of the *aporia*.

No wonder, then, that the discussion concerning truth always revolves in circles and that, throughout its long history, new variants of the concept of truth have arisen time and again. Ontological, logical, empirical, contingent or pragmatic truth are forms of truth which at the same time reflect the progressively increasing sophistication of our understanding of the world.

Table 2.2 The variety of truth concepts and truth theories reflects the context-dependence of the truth problem. The horizon of reflection regarding reality defines the context. Correspondingly, there is no sharp boundary between the different truth concepts and theories. Because the contexts of reflection are intertwined, there must also be an overlap of the truth conceptions. Thus, for example, the consensus and the discourse theory have a degree of overlap. The larger the overall context of philosophical reflection, the greater is the number of truth conceptions that emerge. It becomes clear that there is no ultimate truth covering all basic philosophical questions and problems. Instead, the truth has only a regulative function in the process of cognition. For example, in the empirical sciences, cognition is always subject to the proviso that it embodies hypothetical truth

Truth concepts	Truth theories
Empirical truth (Sect. 1.9)	Correspondence theory of truth (Sect. 2.1)
Pragmatic truth (Sect. 2.1)	Consensus theory of truth (Sect. 2.1)
Ontological truth (Sect. 2.2)	Coherence theory of truth (Sect. 2.1)
Hypothetical truth (Sect. 2.7)	Discourse theory of truth (Sect. 2.9)
Discursive truth (Sect. 2.9)	
Mathematical truth (Sect. 3.2)	
Historical truth (Sect. 4.6)	
Logical truth (Sect. 5.9)	
Computational truth (Sect. 7.10)	

They are a sign that even our notion of truth is context-dependent (Table 2.2). Since the numerous concepts of truth cannot be brought into line, all attempts to standardize the variety of these concepts are also inevitably doomed to fail.

After making our way over the confusing terrain of truth, we are still at a loss. With Nietzsche, one would like to ask: What are the truths of man? Are they just the "irrefutable errors" of man, as he had mocked [52, p 518; author's transl.]? Are the supposed truths of man nothing more than truths without truth? Is truth, as Nietzsche wants us to believe, only to be grasped as the difference between errors? Or, expressed positively: Is the actual core of the truth just the difference between all human truths?

I think it is better to conclude our discussion here. We seem to have lost ourselves hopelessly in the maze of "truth". It has become evident that we will never attain absolute certainty concerning truth. Not even untruth can be demonstrated with absolute certainty, as any refutation, like any assertion, is subject to uncertainty. In our search for the truth, we seem to be like the diver who reaches for a pearl on the seabed and returns with a handful of mud instead. In the end, truth remains just a regulatory idea, which however still appears to be indispensable—as, without it, the methods of acquiring knowledge would lack both an external goal and internal consistency.

References

1. Nietzsche F (1980) Aus dem Nachlaß der Achtzigerjahre. In: Schlechta K (ed) Friedrich Nietzsche. Werke in sechs Bänden, Bd VI. Hanser, München
2. Tarski A (1944) The semantic conception of truth. Philos Phenomenol Res 4 (3):341–376
3. Aristotle (1975) Categories and De interpretatione (transl: Ackrill JL). Oxford University Press, Oxford [Original: Peri hermeneias, 350 BC]
4. Patzig G (1981) Sprache und Logik. Vandenhoeck & Ruprecht, Göttingen
5. Crivelli P (1999) Aristotle on the Truth of Utterances. In: Sedley DN (ed) Oxford Studies in Ancient Philosophy, vol XVII. Oxford University Press, New York, pp 37–114
6. Peirce CS (1998) The Essential Peirce: Selected Philosophical Writings, vol 2 (1893–1913). Indiana University Press, Bloomington
7. Popper KR (1998) The World of Parmenides. Essays on the Presocratic Enlightenment. Petersen AF (ed). Routledge, London, New York
8. Diels H, Kranz W (1974) Die Fragmente der Vorsokratiker, Bd. 1–3. Weidmann, Berlin
9. Nietzsche F (1980) Götzen-Dämmerung (1889). In: Colli G, Montinari M (eds) Friedrich Nietzsche Sämtliche Werke: Kritische Studienausgabe Bd 6. De Gruyter, Berlin
10. Heisenberg W (1983) Schritte über Grenzen. Piper, München
11. Descartes R (1965) Discourse on Method, Optics, Geometry, and Meteorology (transl: Olscamp PJ). Bobbs-Merrill, Indianapolis [Original: Discours de la methode pour bien conduire sa raison, & chercher la verité dans les sciences: plus la dioptrique, les meteores, et la geometrie, qui sont des essais de cete methode, 1637]
12. Weizsäcker CF von (1980) The Unity of Nature. Farrar, Straus & Giroux, New York [Original: Einheit der Natur, 1971]
13. Locke J (1894) An Essay Concerning Human Understanding (1690), 2 vols, Campbell Fraser A (ed). Oxford University Press, Oxford
14. Leibniz GW (1982) New Essays on Human Understanding (transl: Remnant P, Bennett J). Cambridge University Press, Cambridge [Original: Neue Abhandlungen über den menschlichen Verstand, 1704]

15. Russell B (2004) History of Western Philosophy. Routledge, New York

16. Kant I (1998) Critique of Pure Reason. Cambridge University Press, Cambridge [Original: Kritik der reinen Vernunft, 1787]

17. Schelling FWJ (1990) Über den wahren Begriff der Philosophie und ihre richtige Art Probleme aufzulösen (1801). In: Schelling. Schriften von 1799–1801. Wissenschaftliche Buchgesellschaft, Darmstadt

18. Schelling FWJ (1980) Ideen zu einer Philosophie der Natur als Einleitung zum Studium dieser Wissenschaft (1797). In: Schelling. Schriften von 1794–1798. Wissenschaftliche Buchgesellschaft, Darmstadt. English edition: Schelling FWJ (1988) Ideas for a Philosophy of Nature: as Introduction to the Study of this Science (transl: Harris EE, Heath P, introduction Stern R) Cambridge University Press, Cambridge

19. Schelling FWJ (1980) Von der Weltseele (1798). In: Schelling. Schriften von 1794–1798. Wissenschaftliche Buchgesellschaft, Darmstadt

20. Goethe JW von (1988) Letter to Wilhelm von Humboldt, 22 Aug 1806. In: Mandelkow KR (ed) Goethes Briefe, Bd 4. Beck, München

21. Schlegel F (1803) Europa. Eine Zeitschrift, Bd 1, Heft 1. Wilmans, Frankfurt am Main

22. Ritter JW (1800) Beyträge zur nähern Kenntniß des Galvanismus und der Resultate seiner Untersuchung, Bd 1. Frommann, Jena

23. Goethe JW von (1981) Goethe. Werke, Bd 13. Beck, München

24. Daiber J (2001) Experimentalphysik des Geistes. Novalis und das romantische Experiment. Vandenhoeck & Ruprecht, Göttingen

25. Ostwald W (1902) Vorlesung über Naturphilosophie. Veit & Comp, Leipzig

26. Peters CAF (ed) (1862) Briefwechsel zwischen CF Gauß und HC Schumacher Bd 4. Esch, Altona

27. Hegel GWF (1986) Werke, Bd 9. Suhrkamp, Frankfurt am Main. English edition: Hegel GWF (1970) Hegel's Philosophy of Nature, 3 vols (transl: Petry MJ). Allen & Unwin, London [Original: Enzyklopädie der Philosophischen Wissenschaften im Grundrisse. Zweiter Teil: Die Naturphilosophie, 1830]

28. Liebig J (1840) Über das Studium der Naturwissenschaften und über den Zustand der Chemie in Preußen. Vieweg, Braunschweig

29. Wöhler F (1901) Briefwechsel zwischen J. Berzelius und F. Wöhler, Bd 1. Engelmann, Leipzig

30. Schleiden MJ (1863) Über den Materialismus der neueren deutschen Naturwissenschaft, sein Wesen und seine Geschichte. Engelmann, Leipzig

31. Mises R von (1968) Positivism. A Study in Human Understanding. Dover Publications, New York [Original: Kleines Lehrbuch des Positivismus, 1939]

32. Wittgenstein L (2001) Philosophical Investigations (transl: Anscombe GEM). Blackwell, Oxford [Original: Philosophische Untersuchungen, 1953]
33. Carnap R (1945) On inductive logic. Philos Sci 12:72–97
34. Popper KR (1992) The Logic of Scientific Discovery. Routledge, London, New York [Original: Logik der Forschung, 1935]
35. Popper KR (1974) Unended Quest. An Intellectual Autobiography. Fontana-Collins, London, Glasgow
36. Küppers B-O (1983) Molecular Theory of Evolution: Outline of a Physico-Chemical Theory of the Origin of Life. Springer, Berlin, Heidelberg
37. Popper KR (1972) Objective Knowledge: An Evolutionary Approach. Oxford University, Oxford
38. Quine WVO (1960) Word and Object. MIT Press, Cambridge, MA
39. Kuhn TS (1962) The Structure of Scientific Revolutions. University of Chicago Press, Chicago
40. Lorenz K (1977) Behind the Mirror: A Search for a Natural History of Human Knowledge (transl: Taylor R). Harcourt Brace Jovanovich, New York [Original: Die Rückseite des Spiegels, 1973]
41. Haeckel E (1905) The Wonders of Life. A Popular Study of Biological Philosophy. Harper & Brothers, New York/London 1905 [Original: Die Lebenswunder, 1904]
42. Eigen M (2013) From Strange Simplicity to Complex Familiarity. Oxford University Press, Oxford
43. James W (1907) Pragmatism: A New Name for Some Old Ways of Thinking. Longmans, Green & Co., London/New York
44. Wolenski J (2004) *Aletheia* in Greek thought until Aristotle. Ann Pure Appl Log 127(s 1–3):339–360
45. Habermas J (1984, 1987) Theory of Communicative Action, 2 vols [Original: Theorie des kommunikativen Handelns, 1981]
46. Austin JL (1962) How to Do Things with Words. Harvard University Press, Cambridge, MA
47. Searle J (1969) Speech Acts: An Essay in the Philosophy of Language. Cambridge University Press, Cambridge
48. Wittgenstein L (1967) Zettel (eds: Anscombe GEM, Wright GH von). Basil Blackwell, Oxford
49. Wittgenstein L (1999) Tractatus logico-philosophicus. Dover Publications, New York [Original: Logisch-Philosophische Abhandlung, 1921]
50. Rorty R (1989) Contingency, Irony, and Solidarity. Cambridge University Press, New York

51. Rorty R (1982) Consequences of Pragmatism. Essays 1972–1980. University of Minnesota Press, Minneapolis.
52. Nietzsche F (1980) Die fröhliche Wissenschaft (1882). In: Colli G, Montinari M (eds) Friedrich Nietzsche Sämtliche Werke: Kritische Studienausgabe Bd 3. De Gruyter, Berlin

3

Methods: Ways of Gaining Knowledge

3.1 Implicit Knowledge

In answer to the question "What, then, is time?" Augustine, the early Christian teacher, once gave the mysterious answer: "If no one asks me, I know; if I want to explain it to someone who asks me, I do not know." [1, chapter XIV, p. 295] It seems as though Augustine wanted to distinguish here between two forms of knowledge: an "explicit" and an "implicit" knowledge. Explicit knowledge is knowledge that can be expressed in words and communicated to others, while implicit knowledge is a hidden, a silent knowledge. The latter is the knowledge that Augustine had of time, which he was unable to express in words. If one takes the idea of implicit knowledge seriously, it means nothing less than that "we can know more than we can tell" [2, p. 4].

There sees to be some evidence for the existence of hidden or, as the physicist and philosopher Michael Polanyi called it, "tacit" knowledge. Even in our daily life, we often rely on knowledge that is not immediately comprehensible. Thus, we seem to be able to distinguish between objects and appearances in our world without being able to name the differences explicitly. For example, hardly anyone will be able to say precisely what "mind" or "consciousness" is. Nevertheless, even today we have no trouble in distinguishing a human being from a soulless robot, no matter how artfully the robot is constructed. In general, it is easy for us to differentiate between something animate and something inanimate, without having even a halfway exact definition of the difference.

Somehow, we seem to know inwardly what mind, consciousness, life, time and so forth are, even if we cannot adequately express this knowledge. Without such an implicit understanding of objects and appearances, we would probably have no reasonable access to the world at all—a thought that is reflected in Kant's ideas of the *a priori* forms of cognition (Sect. 2.5).

© Springer Nature Switzerland AG 2022
B.-O. Küppers, *The Language of Living Matter*, The Frontiers Collection,
https://doi.org/10.1007/978-3-030-80319-3_3

Even at the beginning of theory formation in science, there is always an implicit, if only preliminary, understanding of the objects of cognition. Only as a theory is progressively shaped does the tacit pre-understanding finally recede more and more behind increasingly explicit forms of objective knowledge. At the same time, the basic concepts borrowed from everyday language take on ever-sharper contours, until they ultimately merge into the "theoretical" terms of scientific language. This act of gradual specification is excellently illustrated by the development of fundamental physical terms such as force, energy and the like (cf. [3]).

The origin of explicit knowledge can probably not be understood without assuming the existence of implicit knowledge. Of course, experience has a knowledge-forming function as well, but it cannot be the sole source of knowledge. Even the idea that there could be a pure, i.e. unmediated, experience proves to be problematic: the moment we focus our attention on an object of knowledge, our perception is already guided by inner states of knowledge.

Moreover, experience can only unfold its regulative and corrective function if it relies on background information against which its contents can be ordered and evaluated. This background information is, likewise, an item of knowledge which, as a rule, has emerged from experience itself. However, it must also contain elements of implicit knowledge if it is not to end up in an explanatory circle. This difficulty was also the point at which, in the eighteenth century, the conflict between rationalism and empiricism ignited (Sect. 2.4).

The idea of implicit knowledge seems indispensable for a consistent, self-contained theory of knowledge. The history of this thought is equally old. Already Plato's doctrine of the soul's remembrance (*anámnēsis*) of previously contemplated ideas is in principle a doctrine of implicit knowledge. It was intended to make understandable the notion hat insights into general contexts cannot be acquired through sensual experience. From Plato's theory of ideas, in turn, a direct path leads to the neo-Platonic notion of the "archetype" as the original form of all knowledge.

In the early seventeenth century, the astronomer and mathematician Johannes Kepler returned to the antique idea of the archetype and made it the starting point of his doctrine of the construction of the world. Kepler believed that specific patterns were implanted in the depths of the human soul, and that these would enable mankind to fathom the structure of the world by comparing sensually perceived objects with these inner archetypes. For example, Kepler considered geometry to be the archetype of beauty and mathematical proportions to be the archetype of harmony—a world-view that subsequently attracted many mathematicians and physicists.

In the twentieth century, Carl Gustav Jung's idea of archetypes in depth psychology led to a renaissance of this concept. Sigmund Freud had believed that the unconscious comprised items that formerly belonged to the realm of the conscious, but then, for various reasons, has been repressed out of it. In contrast to Freud, Jung posited that repressed memories make up only a small and rather insignificant part of the unconscious. Instead, he claimed, the unconscious consists mainly of archetypical, collective patterns, the "archetypical ideas". Among these he ranked the ideas of "father", "mother", "son", "God" and the gender-related images of the human spirit *animus* and *anima*.

According to Jung, the archetypical ideas in turn point to an unclear, empty basic form, the archetype in itself. By this, Jung understood a purely formal element of the unconscious that itself is incapable of consciousness but has a formative and regulatory function for the archetypical ideas. Thus, for example, Jung compared the archetype to "the axial system of a crystal, which, as it were, preforms the crystalline structure in the mother liquid, although it has no material existence of its own" [4, p. 79]. Following Jung, the unconscious must be seen as a hidden reality *par excellence*, which has never before been in the consciousness of man and which is fundamentally inaccessible to consciousness. He thus regarded any attempt to lift the unconscious into consciousness as a highly contradictory undertaking.

The notions of innate ideas, archetypes and the like have one thing in common: they should make us understand how we attain knowledge of the real world and how we create connections in our knowledge base before we become aware of this knowledge. However, as little as the unconscious can be raised into consciousness in the Jungian sense, just as little can implicit knowledge be made explicit. One even has to ask whether it is possible to talk meaningfully of implicit or "tacit" knowledge at all without becoming entangled in contradictions.

Nevertheless, Polanyi tried to clarify the problem of tacit knowledge [2]. He attached the utmost importance to this kind of knowledge, as he was convinced that we use this knowledge in the same way as we exercise our bodily functions. Just as our body serves us as a "probe", through which we gain knowledge of the outside world, tacit knowledge serves us as a base from which the entire internal world of our thinking emerges—starting with the forms of primitive knowledge that we have of sensually perceived external objects, up to the knowledge of premonitions, problems, skills and the like. According to Polanyi, it is impossible to establish the essence and validity of our knowledge without encountering its underlying forms of dependence and conditionality.

However, this would mean not only that implicit knowledge is a prerequisite for any objective knowledge, but also that the structures of tacit knowledge remain hidden for all time in the dark zone surrounding our consciousness. Polanyi inferred from this that one encounters a paradox here, one that questions fundamentally the knowledge ideal of modern science: If the declared goal of science is strictly objective knowledge, then, according to Polanyi, it must contain no trace of tacit, non-objective knowledge. Ideally, tacit knowledge should be absorbed entirely into explicit knowledge. Conversely, if implicit knowledge is the indispensable basis of all explicit knowledge, then the idea of objective knowledge would, in fact, amount to what Polanyi called the "destruction" of all knowledge. Accordingly, the knowledge ideal of modern science could ultimately prove to be a fallacy, since it would be inherently contradictory. Such considerations, however, amount to little more than a mental game; there are sound reasons to believe that, in reality, complete knowledge can never be achieved anyway (Sect. 5.4).

We must therefore, like Polanyi, presume that implicit knowledge will always remain a blind spot in our understanding of the world. However, this does not necessarily mean that there is no access whatsoever to implicit knowledge. The mere fact that this form of knowledge appears to be a necessary model of thought already assigns it a place among our explicit ideas. If, conversely, implicit knowledge were to belong to a completely different world of the imagination, this would mean a fundamental break in our conception of what we understand by knowledge.

Consequently, the demand that the world of our ideas must be self-consistent means that the conceivable forms of implicit knowledge are already delimited and pre-structured. For example, if we take as our starting point the idea that the world is recognizable because it possesses the structure of a language, then we are compelled, for the sake of consistency, to transfer this idea to implicit knowledge. Against this background, the famous dictum of psychoanalyst Jacques Lacan [5] that "the unconscious is structured like a language" gains a profound meaning, one that is in line with our considerations in the first chapter of this book.

3.2 The Aesthetics of Cognition

Just as enigmatic as implicit knowledge is the creative process in which the structures of our explicit understanding emerge. A first hint of the underlying mechanism may be the fact that the fundamental structures of our explicit knowledge—despite all the enrichment that they have accrued in the long cultural history of mankind—remain largely the same. In Chap. 2 we interpreted this with the idea that the structures of our knowledge go back to a limited number of figures of thought, which, under the influence of accumulating experience, are repeatedly recombined with each other to yield new ideas.

A similar mechanism underlies human language. The richness of linguistic forms of expression also arises from the recombination of basic elements—those of language. Analogously, one could regard the primary figures of thought as elements of a "superlanguage" from which the complex structures of human knowledge emerge. In the creative act of the cognition process, the known elements of our knowledge are turned back and forth and recombined time and again until they suddenly assemble into new cognition. We describe this process metaphorically as a "flash of thought".

It is remarkable that biological evolution is also based on this creative principle. Here, sexual reproduction serves as a tool to recombine continually the genetic material of organisms. In this way, new forms of expression of genetic information emerge again and again, and these are then tested for their biological efficiency by natural selection. The molecular geneticist François Jacob [6] accurately described the creative process of natural evolution as a constant "tinkering" with existing genetic structures. Accordingly, evolutionary change takes place through a continuous rearrangement and redesign of structures of living matter that already exist.

Let us return to the evolution of knowledge. Arthur Koestler [7] described the sudden leap of creative imagination, which combines two previously unconnected ideas, observations, perceptual structures or thought universes in a new synthesis, as "bisociation". With this term he wanted to characterize the creative act as a process which, in contrast to "association", connects elements that come from entirely different domains of human thinking.

The creative play of thought, however, would lose itself in the infinite vastness of thinking if our background knowledge did not at the same time have an organizing function. The harmonization of the new with what is familiar to us—Koestler speaks of the dissolution of an irritating problem, perceived as dissonance—may also evoke the feeling of beauty that scientists have always reported in connection with their discoveries. Accordingly, Koestler considered the creative act of bisociation as a characteristic common to both scientific and artistic creativity.

In fact, the parallels between these two forms of creativity cannot be dismissed. Physicists and mathematicians have repeatedly emphasized the common features of the sciences and the arts. Werner Heisenberg [8], for example, saw in the fresh awareness of new ideas an event of "artistic vision" and "semiconscious premonition" rather than a process of rational cognition. Carl Friedrich von Weizsäcker held the same opinion. He noted: "The truly productive, truly eminent researcher is distinguished by an instinct for, a feel for, a not quite analyzable perception of interrelations that is deeper than that of most other people; this accounts for his being the first to arrive a particular truth. A scientific truth is almost always surmised, then asserted, then fought over, and then proven. This is essential, it is the nature of science, it cannot be any different." [9, p. 99]

It is the precursors of rational cognition—such as inkling, intuition, inspiration and the like—that bring the sciences close to art. At the same time, however, they also surround science with the mantle of mysteriousness. François Jacob [10] went so far as to describe this side of science as "night science", which precedes the actual "day science" because it is based not on sound and unbroken argumentation, but on instinct and intuition. Night science is for Jacob "a sort of workshop of the possible where what will become the building material of science is worked out. Where hypotheses remain in the form of vague presentiments and woolly impressions. Where phenomena are still no more than solitary events with no link between them. Where the design of experiments has barely taken shape. Where thought makes its way along meandering paths and twisting lanes, most often leading nowhere. At the mercy of chance, the mind thrashes around a labyrinth, deluged with signals, in quest of a sign, a nod, an unexpected connection." [10, p. 126]

Intuition and inkling play an essential role even in such a rigorous science as mathematics, which is supposed to adhere to strict, logical reasoning. Famous examples of this are the mathematical "conjectures". These are assertions that are based on thought and reflection but cannot yet be proved. The fact that it has sometimes taken centuries to find proofs of these underlines the importance of instinct and intuition for creativity in science.

Some mathematical conjectures have become famous far beyond mathematics, such as Pierre de Fermat's seventeenth-century conjecture, which was proved only a few years ago, or the still unproven conjectures in number theory by Christian Goldbach (1742) and Bernhard Riemann (1859). Such conjectures mostly result from the generalization of one or more individual cases, raising the question of whether properties verifiable in individual cases are also displayed by an infinite number of cases.

Fermat's conjecture, for example, takes up the well-known theorem of Pythagoras and asserts that this theorem cannot be generalized. Another famous conjecture, made half a century ago by Yutaka Taniyama and Goro Shimura, is based on an analogy, linking the two very different subjects of topology and number theory. Incidentally, the proof of the Taniyama-Shimura conjecture by Andrew Wiles in 1993 also proved Fermat's conjecture at the same time. Since then, both conjectures have been regarded in mathematics as irrevocably true theorems.

Another fascinating topic is that of the aspects of scientific cognition that are guided by aesthetic perception. Mathematicians and natural scientists in particular are often inspired by a sincere belief in the beauty and harmony of Nature and are convinced that the perception of beauty opens up access to scientific truth. One of the most impressive works of modern science, the *Harmonices Mundi* (The Harmony of the World) published in 1619 by Johannes Kepler, took up in its title the leading idea of the harmonious construction of the world [11].

In the context of the connection between beauty and truth, another legacy of Platonism comes to light. Plato believed that the idea of beauty as the epitome of highest perfection, together with the idea of truth and the idea of the good, forms a unity that tops the hierarchy of all ideas. The unity of these ideas was, for Plato, the highest level of the spiritual-divine being. In the neo-Platonism of the third century, this view finally took on a mystical character. Plotinus, for example, believed that the splendor of the "absolute One" shone through all material appearances and was reflected in every part of the universe. Consequently, the world was already to be considered beautiful, since it had arisen from divine creation and mirrored the fullness and glory of the Creator. At the same time, the beauty of creation had to be an absolute truth, because it could only be so, and not otherwise.

With the Age of Enlightenment and the beginning of modern science, the contemplative admiration of creation gave way to a sober analysis of reality. The phenomenon of beauty and its function as a guide of cognition also became the subject of rational scientific thought. A treatise by Alexander Gottlieb Baumgarten published in 1750/58 under the title *Aesthetica* was ground-breaking in this. Baumgarten had derived the artificial word "Aesthetica" from the Greek word *aisthesis*, meaning perception or sensation. In the broadest sense, it denotes the doctrine of the manifestly beautiful, and in the narrower sense the science of sensual cognition and representation.

Baumgarten believed that sensual cognition is a form of cognition that assists logically structured cognition. He also denoted sensual cognition as "the younger sister" of logic [12, vol. 1, §13; author's transl.]. Consequently, Baumgarten attempted to supplement (the in his time prevailing) rationalism with the elements of a theory of sensual cognition. However, despite all philosophical efforts to render the idea of beauty more precisely, it has remained somehow dark and vague. Like the notion of truth, the idea of beauty has disintegrated into a number of diverse concepts. "Subjective beauty", "objective beauty", "existential beauty" and "natural beauty" are only some examples of the variety of concepts that have emerged around the idea of beauty in the course of time.

Let us consider the concept of natural beauty. Since antiquity, natural beauty has been regarded as the epitome of the lawful order of Nature and thus as one of the sources from which we draw our knowledge about Nature. However, what about the alleged "beauty" of the abstract theories that enthuse mathematicians and physicists? What can be attractive about a dry-as-dust physical theory like the kinetic theory of gases? Ludwig Boltzmann, the pioneer of statistical physics, could go into raptures about the elegance and beauty of mathematical formalisms. For him, the intensity of the aesthetic experience of formal representation seemed to be comparable only to artistic experience.

The following passage provides an example of Boltzmann's enthusiasm for the beauty of science: "Even the Pythagoreans recognized the similarity between the most subjective and the most objective of the arts.—Ultima se tangent". And how expressive, how wonderfully exact mathematics is. Like the musician who recognizes Mozart, Beethoven, Schubert after the first few bars, the mathematician can distinguish between his Cauchy, Gauß, Jacobi and Helmholtz after a few pages. The highest external elegance, sometimes a weak skeleton of the conclusions, characterizes the French; the greatest dramatic impact the Englishmen, above all Maxwell. Who does not know Maxwell's dynamic gas theory? At first, the Variations of velocity develop majestically. Then, from one side, the equations of state come in, and from the other side the equations of central motion. The chaos of formulae rises ever higher. Suddenly, four words resound: Put n equal to 5. The evil demon V disappears, falling silent like, in music, a wild figure of the basses that undermines everything; at a magic stroke, everything that seemed intractable suddenly falls into place. There is no time to say why this or that substitution was made; if one doesn't sense it, then one should put the book aside. Maxwell is not a program musician who has to annotate his score with explanations. The formulae now churn out one result after another until, surprisingly, as a final effect, the thermal equilibrium of a heavy gas is attained and the curtain falls." [13, p. 50 f.; author's transl.]

Of course Boltzmann is parodying the ideal of beauty in mathematical physics. Boltzmann was not only a brilliant physicist, but he also possessed a sense of humor that he could have gone on stage with. Nonetheless, one cannot fail to see that his parody is at the same time a eulogy on the aesthetics of scientific cognition. It expresses in an inimitable way what many mathematicians and natural scientists believe in: that there is a deep connection between formal beauty and scientific truth.

Henri Poincaré, for example, even pointed out explicitly that his discoveries had been influenced very much by a sense of mathematical beauty. He described this as "a true aesthetic feeling" for the harmony of numbers, forms and geometric elegance that "all real mathematicians know" [14, p. 391]. Even more downright was the confession of Godfrey Hardy concerning the beauty of mathematics. He asserted apodictically: "Beauty is the first test; there is no permanent place in the world for ugly mathematics." [15, p. 85] For this very reason, many mathematicians today reject evidence obtained with the help of computers. To them, computer-aided proofs do not seem original, because they often require a significant amount of computational effort, which lacks the elegance and coherence of a simple and thus aesthetically satisfying representation—quite apart from the fact that computer proofs are often opaque and resistant to checking.

Not only mathematicians, but also physicists (see above) have occasionally confessed profusely to the ideal of scientific beauty. The theoretical physicist Paul Dirac, for example, once claimed that "it is more important to have beauty in one's equations than to have them fit experiment" [16, p. 49]. One may wonder whether such overt confessions are still objectively justifiable or whether they can only be seen as excesses of abstract science, one that stylizes the very process of acquiring knowledge into an art form and sacrifices its claim to truth uncritically on the altar of formal beauty.

This aspect of science, however, is not new. Even in antiquity, models had been developed that attempted to explain the structure of the world by referring to the ideal of formal beauty and harmony. Here the Pythagoreans' number theory and Plato's natural philosophy probably had the most

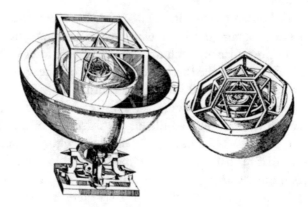

Fig. 3.1 Kepler's model of the planetary system. In his *Mysterium Cosmographicum*, published in 1596, Kepler [17] tried to give an explanation of our planetary system based on the teachings of Copernicus. He supposed that the six then known planets (Mercury, Venus, Earth, Mars, Jupiter, Saturn) moves on spherical shells into which the five regular platonic solids (dodecahedron, tetrahedron, hexahedron, icosahedron, octahedron) are inscribed. Each platonic body is touched by two neighboring planetary spheres and thereby determines the distance between the outer and the inner sphere. In the figure (left) the spheres of Saturn, Jupiter and Mars are highlighted (from the outside to the inside). Between the planetary spheres of Saturn and the neighboring sphere of Jupiter, a hexahedron is inscribed, and between Jupiter and Mars a tetrahedron. The earth, in turn, was supposed to move on a shell with an "inner" icosahedron (shown on the right)

significant influence on later thinking. In particular, Plato's thought that the world is built on regular structures, the Platonic solids, has inspired the imagination of natural scientists for centuries (Sect. 2.2).

Kepler, in particular, used the idea of the Platonic solids to provide a "harmonious" explanation for the structure of our planetary system. He believed that the planets move on spherical shells, in which the five regular platonic solids are inscribed (Fig. 3.1). Moreover, Kepler had made the astonishing discovery that the difference between the maximum and minimum angular speeds of a planet in its orbit approximate to a harmonic proportion. In the *Harmonices Mundi*, Kepler even compared the movements of the planets around the sun to the vibrations of a string, generating the "music of the spheres". Although later astronomical discoveries did not fit this model and brought some dissonance into Kepler's harmony of the planetary motions, Kepler stressed unflinchingly his admiration for the perfection of the divine creation: "I feel carried away and possessed by an unutterable rapture over the divine spectacle of heavenly harmony." (Quoted from [18, p. 267])

For a long time, the platonic solids were regarded as symbols of the symmetry principle discovered by the Greek philosophers of Nature. However, in contrast to the antique conception, according to which symmetry was perceived as a harmonious balance between a whole and its parts, the modern concept of symmetry has largely lost its contemplative character. In physics, for example, one speaks of the "symmetry" of objects, states, or laws of Nature if these remain invariant under the application of specific mathematical transformations. That is the case, for example, when objects— such as the regular solids—maintain their shape when they are rotated or reflected.

The idea of symmetry has become an integral part of our understanding of Nature. It is, as the mathematician Hermann Weyl put it, an "idea by which man through the ages has tried to comprehend and create order, beauty and perfection" [19, p. 5]. In fact, symmetry principles can be found in almost all areas of inanimate and animate Nature. The most spectacular form of (higher) symmetry is the so-called supersymmetry, which nowadays forms an important explanatory basis for the physics of elementary particles.

Sometimes, symmetry principles also have a guiding function for modern research of Nature. When a few decades ago chemists started to search more intensively for symmetric molecules in their laboratories, they discovered not only "dodecahedrane", which is made up of carbon and hydrogen in the form of a dodecahedron, but in doing so they also improved considerably the preparative and analytical techniques of chemistry. In recent times, the search for highly symmetrical molecules has led to spectacular syntheses. For example, the chemists Pierre Gouzerh and Achim Müller succeeded in synthesizing a giant molecule consisting of 700 atoms with perfect symmetry [20]. It looks like a large sphere with an inscribed icosahedron (Fig. 3.2). The shape of this molecule bears a striking resemblance to Kepler's amazing model of the planetary spheres with inscribed platonic solids. Consequently, giant molecules of this kind has been named "Keplerates".

We now see that beauty is spoken of in the sciences in two senses. One refers to the beauty of Nature as it presents itself to sensual perception. The other refers to formal elegance, which is generally regarded as a kind of order and harmony inherent in mathematical thought structures. If, however, the idea of beauty is indeed a true idea that underlies natural events and points the way to scientific truth (as Werner Heisenberg asserted [8]), then there must necessarily be a close connection between natural beauty and formal elegance.

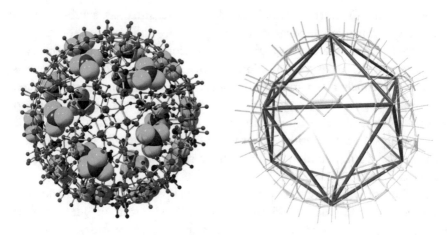

Fig. 3.2 A giant "Keplerate" molecule. The molecule's outer shell is a sphere of 132 atoms of molybdenum (Mo). Another sphere of 60 molybdenum atoms is enclosed within the molecule. The total number of atoms (including other elements) is 700. Despite its large number of atoms, the outer sphere has an icosahedral structure. It is composed of 12 {Mo_{11}} fragments in such a way that the five-fold rotational axes remain in the resulting shell. Since one can discern an icosahedron inscribed into a sphere, this molecule is structurally similar to Kepler's shell-model of the cosmos. [Images: courtesy of Pierre Gouzerh and Achim Müller]

With the current state of knowledge, we can only speculate about the nature of such a connection. Perhaps the following train of thought can point in the right direction (cf. [21, chapter 6]): By perceiving something as beautiful in Nature, we recognize in the complexity of its underlying phenomena something that is harmoniously ordered because it exhibits regular features. Such regularities, in turn, point to lawful patterns that can be formulated as objective characteristics of Nature. Following this thought, we may conclude that our aesthetic perception serves as an instrument that filters and checks the perceived complexity of Nature, looking for regularities. In this way, natural phenomena finally become transparent from a formal perspective and, thus, potentially calculable.

From this standpoint, it would appear understandable why regularity, symmetry, proportion, coherence, elegance and the like—in short, aesthetic aspects—play such an essential part in the acquisition of scientific knowledge. It is precisely these characteristics that go hand in hand with a law-like description of Nature which categorizes, and thus reduces, the complexity of natural appearances. In other words: the beauty of Nature and formal beauty can be brought together under the mantle of simplicity, beneath which the idea of beauty exercises its function as a guiding idea in gaining cognition—in accordance with the motto that once adorned the entrance to the physics auditorium of the University of Göttingen: *Simplex sigillum veri*—Simplicity is the seal of truth.

Science and technology have also influenced contemporary music. For example, the composer Iannis Xenakis has included elements of probability theory, number theory and set theory in his extremely formal compositions. Beside, the intervals and tone lengths in his compositions are often built up according to the pattern of a geometric series or the proportions of the Golden Section—stylistic elements that have been commonplace in the fine arts and architecture since Greek antiquity. Similarly, György Ligeti was inspired by scientific techniques in his mechanical and electronic compositions.

The close relationship between science and art is evident not only in the aestheticization of science but also in the scientization of art. For centuries, the visual arts have repeatedly turned to pseudo-scientific and scientific doctrines such as Pythagorean number theory, the doctrine of the Golden Section or the principles of representative geometry. Especially in modern painting, there have been attempts to bring the conditions of scientific objectivity into the subjective world of art. Paul Cézanne's doctrine of the basic geometric elements of Nature—sphere, cylinder and cone—bears witness to this, as do Wassily Kandinsky's writings on the formal foundations of abstract painting.

In the course of modern genome research, the idea came up to transcribe the sequence pattern of genes into sequences of sounds. Figuratively speaking, this would mean to transliterate the language of the living into the language of music, as it were into a "symphony of the living". Some decades ago, a similar experiment was started, transcribing the harmonic proportions of the motions of the planets into a sequence of sounds which Kepler believed to reflect the harmony of the world. In the end, however, it turned out that Kepler's "music of the spheres" was just what it always has been: music for the intellect, but not for the ear [22].

Scientific techniques were first used in arts at the end of the nineteenth century, when the development of photography made possible new forms of artistic expression. Nowadays, the techniques in primary use are the so-called digital media, first and foremost the various computer-aided techniques. An important branch of this is the so-called computer art, with its diverse styles of production and presentation ranging from image-, sound- and video-installation to the generation of art works with the help of creative algorithms (Fig. 3.3).

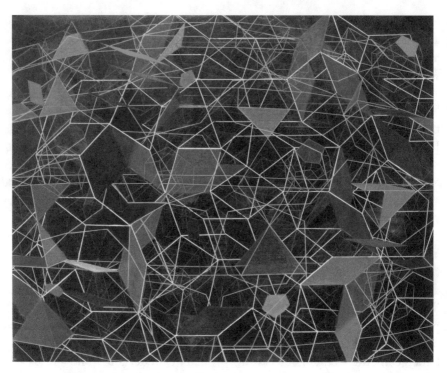

Fig. 3.3 Computer art. An attempt to visualize four-dimensional geometry. Computer artist Tony Robbin generated this painting by superimposing five sheets, each representing a specific geometric pattern (for details see [23]). This technique generates an impression of multi-dimensional space. In this way, art may fill a gap in our limited intuition and build a bridge to the abstract theoretical concepts of physics. Importantly, higher-dimensional spaces play a fundamental role in a theoretical understanding of the origin of genetic information (Sect. 6.10). In the artistic composition shown above one may also perceive a certain relation to the Platonic idea of the construction of the world from geometric archetypes (Sect. 2.2). [Tony Robbin, Acrylic on Canvas, 56 × 70 inches. Collection of the Artist, 2006]

An impressive example of the scientization of the arts—or perhaps the aestheticization of science—is the visualization of the so-called Mandelbrot set by modern computer techniques. The Mandelbrot set, named after the mathematician Benoît Mandelbrot, is a set of points in the complex number plane (Fig. 3.4). It is generated by the application of a "recursive" algorithm. This is a calculation rule using the result of a calculation as the initial value for the next, similar, calculation (for details see the figure caption).

◄**Fig. 3.4** Mandelbrot set and details of its boundary at increasing magnification. Mathematical algorithms may have a complex deep structure which, however, can only be made visible with the help of a computer. This example shows the boundary of the Mandelbrot set, which possesses an inexhaustible wealth of details. Mathematically speaking, the Mandelbrot set is a point-set in the complex number plane. It is calculated by using a simple (recursive) algorithm. In a first step, a real number z (starting with $z = 0$) is multiplied by itself, and the resulting number z^2 is increased by adding a complex number c. The result is a new value for z. In the subsequent steps, this procedure (squaring z and adding c) is repeated again and again, each time using the value of z that was calculated in the preceding step. All numbers that remain finite after repeated application of the algorithm constitute the Mandelbrot set. In contrast, all numbers outside the Mandelbrot set increase toward infinity. The most exciting part of the Mandelbrot set is its edge. Here, we see many tiny figures that each have the shape of the Mandelbrot set. If one looks more precisely at the border (zoomed in toward by the computer), one sees details of infinite variety which have a strong aesthetic quality. However, the pictures shown have objective and subjective features: the deep structure of the graphic is determined by the mathematical formalism, while the coloring of the graphic is subject to the scientist's (or artist's) creative freedom. In the example shown, all the points outside the Mandelbrot set are colored, with the depth of the color depending upon the speed with which the iteratively calculated number z increases toward infinity. [Images: Wolfgang Beyer, Wikimedia Commons]

The Mandelbrot set does not only allow a greater depth of insight into the essence of formal beauty; it also demonstrates that even simple algorithms can have creative power and lead to highly complex structures. In the present case, the never-ending complexity becomes visible if one zooms into the boundary of the Mandelbrot set. Such computer-assisted methods do not only reveal the aesthetics of mathematical structures; they have also inspired the new research field of experimental mathematics (cf. [24]). Moreover, the findings have implications for our physical and chemical understanding of Nature, because the algorithms investigated also underlie numerous natural phenomena. Algorithms have even been detected that generate, with deceptive accuracy, artificial structures resembling biological patterns. They have opened the door to a virtual world, in which the distinction between natural and artificial, living and non-living is becoming ever more unclear (cf. Chap. 6).

3.3 Is "Understanding" a Method?

It is no coincidence that there are so many parallels between science and art. Both seek to uncover aspects of reality that are beyond the sphere of immediate perception and thus beyond everyday experience. Furthermore, both make extensive use of symbols. However, with all the parallels that exist between science and art, we must not lose sight of what separates them. First of all, there are differences in the information content of science and art. The one is objective, the other subjective. The one is clear and general; the other is ambiguous and depends on interpretation. Furthermore, science relies on logical conclusions, art on sensual perceptions. Finally, in contrast to science, there is no methodology to which artistic creativity is submitted: quite on the contrary, an existing work of art can be replaced at any time by the creation of something new. Innovativeness as such is regarded as a sign of artistic quality, as it suggests originality and creative individuality. In contrast, in science, tried and tested methods and concepts are retained for as long as possible: if something, once considered proven, has one day to be abandoned and to be replaced by something entirely new, then in science this is regarded as revolutionary.

The differences between science and art become evident as soon as one tries to apply the methods of the exact sciences to art. This is not to say that works of art could not be investigated scientifically. On the contrary, it is quite possible to check, for example, the authenticity of works of art—their dating, the color composition, or the possible overpainting of a picture—by applying physical methods. However, scientific investigations of this kind can at best only reveal certain material aspects of a work of art, and they cannot provide any information about its artistic content.

Even if works of art are often subjected to formal criteria, this does not exhaust their aesthetic content. For this reason, the philosopher Max Bense [25] and others were doomed to fail in their attempt to find—by applying information theory—a quantitative measure for aesthetic aspects of natural and artistic objects. Just as one cannot deduce the content of a message from its mere form, one cannot infer from the form of a particular art work its aesthetic substance. In short, there is no definite, law-like link between form and meaning. Instead, each interpretation of an work of art is highly context-dependent. The context includes all relationships to the world that are induced by a specific art work and which are reflected in the unlimited variety of its possible interpretations. The analysis of the meaning content of a work of art thus goes far beyond the mere study of its form. In the simplest case, it leads to basic questions of what the work represents and what the artist wanted to express by this. Such questions, aimed at the artist's intentions, are supposed to make the substance of his work understandable in an objective sense. However, even such superficial aspects of meaning already go beyond the scope of scientific explanations, because the understanding of art cannot be reduced to the history of its genesis. A work of art always affects our inner experience. This, however, cannot be squeezed into the simple form of a causal explanation. Seen in this way, "understanding" is always more than "explaining".

Needless to say, the problem of understanding is not limited to art. The same problem arises in all areas of life, starting with everyday questions about the meaning of a gesture or the sense of an action up to profound and final questions regarding the meaning of human existence. All these items of meaning are interrelated, and their connecting element is what constitutes our historical and cultural world. This world is essentially handed down to us in documents. That is why the interpretative understanding of reality is, amongst other things, the understanding of what the records tell us. Art works are just as much documents as are writings, archaeological pieces of evidence and other relics and traces of human activity. As mentioned in Sect. 1.10, the interpretation of such records is the central concern in the field of hermeneutics. However, our considerations show that the analysis of an art work is a highly subjective form of understanding, which seems to remain largely impervious to strict methodology, unless one could put hermeneutics on a scientific fundament.

In fact, as early as the seventeenth century, Baruch de Spinoza had already considered how hermeneutics as a method of text interpretation could be shored up in such a way that it guarantees the highest degree of clarity and unrestricted validity for the understanding of texts. Spinoza referred to the rigor of the rationalist method of cognition, which he also recommended as a model for the technique of interpretating texts. From the perspective of his philosophy, which started from the absolute identity of Mind and Nature, this methodological approach is entirely understandable, for, if Mind and Nature—as Spinoza thought—are two attributes of the same essentiality, namely the divine being, then it followed that the understanding of books, created by humans, should be approached in the same way as that with which one reads the "Book of Nature". In his time this was the generally accepted, ideal path to discovery, referred to as rationalism (Sect. 2.4).

In the nineteenth century, the philosopher and psychologist Franz Brentano stepped up the claim to accuracy when he stated succinctly that "the true method of philosophy is none other than that of the natural sciences" [26, pp. 136/137; author's transl.]. Almost at the same time, Friedrich Albert Lange published his book, at the time widely acclaimed, on the *History of Materialism and Critique of its Meaning in the Present*. Lange took a path similar to Brentano's. He warned against an ideological overstretching of the materialism of natural sciences, but he advised at the same time that linguistics and other disciplines of the humanities should adhere to the methodology of the natural sciences if they intend "to promote true and lasting knowledge, in whatever field, even by a single step" [27, p. 829; author's transl.].

Centering on its methodology is characteristic of the phase of upheaval in the science of the nineteenth century in which—a consequence of the stormy development of natural science and technology—almost all branches of science seemed to get caught under the wheels of a materialistic world-view. At the same time, there were growing doubts as to whether historical and philological research can even claim to be a science, as measured against the methodological standards of the natural sciences.

However, resistance soon formed against the predominance of the natural sciences. The philosopher Wilhelm Dilthey and the historian Johann Gustav Droysen played a leading part in this. Dilthey, for example, was concerned with putting the methodology of natural science into perspective and creating a methodologically robust free space for the humanities. In doing so, he relied on a fundamental idea of his *Lebensphilosophie*. According to this, man can only gain access to his historical and cultural world through an understanding of his inner experience. The latter, Dilthey claimed, would justify the decisive difference between the natural sciences and the humanities. While natural science found the objects of its cognition independently of all expressions of human life, the objects of cognition in the humanities were those very life expressions themselves. In other words: contrary to the material phenomena that belong to the outer world, historical and cultural phenomena were assigned by Dilthey to the inner world of mental experience.

According to Dilthey, the task of the humanities is to understand the mental world from the inside, by "re-living (*Nacherleben*)" mental phenomena. Only in this way, he argued, would the phenomena of our historical and cultural world finally come to the fore that are manifested in the symbol-systems and symbol-rules of language, action, and creative activity. In the act of "re-living" mental phenomena, Dilthey believed that he recognized the most general trait of the structure of the humanities, which "leads us from the narrowness and subjectivity of experiencing into the region of the whole and the general" [28, p. 164]. Dilthey thus classified the sciences according to the form of the items of their cognition. The items of the humanities, he believed, are initially formed by inner experience, while those of the natural sciences are already given *a priori* by the objects of the external world.

This understanding of science developed into a stereotypical idea of two forms of science, irreconcilably opposed in their major objectives: The natural sciences deal with the material order of reality. They reveal the causal relationships between the phenomena of the outside world, which allow one to explain them in a law-like manner, to calculate them and, if possible, also to predict them. However, they have to pay dearly for their precision and mathematical superiority by the fact that they can never advance to the essence of things. In other words, the natural sciences merely "explain" phenomena, but they do not allow one to "understand" them. In contrast, the humanities' gaze is directed towards the mental existence of man and thus towards a higher level of reality. In their aim to "understand" man and his mental world, the humanities, according to Dilthey, seem to come closer to the truth. Unlike the natural sciences, which do not rise beyond the purely physical level of natural givenness, the humanities are focused on the historical and cultural world as the existential foundation of man. Thus, it seems as if the humanities have a more profound access to the world than do the natural sciences. Is that true?

3.4   Fundamentalism of Understanding

The doctrine of understanding, as proposed by Dilthey, far exceeds the objectives of traditional hermeneutics, a discipline that initially saw itself as a pure art of interpretation. Yet can a form of knowledge that is based on inner experience meet the demands of scientific objectivity at all? One may rightly doubt it. An understanding of reality that emerges from the inner perception of an individual cannot be separated from the individual. When the actual process of understanding, as Dilthey thought, is the "re-living" of what needs to be understood, then the object of understanding arises in the first place by the act of understanding itself. However, how can objective understanding be attained under these conditions? After all, the process of objectification presupposes that the individual steps aside from the object of cognition. According to Dilthey, however, the process of understanding in fact enforces the unity of subject and object.

It is no wonder that Dilthey's methodology was accused of violating the most elementary principles of objective thinking. In the twentieth century, this criticism finally led to a reorientation of hermeneutics, to overcome the subjectivity of understanding and to reinterpret understanding as a supra-individual concept. In contemporary philosophy, Jürgen Habermas, Karl-Otto Apel and others pursued this path by placing inter-subjective forms of understanding—such as dialog, communicative action and the like—at the center of hermeneutic thought.

Today, philosophical hermeneutics has become a powerful intellectual current, which often strides onto the scene with the bold claim of being a binding doctrine of "correct" world understanding. These overblown pretensions are ultimately rooted in the philosophy of Martin Heidegger, who interpreted the mode of understanding as an existential—i.e., as the most elementary—form of human existence. Heidegger's approach was pursued further by his student Hans-Georg Gadamer, who placed human language at the center of world understanding, because language, as he emphasized, is always associated with the entirety of being.

In respect of the existential interpretation of understanding, the approaches of Heidegger and Gadamer can largely be read in parallel. When Heidegger in his major work *Being and Time* [29] elaborates "Being-in-the-world" as the most primordial characteristic of human life, and understanding as the most primordial form of fulfillment of existence, Gadamer follows him by saying that the first prerequisite of understanding is the rootedness of Being in the experience of life and that this rootedness is connected with the experience of the *you*, the *other* and the *foreign*. The pre-scientific conditions of human experience are specified at the same time. According to Heidegger, these include the "fore-structure of existence", i.e. the "fore-having, fore-sight and fore-conception of understanding" [29, p. 195], or, as Gadamer calls it, "prejudices as conditions of understanding" [30, p. 289 ff.].

Let us take a closer look at the concept of prejudice. Usually, we associate this term with an unfounded, mostly false judgment. Gadamer attributes this negative connotation to the thinking of the Enlightenment, which would have seen in prejudice the expression of an unenlightened "prejudice against prejudice itself, which denies tradition its power" [30, p. 283 f.]. According to Gadamer, prejudices are, strictly speaking, cognitively neutral because they can be both true and false until a final examination can be made of all the elements that determine a situation.

If we were to pursue this thought consistently, we would have to conclude that we can never achieve a definitive examination of our prejudices. This is because our prejudices are always interwoven with tradition, that is to say with the whole of our historical existence. That in turn means that we will *de facto* never overcome our prejudices. Since our prejudices are always epistemologically neutral, they can never contribute to our understanding of the world. Instead, Gadamer considered, all prejudices prove to be empty in the end, depriving hermeneutic thinking of its foundation.

That is why Gadamer also had to admit that we must somehow distinguish "the true prejudices by which we *understand*, from the *false* ones, by which we *misunderstand*" [30, p. 309]. However, since for Gadamer prejudices themselves are the unassailable basis of all understanding, the "temporal distance", and the "historical consciousness" that feeds on it, are the only means of confirmation or refutation that he is able to specify. Once again, the whole appears here as the truth that in an obscure manner filters out, in the constantly changing world of our prejudices, those judgments that are supposed to guide our understanding.

All in all, this means that a universal and ultimate concept of understanding, to which our entire knowledge of the world is ideally related, must already presuppose the concept of understanding itself as understood. Here we come across a figure of thought, which is also referred to in the philosophical literature as the "hermeneutic circle". The question arises as to whether this is a vicious circle—i.e., a logical circle in which what is to be proved is already contained in the preconditions—or whether this circle genuinely expresses the true character of human understanding, as is claimed by hermeneutic philosophy.

Referring to Heidegger, Gadamer argues that "in the domain of understanding there can be absolutely no derivation of one domain from the other, so that here the fallacy of circularity in logic does not represent a mistake in procedure, but rather the most appropriate description of the structure of understanding" [30, p. 159]. In reality, Gadamer says, the hermeneutic circle aims at "the structure of the Being-in-the-world itself" to overcome the transcendental problem of the "subject-object bifurcation" (see Chap. 2). Accordingly, the hermeneutic circle is not the expression of a logically contestable doctrine of understanding, but the epitome of a true structure of understanding. This structure, according to Gadamer, is represented precisely in the absoluteness of understanding, because understanding is—as the most elementary form of "Being-in-the-world"—related in a genuine way to reality in its entirety.

However, against the background of the thesis that every understanding refers to some pre-understanding, the "true" structure of understanding could also be interpreted in a completely different way from that intended by Gadamer. According to this, true understanding is not articulated in its absoluteness, but rather in its context-dependence and thus in its relativity. This would also be a conceivable reading of the hermeneutic circle, if one does not exaggerate the essence of understanding existentially, but interprets it merely as a more or less fleeting moment in man's reflection on his being and his world. Admittedly, the concept of understanding then loses its fundamental interpretation, since contextual understanding never reaches a final point in its relative relations and therefore could never fulfill the hermeneutic claim to totality.

Such an interpretation, however, would completely contradict Gadamer's philosophical intentions. His interpretation aimed at the totality and absoluteness of understanding, as deduced from the existential unity of life and understanding. Since the act of understanding, as Gadamer asserted, is deeply rooted in life, and since life itself becomes manifest in history and as history, the historicity of understanding becomes irrefutable for Gadamer's hermeneutics. Consequently, Gadamer demanded that "hermeneutics adequate to the subject matter would have to demonstrate the reality and efficacy of history within understanding itself" [30, p. 310 ff.]. This, Gadamer continued, can only succeed if the horizon of understanding merges with the horizon of what is to be understood. Understanding, therefore, in its innermost essence, proves to be a fusion of horizons and as such constitutes a process of enrichment, in the same way as we experience a constructive conversation as gain.

In contrast to Dilthey, who—as Gadamer criticized—had "reprivatized" history by trying to bind the process of understanding to inner experience, Gadamer shifted the proper understanding back from inner to outer history. "In fact," he wrote, "history does not belong to us but rather we to it. Long before we understand ourselves through the process of self-examination, we understand ourselves in a self-evident way in the family, society, and state in which we live. The focus of subjectivity is a distorting mirror. The self-awareness of the individual is only a flickering in the closed circuit of historical life. That is *why the prejudices of the individual, far more than his judgments, constitute the historical reality of his being.*" [30, p. 288 f.]

How much hermeneutics differs from the analytical thinking of modern sciences becomes particularly evident when we examine the role of the object of cognition. For example, Gadamer asserted that real historical thinking does not pursue the "phantom of a historical object", but rather "learn[s] to view the object as the counterpart of itself and hence understand both". "The true historical object", he pointed out, "is not an object at all, but the unity of the one and the other, a relationship that constitutes both the reality of history and the reality of historical understanding." [30, p. 310]

In other words, Gadamer's hermeneutics abandon the objectivity principle of modern research according to which it is the task of the recognizing subject to strive, as far as possible, for an exact representation of the object of cognition. Richard Rorty [31] even regarded Gadamer's alternative draft as a necessary step of hermeneutics, since it is only in the renunciation of the elaboration and constitution of the object of cognition that the path toward the actual goal of hermeneutics is smoothed out, namely to approach the whole in a dialogical way. Hermeneutics thus turns out to be a variant of a "non-representational" philosophy whose cognitive aim is no longer represented by the object of cognition but instead merges into the unity of subject and object.

At the same time, a philosophy that so fundamentally abandons the idea of objectivity throws all principles of objective and critical thinking overboard. Beside, the hermeneutic concept of history takes on a somewhat peculiar character. When Gadamer asserts that any understanding is "being situated within an event of tradition" [30, p. 320], and thus joins in with the historical and cultural tradition, then history is not only the prior meaning-context for all understanding, but it also possesses a normative power. In a more precise sense, it is claimed, history is "impact history" (*Wirkungsgeschichte*) that, in the truest sense of the word, has an impact on us. Thus history makes the final decision as to what we consider to be remarkable, meaningful or questionable.

In fact, for Gadamer, the goal of a hermeneutics of history is to draw up the norms that are obligatory upon us from the events of the past. It is the same task that faces the jurist when he tries to discern from the jungle of legal rules and laws the intent of the lawmaker, or the theologian when he attempts to deduce God's will from the Bible. In this sense, Gadamer claims, that "we have the task of *redefining the hermeneutics of the human sciences in terms of legal and theological hermeneutics*" [30, p. 321].

All this sounds suspiciously like a higher form of truth, which has to be opened up hermeneutically. And indeed, in Gadamer's case, one encounters the suggestive remark that "human sciences are connected to modes of experience that lie outside science" and "in which a truth is communicated that cannot be verified by the methodological means proper to science" [30, p. xxi]. However, a doctrine of understanding that refers to the individual's experience within the whole experiential horizon of the world eludes inter-subjective control. For this reason, there is no objective yardstick that could be applied to the claim of such a doctrine to be true. Therefore, Gadamer has no other course but to appeal to the whole itself as the truth.

Scholars of the humanities who are committed to philosophical hermeneutics often display a certain arrogance when encountering natural scientists, when it comes to the prerogative of interpretation of scientific knowledge. In this, Gadamer himself was a forerunner, in that he deprecated the "methodical nurturing of certainty" of modern science and tried to out-bid this with the hermeneutic certainty of life: "In the natural science one speaks of the 'precision' of mathematizing. But is the precision attained by the application of mathematics to living situations ever as great as the precision attained by the ear of the musician who in tuning his or her instrument finally reaches a point of satisfaction? Are there not quite different forms of precision, forms that do not consist in the application of rules or in the use of an apparatus, but rather in a grasp of what is right that goes far beyond this?" [32, p. 5]

For Gadamer, it was the hermeneutic approach that gives us true access to the world and lays the foundations of the humanities. However, which scientific concept underlies this? What can the humanities actually bring to bear against the claim of the exact sciences to *explain*? Gadamer responded to this question by stating: "What makes the human sciences into sciences can be understood more easily from the tradition of the concept of Bildung than from the modern idea of scientific method." [30, p. 17]

Gadamer's concept of *Bildung* ties in with the humanistic tradition in which man, his language and his history are at the center of any understanding of the world, and which is carried by the ideal of a comprehensive intellectual and artistic education of man. Just as in the humanist tradition, Gadamer's idea of *Bildung* is also based on the claim of a normative understanding of history: namely to be the binding benchmark for moral integrity. Here, the circle of Gadamer's arguments against modern science closes: The exact sciences do not meet the requirements of the humanistic ideal, because their methodological prerequisites make them lose sight of the human being and his historical and cultural tradition and instead reduce the reality of life to mathematical precision and technical availability.

Gadamer was convinced that the humanistic tradition would gain a new significance "in its resistance to the claims of modern science" [30, p. 17]. This remark, however, expresses an extremely one-sided view of humanism. The humanist of the 14th and 15th centuries could not, of course, yet know the options that modern science would open up for human life and actions. He inevitably orientated himself toward the ideal of ancient humanism, which placed the virtues of measure, justice, aesthetic sensibility, harmony with Nature and the like at the center of world-understanding. This is obviously the image that Gadamer, too, had of humanism.

However, in the sixteenth century, at the height of the Renaissance, humanism gained a new dimension through the awakening sciences. It is remarkable that at that time science and its ideal of mastery of Nature were not regarded as in conflict with humanism but, on the contrary, they were seen as an enrichment which made humanism flourish in the first place. Francis Bacon and René Descartes are the impressive representatives of this modern humanism, the primary objective of which was the wellbeing of humankind. They provided important stimuli for a humanism based on science and technology (Sect. 1.2). It appears that, in Gadamer's hermeneutics, humanism and humanity fall apart. In his image of humanism, science and technology are not conceived of as a form of emancipation of man from the arduousness of his life-world, providing man with a livable existence, but as ways of thinking and acting that bypass "true" life.

The philosophical thrust of hermeneutics is the same as that of German Idealism. In both, the unity of subject and object, of man and the world, is the pivotal point of philosophical reflection, and both refer to a higher truth to which the analytical thinking of the exact sciences is claimed to have no access (Sect. 2.6). Thus, the two philosophical currents meet in their fundamentalistic claim to understand the world—entirely, absolutely, truly and cohesively—a claim that stands in irreconcilable contrast to the ideal of critical cognition in modern science.

Can, under these conditions, hermeneutics ever be the basis of a science, for example, of the humanities? On the part of hermeneutics, the contradiction emerging here is usually counteracted with the argument that hermeneutics ultimately does not claim to be a scientific method, or a method analogous to the scientific procedure. In fact, hermeneutics does not aim for conceptual constructions or theoretical foundations. Gadamer, too, did not regard hermeneutics as a scientific methodology in the narrow sense, but rather as a fundamental insight into what thinking and recognition mean to mankind in practical life, even if one is using scientific methods.

One could conclude from this that hermeneutics understands itself as a kind of "metamethod", supposed to accompany human thinking and cognition. However, even if one grants hermeneutics in this sense a cognition-leading function, as a doctrine of "correct" understanding it must be able to differentiate itself from other, competing conceptions of understanding. Yet this is only possible with the help of a suitable procedure for confirmation and refutation, which is oriented toward the methodological standards of critical acquisition of knowledge.

Gadamer, however, refused to acknowledge such control procedures for understanding, because his universal doctrine of understanding "goes beyond the concept of method as set by modern science" [30, p. xx]. Whether the critical instrument of fallibilism (Sect. 2.7), as has been suggested, can also be made useful for hermeneutics must appear more than questionable in the light of the self-understanding of philosophical hermeneutics. The dogmatism of a doctrine of understanding aimed at the world as a whole does not permit confirmation and refutation procedures in which knowledge is examined following the pattern of analytical sciences and, if necessary, restricted in scope or even abandoned.

It is inevitable that, in the hermeneutic discourse, the truth is occasionally lost; this is because every interpreting individual can, and even must, appeal to the totality of his or her own, particular world-experience. This is the flip side of a philosophy that is dogmatically fixated on the whole. One may also have doubts as to whether a doctrine that makes human history the absolute yardstick of our understanding of the world can satisfy its universal claim at all. In the face of human evolutionary history, such a belief may be an error because of the mere fact that human history itself is only a late and fleeting episode in natural history.

By inadmissibly reducing the world as a whole to the perspective of human history, universal hermeneutics forces itself into a dilemma of reasoning from which it cannot escape without abandoning its foundation in the philosophy of history. For hermeneutics, the question arises at the end, albeit involuntarily, of the specification of the whole. Is the world as a whole, to which all understanding is related, only the history of man, or is it the overarching history of Nature? If the history of Nature is to be the reference point of our world-understanding, then the roots of human understanding must also lie in Nature. This, in turn, can only be understood with the help of the "explanatory" methods of natural sciences.

3.5 Facets of Holism

In daily life, "understanding" and "explaining" are more closely intertwined than in academic debate. This is already visible in everyday language, where we often use the two terms synonymously. For example, one uses the word "explanation" in quite different contexts: the description of a route, the justification of a particular behavior, or the disclosure of financial conditions. The usage of the term "understanding" is just as vague. We not only say that we understand gestures, images, poems, actions and so forth but also the function of a cash machine or the instructions for using a kitchen appliance. In all these examples, the terms "understanding" and "explaining" are used as if they were more or less identical in meaning.

Any attempt to separate "understanding" from "explaining" must, therefore, appear amply artificial. Perhaps understanding is a mode of cognition which, by its very nature, includes a causal explanation. This is evident, for example, in all situations in which the wish to understand something is expressed in the question of "why" or "how". Since adequate answers to these questions always require an appeal to causes, they automatically fall into the category of causal explanations.

The argument that every understanding, if it is not limited to a mere interpretation of texts, images and the like, is essentially an explanation, can only be countered by taking "understanding", as Dilthey did, as an act of inner perception or contemplation. In this case, however, the opaque veil of mental experience lays itself over the concept of understanding which analytical science is no longer able to uncover. Conversely, if one comprehends "understanding", as Heidegger and Gadamer did, as an *"Existential"* by placing it in the overall context of man's existence, then the concept of explanation becomes lost in the totality of man's relationships to the world. Such a world-interpretation, which claims to have higher, namely existential, access to reality, remains inaccessible to any analytical thinking. It does not try to uncover the causal structure of reality; rather, it argues directly or indirectly in favor of a holistic approach to the world.

Holism has a long tradition in philosophical thinking, going back to the ontology of the Eleatic school of philosophy (Sect. 2.2). In modern times, holism has become known above all as a philosophical countercurrent to the evolutionism of the late nineteenth century. At that time, evolutionism had attracted a great deal of attention, especially in the social sciences and ethnology, because it claimed to be able to describe and explain the development of peoples and their cultures by the idea of evolution, borrowed from biology. Thus, evolutionism advocated a world-view according to which reality gradually evolves from simple preliminary stages into a rich and sophisticated, elaborate reality. Holism now contrasted the mechanistic, progressive thinking of evolutionism by offering an understanding of the world in which the world is represented as the development of entities whose elements explain themselves on the basis of their associated entireness.

In the first half of the twentieth century, holism was propagated above all by the physiologist John Scott Haldane and the biologist Jan Christian Smuts. The latter was first and foremost known as a politician, as for a while he headed the South African Union as prime minister. While Haldane expanded certain holistic concepts of biology into a methodological holism, Smuts gave an ideological direction to the holistic understanding of reality, aimed not least at social politics. He stressed that the organism is "a little living world in which law and order reign, and in which every part collaborates with every other part, and subserves the common purposes of the whole, as a rule with the most perfect regularity" [33, p. 82]. Finally, he came to the conclusion: "The holistic nisus which rises like a living fountain from the very depths of the universe is the guarantee that failure does not await us, that the ideals of Well-being, of Truth, Beauty and Goodness are firmly grounded in the nature of things, and will not eventually be endangered or lost." [33, p. 345]

Holism, in the form outlined by Smuts, regards itself primarily as a doctrine of order, which claimed to justify the order of reality from above, i.e.

from the higher levels of being. The world was seen as a hierarchically ordered, organic whole in which the higher levels of being are the basis for explaining the levels beneath them. Later, the philosopher Adolf Meyer-Abich [34] developed a similar theory of order, according to which the "world organism" comprises four levels of being: social wholeness, spiritual-mental wholeness, organic wholeness and inorganic wholeness. According to Meyer-Abich, these different levels form a gradient of order, with the "lower" levels of being participating in each of the "higher" levels, as, according to Plato's doctrine of ideas, real things participate in the higher world of ideas. Even the wording, following Aristotle, according to which "the higher is always the 'prime mover' of the lower" [34, p. 356; author's transl.], reveals the influence of antique thinking on the holistic doctrine of order.

Although holism is deeply rooted in the tradition of occidental philosophy, it never acquired any significance in science, either as a philosophical doctrine or as a counterweight to the analytical world-understanding. However, the ideological influence of holism on society and politics was stronger and, in consequence, all the more dangerous. A view according to which the order of reality is determined from "above", namely from the social sphere, naturally holds the danger of being misused for authoritarian thinking. In fact, in the first half of the twentieth century holistic ideas, in various forms, found their way into the ideologies of totalitarian systems (see [35]).

Given its far-reaching consequences, the ideological branch of holism must ultimately be seen as an aberration of thought. Nevertheless, its influence is still present in the public debate on the scope of science and technology. Thus, for example, around holism, so-called "alternative sciences" have emerged which claim to have new access to the complex issues of Nature and society. They often argue with catchwords such as "systemic thinking", "network thinking", "integral thinking" and the like, to underpin the importance of their "new thinking" for an adequate understanding of our life-world (see Sect. 7.4).

However, what is "holistic thinking", in contrast to analytical thinking, actually supposed to be? This question is crucial, because attempts to answer it have created and continue to create, time and again, much debate about the explanatory power of science. Science must often defend itself against the accusation that its analytical way of thinking is fundamentally unable to grasp the essence of reality. Each causal analysis, the critics argue, is always preceded by the systematic preparation of the object of research by removing it from its original and authentic context.

This methodological approach to reality is indeed unavoidable, as the primary causes and effects in natural events can only be found when reality is first broken down into single relationships of cause and effect. However, in this way, according to the critics, reality is reduced to certain methodological fragments, a process that makes impossible an adequate overall understanding of reality. Consequently, any scientific approach to reality is inevitably fragmentary and, therefore, incomplete.

This criticism of analytical thinking seems to confirm the allegation often heard in public debate that the scientific approach to reality is inevitably accompanied by a loss of reality (Chap. 7). In particular, the natural sciences are accused of looking at our natural environment exclusively from a mechanistic point of view. From this perspective, Nature appears as a giant machine in which all processes take place as if they were part of some huge clockwork. This, in turn, would mean that the actual complexity of natural events is reduced, inadmissibly, to a conglomeration of simple cause-effect mechanisms.

The debate is further fueled by the accusation that the mechanistic understanding of Nature promotes the disastrous attitude of modern industrial society toward the intervention, for any arbitrary purpose, into the natural basis of our life. The consequence, it is asserted, is the ruthless exploitation of Nature. Here, criticism of the analytical research strategy flows over into a general critique of modern science and technology. As an alternative concept of Nature, intended to complement or even replace the mechanistic understanding of Nature, a holistic approach to Nature is demanded, one that takes account of the need to preserve our natural resources.

However, what does the call for holistic thinking mean at all? A simple consideration shows that this demand is quite irrational. Whatever a "whole" may be, it must have an internal structure if the it is not to be a mere empty shell. Only its internal structure gives a whole entity a character at all—quite independently of the question of its relationship to its external world.

If one wants to understand the essential core of a given wholeness, one has to break the whole down into its parts. This procedure, in turn, is the guiding principle of analytical thinking. Only after this step can the actual tools of the analytical sciences come into play. These tools include, among others, the methods of abstraction, idealization and simplification. They make possible the explanation of the natural happening by the terms and laws of science. It should nonetheless be mentioned that the analytical procedure of dissection and separation ends where it encounters the smallest known systems, quantum objects; this thought is taken up in the next section.

The question of the whole and its parts was already examined from a logical and philosophical point of view by Gottfried Wilhelm Leibniz. At the beginning of the nineteenth century, Bernard Bolzano embarked on a similar train of thought. At the beginning of the twentieth century, the philosopher and mathematician Edmund Husserl did significant preparatory work for the foundation of an independent scientific discipline, the so-called "mereology", which aims at the formal description of the relationship between the whole and its parts. In 1916, mereology received a further essential impetus from a formal theory set out by the logician Stanislaw Leśniewski [36].

Despite these efforts, however, mereology is still far from opening up new areas of application outside mathematics and logic. It remains to be seen, therefore, whether mereology will one day be able to contribute to the clarification of the epistemological controversy that is raised by the question of the whole and its parts, and which continually re-appears in connection with the adequate understanding of complex phenomena. This could be the birth of a new structural science (Sect. 4.9).

Here, it suffices to emphasize that analytical thinking as such is by no means associated with a loss of reality, as the critics of science fear. On the contrary, since there are many ways of dissecting something, it is only this method that has the capacity to reveal the inherent manifoldness, and thereby the potentiality, of a whole. To risk a comparison: a mountain landscape does not show its full richness when seen from a distance, but only on the various trails that traverse it. In the same sense, one can only comprehend a whole by uncovering its hidden diversity.

In this connection, it is essential to note that every analysis also implies a synthesis. If one wishes to understand the essence of a whole then one will not break it down into arbitrary parts, but only into those parts that are relevant for the function in question. This is particularly clear in biology, because living organisms have many functional levels which, moreover, are hierarchically organized. Thus one will never analyze an organism by blindly dissecting it into all its molecular constituents, since most of its life functions would then also be destroyed. For this reason, the first step of analysis is always directed toward the whole.

The same is true in analyzing a machine. To understand its function, one would never consider "atomizing" the machine, that is, decomposing it into its atomic parts. Instead, the analysis must be carried out in such a way that the parts of the machine still refer meaningfully to the whole. For this reason, the whole is always included in the analysis. In this case, however, the whole is a "synthetic", and not an irreducible, entireness. We make this experience already in our childhood. Anyone who has, out of curiosity, dismantled a device into its components, knows that the urge to understand can only be successful if one bears in mind that the device is a synthetic whole. Anyone who forgets this will most probably fail to put the parts together again successfully. This prompted Johann Wolfgang von Goethe to the remark:

"The essential point that one does not seem to think of in the exclusive application of analysis is that every analysis presupposes a synthesis." [37, p. 52; author's transl.]

In science, the analytical and the synthetic approaches are not two mutually exclusive methods of cognition; rather, they belong together. Already Descartes had attached great importance to this aspect in his methodology (Sect. 3.1). Isaac Newton, the founder of the mechanistic world-view, likewise emphasized the complementarity of the analytical and synthetic methods. In his lectures on *Optics*, he suggested: At first, one should take the path of analysis and induction, to draw general conclusions from observations and experiments. This path would lead from the compound to the simple, from movements to the forces generated by them, from effects to their causes. Subsequently, one should proceed to the synthesis elevating the causes discovered to principles, from which phenomena can be explained.

3.6   The Problem of Non-separability

The analytical and synthetic ways of looking at things do not contradict one another. On the contrary, they are essential features of the scientific method. Every analysis presupposes a synthesis, as an investigation only makes sense if it is guided by the thought that the research object is a synthetic whole. If it is not, the analysis remains a pointless endeavor, since it would have no goal of cognition and would therefore inevitably come to nothing. Richard von Mises stated this succinctly: "We know of no other synthesis than the sum of all we have to say." [38, p. 57]

There is no need for a new kind of thinking in science. The conventional sciences themselves have already developed new disciplines such as systems theory, network theory, synergetics and others, designed to analyze and understand the integrated behavior of complex systems. Above all, biology has enhanced this development. Nevertheless, one may ask whether there are problems that persistently elude a solution within the framework of these modern sciences. Here we will focus upon quantum objects and the related problems of non-separability and non-locality.

The more profoundly science advances into the border areas of tangible reality, the more urgently we are called upon to look at the very basis of our conceptional thinking. The ongoing discussion about the appropriate inter-pretation of quantum phenomena conveys an impression of the enormous difficulties encountered in the endeavour to reconcile modern scientific results with our conventional perceptions of the world.

Quantum theory is the most fundamental theory of physics. Therefore, it is worrying that quantum theory seems to question our common world-view, according to which reality exists "simply as it is", i.e., independently of any human observer (Sect. 1.10). The most challenging problem of quantum physics is that of limited divisibility. If we want to describe reality in physical terms as precisely as possible, we have to make increasingly subtle distinctions regarding physical objects in space and time. In a thought experiment, one can continue this distinguishing process indefinitely. However, ultimately, the concept of a real object will dissolve in the idealized image of a point mass, which is an abstract idea without any intuition content.

In the submicroscopic domain, physical description and experienceable reality necessarily diverge. This problem is manifested in the "strange" properties of quantum objects: non-locality, non-separability, indeterminism and others (see below). The contradiction between quantum theory and our conventional conception of the world is so strong that it has led, for decades, to fierce philosophical controversies about the correct interpretation and classification of its findings.

An example of this is the famous Bohr–Einstein debate on whether quantum theory is already complete or needs further theoretical substantiation. Is quantum–mechanical uncertainty due to a lack of detailed information, or is it of a fundamental nature? Asked differently: Are there hidden, i.e., hitherto unknown, factors or parameters that determine quantum particles' behavior, or is their quantum behavior subject to pure chance and thus inherently undetermined?

Niels Bohr insisted that quantum theory is complete and does not require any extension [39]. On the other hand, Albert Einstein was convinced throughout his life that quantum theory is not complete. In a letter to Max Born, one of the co-founders of quantum mechanics, Einstein wrote: "Quantum mechanics is certainly imposing. But an inner voice tells me that it is not yet the real thing. The theory says a lot, but does not really bring us any closer to the secret of the 'Old One'. I, at any rate, am convinced that He is not playing at dice." [40, p. 88]

Some years later, Einstein developed together with Boris Podolsky and Nathan Rosen a thought experiment, the "EPR paradox", that he believed questioned the Copenhagen interpretation (Sect. 1.10) of quantum theory [41]. At the center of the EPR paradox is a conclusion—following from the Copenhagen interpretation—according to which two quantum particles can be "entangled" so that they appear as a single whole. Measuring one particle's state will instantly fix the other one's state, even if the particles are far away from each other (for physical details see Fig. 3.5).

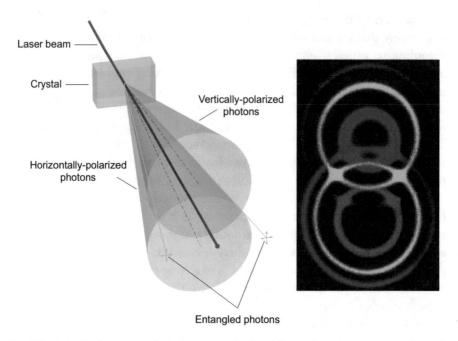

Fig. 3.5 Polarization-entangled photons. Left: If an intense laser beam passes through a nonlinear crystal, some photons of higher energy occasionally undergo a spontaneous "down-conversion" into a pair of photons of lower energy. The resulting photon pairs are emitted along cones, so that one photon belongs to one cone and the other photon to the other cone. The cones' axes (dashed lines in the figure) are arranged symmetrically around the laser beam. The photons moving along one cone are vertically polarized, while those moving along the other cone are horizontally polarized. Along the lines where the two cones intersect, the polarization is undefined. However, if one photon is vertically polarized, the other must be horizontally polarized and vice versa, because the two photons originate from different cones. If both polarizations are possible and neither is specified, then the photons are entangled (image: adapted from [42]). Right: Intersecting beams of down-converted photons shown in different colors. The rings correspond to light with different wavelengths. The figure shows a false-color composite of three images taken with differently colored filters in front of the film. Polarization-entangled photons are obtained by selecting the rays along the green rings' intersection lines by apertures (photo: courtesy of Michael Reck and Paul Kwiat)

Quantum entanglement—a term coined by the physicist Erwin Schrödinger—gives the impression that the states of entangled quantum particles are brought into line by an invisible hand (cf. [42]). However, this idea violates the traditional world-view of physics according to which the principle of "local realism" holds. This principle lays down that that the result of a measurement on object A is unaffected by the measurement on a distant object B with which A has interacted in the past.

Because no long-distance effect can propagate itself faster than the speed of light, Einstein described the conclusion drawn from the EPR paradox as "spooky action at a distance". He did not believe in such a mysterious phenomenon and asserted—in opposition to Bohr—that quantum theory could not provide a complete description of reality. Instead, he argued in favor of the existence of "hidden" parameters, which could remove the incompleteness of quantum theory and restore causality and locality in the submicroscopic domain.

The controversy took another turn in 1964 when the physicist John Stuart Bell formulated a criterion that allows an experimental test of Einstein's idea of hidden parameters [43]. The Bell criterion, which has the mathematical form of an inequality, has the remarkable property of being able to rule out hidden parameters if it is violated in an experimental test. Some years later, when the techniques to carry out the Bell test became available, various experiments were performed. They all violated the Bell inequality and thus argued against the existence of "local" hidden parameters [44, 45].

The realization that "local realism" is violated in the submicroscopic range has massively shaken our traditional world-view. An essential part of this world-view is the assumption that there is a reality, which has pre-determined properties and is independent of the human observer. Therefore, it is not surprising that findings in quantum physics have unleashed a profound discussion about the extent to which quantum physics calls for a completely new understanding of reality.

Approaching the quantum world, we are increasingly forced to abandon the experimental scenario, in which we are merely observers of the world. Instead, by the measurement process we intervene in reality and thereby generate the quantum phenomena that we observe. As shown in Sect. 1.10, the experimental arrangement determines which aspect of light (wave or particle) appears. The experiment acts here like a selection condition acting on the possible states that a quantum object can take on. Therefore, to understand the quantum world, we have to investigate, among other things, the experimental context—that is, the detailed physical configuration or boundary condition under which a quantum phenomenon appears. We will see in Sect. 4.8 that, beside the natural law, the boundary conditions are central for an understanding of physical phenomena. By their specific structure, boundary conditions carry factual information about reality which, together with the applicable natural laws, are the basis of physical explanation (Sect. 4.3).

The world of quantum objects is apparently incongruent with our tangible and experienceable world. Already the epochal discovery, made by Max

Planck, that energy can be emitted or absorbed only in integral multiples of a small amount disproved an idea that since Greek antiquity had been an unquestioned part of man's body of thought: "Nature does not make jumps" (*natura non facit saltus*). In contrast, Planck's quantum of action attributed an essential discontinuity to every atomic process. Heisenberg's uncertainty relation finally demonstrated the peculiar blurring that affects all observations at the submicroscopic level, which, in turn, brought into question the traditional ideas of causality and determinism in physics (Sect. 1.10).

The epistemological problems of quantum physics were triggered by the undecidable question of the nature of an elementary, i.e., ultimately "not further divisible" object. The interpretation that Bohr gave to quantum theory already indicated this dilemma. To take into account the upheaval of physical thinking enforced by quantum physics, Bohr developed the idea of complementarity, which regards the object as the whole of the observations and measurements that can be carried out on it [46]. This means that when the properties of an object as measured by different methods appear to be mutually exclusive, they must be regarded as complementary aspects of one and the same object. Accordingly, wave-particle duality represents two complementary descriptions of light, which only together result in a full picture of light's physical nature.

It was a remarkable coincidence that, almost at the same time as quantum physics emerged, mathematics was also plunged into a deep crisis by a problem of undecidability. Here, it was the finding that formal systems are in principle incomplete, since every formal system contains at least one statement whose truth content is undecidable within that system's framework (see Sect. 5.2). This crisis in mathematics fits well with the question that quantum physics also raised early on: the question of whether a new kind of logic is needed to take adequate account of quantum-physical undecidability. This approach led to the conception of a three-valued logic, in which the classical two-valued logic of "true or false" was supplemented by the third truth-value "undetermined" (Sect. 5.8). Since in quantum physics the statement "undetermined" suggests a dependence on time, classical logic's remodeling would amount to a logic of temporal arguments. However, it soon became clear that the new logic envisaged does not solve the philosophical problems of quantum physics. Instead, quantum logic creates new epistemological issues that concern the fundamentals of logic itself. An overview of these problems can be found elsewhere [47].

What changes in our world-view may we expect from our insights into the submicroscopic world? Since "local realism" (see above) evidently does not apply in the domain of quantum physics, a philosophical reorientation regarding our understanding of reality is thought to be unavoidable. However, in which direction should the renewal go? This question has given rise to a vast amount of scientific literature favoring highly diverse interpretations of quantum physics (cf. [42, 48, 49, 50]).

Most physicists are inclined merely to take note of the fact that the classical understanding of physical reality does not hold in quantum physics. They recognize that non-locality is a feature of the submicroscopic domain. However, they see no need for any further explanation—provided that quantum physics continues to be as successful as ever in theoretical and practical respects. In fact, quantum physics has opened the door to entirely new techniques such as quantum computing, quantum teleportation and quantum cryptography, which are already within reach of technological implementation. To continue along that path we do not necessarily need a philosophical reorientation. One can simply accept the results of quantum physics, taking up a position that is occasionally denoted as "quantum realism". In this case, however, one may risk overlooking something which could be of interest, as pointed out by the physicist Anton Zeilinger [42]. We would leave a fundamental question of our scientific world-understanding without any answer.

However, is there an answer at all that remains on a scientific foundation and does not throw the realism of modern science overboard in favor of speculative idealism? At present, the most promising approach that might contribute to a sound understanding of quantum physics seems to be the concept of information. There is even a historical example that indicates the scope of this approach. This example is the directionality of time, and it refers to a physical problem that once seemed to be more like a mysterious riddle. It follows from the law of entropy, assigning to natural events a direction in time, a feature that does not fit in with the reversible laws of mechanics (Sects. 3.10 and 5.5).

However, Ludwig Boltzmann, who at the end of the nineteenth century gave entropy a foundation based on statistical mechanics, pointed out already then that the riddle of entropy is intimately related to the information we have about a physical system. Thus, Boltzmann considered entropy to be a measure of the loss of information that inevitably occurs when one moves from the microscopic to the macroscopic description of a physical system. Later on, the physicist Leó Szilárd made this idea more specific by equating information with neg(ative) entropy (Sect. 5.6).

The equivalence of information and negentropy underlines information's real character. Information is neither a naturalistic nor an idealistic concept, but is rather a structuralistic one [51]. This finding leads to the collapse of the frequently heard criticism that the natural sciences wrongly regard information as a natural object. This criticism rests upon a fundamental misunderstanding that becomes especially clear in the case of entropy in physics. Although entropy is an object of scientific investigation that can be described by a physical quantity, it is not a natural object in the same sense as stones, trees or any other material entities. Instead, entropy is only a mathematical function providing information on the distribution of the system's energy states. This explains the structural equality of information and entropy.

The concept of information belongs to the rapidly expanding branch of the so-called structural sciences. The structural sciences describe the comprehensive, abstract structures of reality, independently of where we encounter them and whether they characterize non-living or living material, natural or artificial systems (for more details see Sect. 4.9).

With this in mind, we may ask whether the concept of information can contribute to the foundations of quantum physics. Could information become the missing link between classical physics and submicroscopic physics? Is information, alongside matter and energy, becoming the third pillar of physics? A first step in this direction was the attempt by Carl Friedrich von Weizsäcker to reconstruct quantum theory by means of the concept of ur-alternatives, thought of as "atoms of information" (Sect. 1.10). However, von Weizsäcker did not develop this idea further, which may explain why his approach did not attract much attention among physicists.

Von Weizsäcker was guided by the notion that an ultimate justification of quantum physics should reveal the conditions under which cognition of the submicroscopic domain is possible at all. This thought follows Heisenberg's belief that "the laws of nature which we formulate in quantum theory are no longer about the elementary particles themselves, but rather about our knowledge of them. The question of whether these particles exist 'per se' in space and time can therefore no longer be asked in this form, since we can only ever talk about the processes that take place when the elementary particle interacts with another physical system, for example, the measuring apparatus intended to reveal the behavior of the particle. The idea of the objective reality of elementary particles has thus vanished in a strange way—not into the fog of a new, vague or misunderstood notion of reality, but into the transparent clarity of a mathematics that no longer represents the behavior of the elementary particle, but rather our knowledge of this behavior." [52, p. 39; author's transl.]

Heisenberg considered the results of quantum physics from an epistemological point of view. He asked what we can know in principle about submicroscopic reality, but he made no statement about its essence. In no way did he contest the existence of objective reality. In contrast, some physicists have turned the idea of realism—the conviction that reality exists regardless of whether we observe it or not—upside down. They claim that reality exists only through observation—a thesis that must necessarily provoke vehement contradiction by scientists who believe in the reality of their research objects.

The physicist John Archibald Wheeler has expressed the view that reality only exists by humans interacting with reality in the slogan "it from bit" (instead of "bit from it") [53, p. 310 ff.]. Such an interpretation goes far beyond the horizon of human experience. Taking this idea seriously, we would be forced to accept an ontological idealism that has already in the past led to untenable and absurd ideas regarding reality (see Sect. 2.6).

No matter from which side we look at the application of the concept of information to quantum physics: information, by its very nature, always remains a structural concept that is unable to grasp the "essence" of reality. However, the idea of information can very well contribute to the clarification of general epistemological problems. The application of information theory to living matter is an example of this [54]. Information-theoretical issues that may be of interest in quantum physics are the contextuality of quantum objects (Sect. 1.10), the significance of physical boundary conditions as carriers of information (cf. Sect. 4.8) and, last but not least, the constitution of the meaning content of objects (cf. Sect. 6.8). In this book, however, we will face these issues of information mainly from the perspective of living matter.

3.7 The Mechanistic Paradigm

One cannot overlook the fact that the popular view of holism and the desire for alternative forms of science have their roots in a romantic understanding of Nature. We remember that in the romantic period, too, Nature and man were regarded as an organic entirety—as it were, the "original state" of man and world. Even at that time, the technological mastery of Nature, which accompanied the scientific progress of the period, was criticized as man's arrogance toward Nature: at that time, too, the idea of a holistic understanding of Nature was opposed to the mechanistic concept of Nature. It is worth illuminating some aspects of this time again, because it quickly becomes clear that the specter of mechanism, which the modern apostles of wholeness like to evoke, is, in reality, a nightmare of bygone days.

In the seventeenth and eighteenth centuries, Newtonian physics were at the forefront of the natural sciences. The gravitational law discovered by Newton and the laws of motion that he formulated made it possible for the first time to calculate precisely and to predict the movement of the planets. The fact that the celestial bodies follow simple mechanical laws made the cosmos appear like a gigantic machine in which only natural forces play a part.

However, the Newtonian world-view revived not only antiquity's idea of a world-machine (*machina mundi*), but, at the same time, it also brought the specter of mechanism into the world. Robert Boyle, a contemporary of Newton, believed, for example, that the idea of the world-machine had to be interpreted in the sense of a vacuous and dead mechanism (*mechanismus cosmicus*). Thus, Boyle had unintentionally prepared the ground on which a deep aversion to the mechanistic paradigm could subsequently spread. Moreover, the mechanistic world-view seemed to follow seamlessly the ideas of the ancient natural philosophy developed by Democritus and Epicurus, according to which all natural phenomena can be traced back to the path of the finest material particles, which they called atoms—an idea that gave the mechanistic world-view a firm material basis (Sect. 2.2).

At the height of the mechanistic age, Newtonian physics even gave rise to the idea that in principle all processes of the world can be represented and calculated by a system of simultaneous equations of motion. This prospect eventually led the mathematician and physicist Pierre-Simon Laplace to the self-indulgent assertion: "An intelligence that, at a given instant, could comprehend all the forces by which nature is animated and the respective situation of beings that make it up, if moreover it were vast enough to submit these data to analysis, would encompass in the same formula the movements of the greatest solids of the universe and those of the lightest atoms. For such an intelligence nothing would be uncertain, and the future, like the past, would be open to its eyes." [55, p. 2]

The Universal Mind conjured up by Laplace would have been able to calculate the exact past and future of the world from its present state. For him, all world events were in principle completely transparent, because they take their course like a clockwork mechanism. Laplace is said to have responded to an inquiry by Napoleon Bonaparte that the hypothesis of God's existence was no longer needed because the mechanistic image of Nature was absolutely perfect. Thus, in the nineteenth century, the idea of the Universal Mind finally became the symbol of a hypertrophic science that presumes to be able to calculate and predict everything.

The mechanistic program of physics was indeed able to celebrate great successes in the eighteenth and nineteenth centuries. When the industrial age began, with the construction of heat engines, the mechanistic paradigm finally became the dominant paradigm of science. Eventually, nearly all branches of science fell under the spell of the mechanistic approach.

The physiology of the late eighteenth century is an example of the triumphant advance of mechanistic science. Here, the focus of the investigations was on Luigi Galvani's observations that dissected frogs' legs can be stimulated to move by applying electrical discharges, such as occur in the vicinity of a Wimshurst machine (Fig. 3.6). Such observations seemed to suggest that there is a causal connection between electrical phenomena and the phenomena of muscle or nerve excitation. At the same time, it seemed to confirm a hypothesis developed by Albrecht von Haller in 1750 that considered the irritability and sensitivity of organs and muscle fibers as the essential mark of the living organism. Alexander von Humboldt [56], on the other hand, tried to verify the assumption that "animal electricity" was identical to the long sought-after life force. To this end, he carried out some four thousand galvanic experiments on plants and animals.

Although these and other experiments were still based on the wrong notion that there is a fundamental difference between inanimate and animate matter, they were *de facto* already entirely dominated by a mechanistic science that tried to fathom the essence of the living by applying mechanical, chemical and electrical stimuli to organic matter. Sure enough, this method did not fit into the romantic understanding of Nature at all, because it reduced organic matter to a mechanical stimulus-reaction scheme.

The central element of Newtonian mechanics was the concept of force. Therefore, the discovery of new forces such as the electrostatic and magnetic forces was of the highest importance for scientific discourse at the end of the eighteenth century. Beside, there was increasing evidence that these forces are connected in some mysterious way. Galvani's experiments, for example, seemed to indicate a connection between electrical force, mechanical force and life force. Moreover, the discovery of electrolysis shortly after that suggested that electricity too was a chemical force. With these insights, the question was finally raised as to whether all known forces could be attributed to a uniform force, and as to what the character of this force might be.

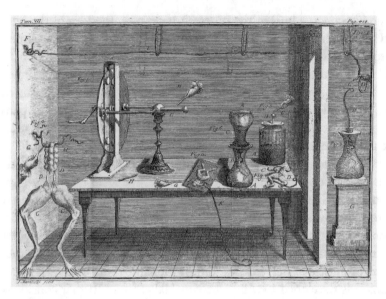

Fig. 3.6 The laboratory of Luigi Galvani (eighteenth century). The picture shows the experimental setup with which Galvani discovered "animal electricity". Galvani was a physician and professor of anatomy. In 1780 he discovered a strange phenomenon. While he was working with dissected frogs' legs in the vicinity of a Wimshurst machine, which generates electric voltage (on the bench, left), one of his assistants accidentally touched the inner thigh nerves of the frog with the tip of a scalpel. At that moment all the muscles in the leg-joints contracted repeatedly, as though they were affected by severe cramps. As it turned out, a second assistant had started up the machine and thereby generated a spark discharge, which in turn led to the stimulation of the muscles' movement. [From: Aloysii Galvani, *De viribus electricitatis in motu musculari*, 1792]

However, behind such questions there still lay the idea of a world formula that could be used to explain all physical events.

Thus, at the end of the eighteenth century, no less a problem than the question of the true basis of the unity of reality was on trial: Can natural phenomena be derived primarily from mechanical principles, or are there superordinate principles of Nature, from which the mechanical principles only emerge as special cases? Can the order of Nature in general, and the order of the organism in particular, be adequately understood by holistic principles alone, or can both be explained by reducing them to mechanisms? These questions sparked heated controversy over what comprised adequate paths toward understanding Nature, and this finally led to the emergence of the romantic natural science as a counter-movement to the mechanistic sciences.

Criticism of the mechanistic sciences was aimed above all against their method, according to which every natural process must be broken down into causes and effects for detailed analysis. This method was also thought to be appropriate for the study of the complex processes of life. However, this seemed to contradict the real character of organisms, because an organism is clearly based on a closed chain of causes and effects in which every cause is at the same time an effect, and every effect is at the same time a cause. How, then, could a mechanistic science ever advance to discover the primary cause and thus to explain organic matter? Should we then not understand Nature, in the same sense, as a living and autonomous entireness that causes itself?

Questions of this kind convey the impression that only the organismic world-view can restore to Nature and man the primary freedom that the mechanistic world-view seemed to take away from them. The rejection of mechanical thinking was reinforced in the nineteenth century when Laplace propagated the, as he claimed, unrestricted capabilities of the mechanistic sciences. However, the disquiet caused by Laplace was based on a misunderstanding, because there was no clear distinction between the total computability of the world that Laplace had spoken of and the law-like constraint that underlies an event. It was this misunderstanding that engendered the fear that, in the mechanistic view of the world, man was merely a tiny wheel in a merciless machinery, a view that placed him under total mechanistic constraints and left him no room for making autonomous decisions.

How far this picture solidified can be seen in the ecology movement of our own times (Chap. 7). Even today, a fundamentalist branch of this movement advocates a romantic image of Nature and often behaves aggressively toward science and technology. Above all, ecological fundamentalism not only castigates the "constraints" imposed upon Nature, man and society by scientific and technical rationality, but also decries the "fatal" consequences that the technical exploitation of Nature has for man and society. Thus, in the age of technical mastery of Nature and global economic dynamics, ecological fundamentalism considers itself as the last bulwark against the usurpation of power by science and technology.

3.8 The Reductionist Research Program

The research program of science relies on the analytical method. Causal analysis serves the aim of reducing the complexity of reality and tracing it back to its essential components and interdependencies. This procedure usually requires two accompanying procedures, namely "abstraction" and "idealization". However, precisely these two methods are often in the focus of criticism, including criticism of the natural sciences as a whole. It is asserted that these disciplines, within their framework of research, improperly simplify complex interdependencies, isolate and examine only individual phenomena and neglect the overall context. It is argued that, in the extreme case, the analytical method of the natural sciences leads to a world-view, called reductionism, according to which reality represents nothing more than a mere accumulation of atoms and molecules.

First of all, it should be noted that the natural sciences in no sense claim that they can map and explain reality to its fullest extent. Instead, their objectives of cognition are from the outset directed only at such aspects of reality that can be objectified and generalized. As a result, their findings—and this applies in general to the analytical sciences—always refer to a methodologically processed reality, which conveys only a reduced picture of actual reality. In this respect, *methodological* reductionism must be clearly distinguished from *ontological* reductionism. Ontological questions, referring to being as such, remain closed to the sciences.

Methodological reductionism is a "research strategy" that determines the objectives of cognition and submits the research process to specific rules. Reductionism could also be described as a "research paradigm", a term coined by the historian of science Thomas S. Kuhn [57]. In the narrow sense, this term denotes the research program of a particular group of scientists who pursue a common cognition goal and make use of the same methods, models and theories.

Within the framework of a research paradigm, it is decided, among other things, which problems are considered relevant and which methods of investigation and solution are admissible. Of course, a research paradigm also includes a fundamental theory, the validity of which is recognized by all. Since no observation is theory-free, the paradigm also determines *how* something is perceived. The insights thus obtained can, therefore, never depict full reality, but only a part of reality filtered according to certain aspects which are predefined by the paradigm. Similarly, the reductionist research program represents a paradigm which determines the methodological procedure in large areas of science.

Let us consider in more detail the reductionist method of simplification and idealization. These procedures abstract from the individual features of the objects of research in order to work out their general aspects. For example, solids differ from each other in shape, volume, composition, color etc. Even by saying that two entities have the same mass, we are already abstracting from all the other qualities that the solids in question may have. Since no object of this world resembles another perfectly, one can only speak of equality or similarity from the perspective of abstraction. Only in this way can we name similarities as such, and form general terms like "mass", "plant", "animal" and so forth.

Apart from this type of abstraction, on which human language relies and which is indispensable for our conceptual thinking, there is a higher form of abstraction on which mathematical reasoning is based. Mathematics mirrors the structural richness of rational thought, independently of the nature of real things to which its findings may be applied. For example, the equation $2 + 2 = 4$ expresses equality regardless of whether one is referring to apples, planets or whatever. With the formation of the concept of numbers, man had already taken the decisive step of abstraction, which leads from the area of direct sensory experience of given structures to the field of rationally comprehensible mathematical structures (see [58]).

Let us now consider the strategy of idealization. This aspect of the reductionist method cannot be separated from abstraction. Idealization refers to characteristics of reality that do not even exist in reality. Therefore, idealizations map reality only approximately. In physics, there is a long list of such idealizations, including the "point mass", the "reversible process", the "elastic impact", the "frictionless movement", the "isolated system", to mention but a few.

The essence of idealization can be well demonstrated by looking at the concept of "reversibility". Physicists call a process "reversible" when it can reverse its direction spontaneously. However, everyday experience teaches us that such spontaneous processes do not exist in reality. If, for example, a drop of ink falls into a glass of water, then we make the common observation that the ink spreads out in a cloud. In the end, the ink is dispersed uniformly throughout the water. We observe a similar effect when a gas emerges from a small volume into a larger one: it will continue to spread out until it fills the larger volume evenly. It seems to be impossible for such processes to run backwards; at any rate, no one has ever seen a gas spontaneously contracting and squeezing itself from a large volume into a smaller one. Likewise, water colored by a drop of ink never loses its color spontaneously because the ink has suddenly concentrated itself back into a single drop, like in a film run backwards. Processes that take place spontaneously have a crucial feature in common: They always lead to increased disorganization of an initially orderly state, and never vice versa. It is precisely this that lies behind what we experience as the temporality of natural events.

Spontaneous changes in the natural world always require an unbalanced behavior of matter. On the other hand, reversible changes, of which physicists speak in an idealized form, must be thought of as a "sequence" of equilibrium states, in which the system being studied never leaves its state of equilibrium. However, such reversible state changes cannot be realized in reality. For example, if one wants to inflate a balloon "reversibly" one would have to proceed in such a way that the air enclosed by the balloon is at all times in a state of equilibrium. This means that there must be no turbulences while the balloon is inflated, i.e. no fluctuations in pressure or density. Therefore, if one wanted to blow up an air balloon reversibly, one would have to do it infinitely slowly. This example shows us that reversible state changes are borderline cases and are only conceivable as "virtual" processes.

What has been described so far applies to the mesocosmic world of our experience. However, at the microscopic level, there is a principle referred to as "microscopic reversibility". This principle does not abolish macroscopic irreversibility, but it explains, for example, why in a chemical equilibrium a detailed balance between the forward and backward reactions is reached (for details see Sect. 3.9 and [21]).

The methods of abstraction and idealization are a kind of cognition raster that we superimpose upon reality and with which we model reality. The physicist Hans-Peter Dürr took up a comparison first used by Arthur Eddington and compared the natural scientist to an ichthyologist, who can only catch and investigate the fishes that are caught in his net—the ones that correspond to the mesh size [59]. Dürr wanted to make it clear that even the natural scientist, with his models and theories, can only study those aspects of reality which his methods are designed from the outset to capture.

This comparison is in some ways correct, but it is also misleading. Without any doubt, the physicist can only ever gain a coarse-grained image of reality. At the same time, however, the image of the "fishing" scientist gives the impression that the choice of fishnet already alienates reality, because only certain fishes are caught in it. Dürr speaks of a "change in quality which reality undergoes through our ideas, and also through our special way of thinking" [59, p. 2; author's transl.]. Indeed, looking at the enigmatic phenomena of quantum physics, an interpretation of this kind seems to be plausible.

On the other hand, the example of reversibility makes it clear that the view described above does not apply generally to science. Thus, with the concepts of reversible physics, the theoretical foundations of steam engines could be described realistically at the beginning of the eighteenth century. In reality, however, no steam engine works according to the laws of reversible thermodynamics. Even though the theoretical concepts of physics often seem unrealistic, they are still very well set up to describe reality.

The reductionist research program is indispensable not only for physics and chemistry, but also for biology. In biology, the reductionist method relies on the thought that all phenomena of life must be—at least in principle—traceable back to physical and chemical processes. This methodological guiding line is also referred to in biology as "physicalism".

Physicalism has undoubtedly celebrated its greatest triumph with the development of molecular biology, which, in turn, has become the basis of the whole of biology. Nevertheless, the question remains whether physicalism is already a sufficient basis for explaining living matter. To anticipate the answer: So far, no problem of biology is known that might question the universal validity of physicalism. This statement applies not least to human consciousness, even though one must concede that we are still a long way from an adequate understanding of the phenomena of consciousness.

The way we have speaking so far may evoke the impression that reductionism and physicalism are equivalent terms. However, this equivalence has been challenged, in particular by the contemporary philosophy of mind. Thus, Donald Davidson [60], Jaegwon Kim [61] and others favor a model of consciousness according to which the physical states of the brain *determine* states of consciousness but do not *cause* them. If, however, there is no causal connection between the brain's physical states and psychological phenomena, then there can be no psychophysical laws which could explain mental phenomena on the basis of the physical states of the brain. In that case, physicalism would be applicable to the phenomena of consciousness, but these would not be reducible to the material processes.

The concept of "non-reductive" physicalism (or materialism) coincides in an essential point with the conventional notion of psychophysical monism, according to which psychological properties are a pure expression of matter. This is equivalent to saying that every mental state corresponds to a physical state. Thus every mental state has a corresponding material state. If two individuals differ in their mental states, they also differ in their physical states. This correspondence is referred to as "determination".

However, unlike traditional psychophysical monism, non-reducing physicalism claims that material states do not cause mental states in the sense of a causation from bottom to top. In other words: Mental qualities are determined by the physical properties of the brain, but they cannot be reduced to the determining level by means of a law-like explanation. This philosophy of consciousness is also known as "anomalous" monism, because it rests on an anomaly of explanation, i.e., an explanation that lacks the element of causal relationship.

The model of non-causal determination of mental states by their physical states seems to be attractive not least because it is monistic and dualistic at the same time and thus mediates between two seemingly incompatible explanatory models of the philosophy of the mind. Nevertheless, at first glance, the model looks contradictory; how can something be determined without any causality? In fact, does the model not leave a fundamental explanation gap?

The following example may help to clarify this question: Imagine a Greek statue, which we generally think to be the epitome of beauty. Apart from all subjective assessment, the beauty of a statue is entirely "determined" by its material features and its shape. However, beauty is not "caused" by the statue's form. There is no balanced relationship between form and beauty. The beauty of a statue does not necessarily impose a definite shape upon the statue. Other configurations can also be perceived as beautiful. Nevertheless, the property of the form is more fundamental than that of beauty, because beauty does not exist without form. Conversely, however, there are plenty of forms that exist without displaying beauty.

The idea of non-causal determination is at the center of a philosophical concept that is termed "supervenience". This philosophy is based the thesis that between two properties A and B a physical dependence may exist in one direction $(A \rightarrow B)$ but not in the other direction $(B \rightarrow A)$. One thus says that feature B "supervenes" feature A at the lower level. Consequently, B cannot be reduced to A. Applied to the previous example, the property of beauty of a statue supervenes the material property of its form. In other words: Beauty is a quality of the statue which cannot be reduced to its material basis.

The conception of supervenience plays a vital role in contemporary philosophy because it seems to provide a solution for the so-called *qualia* problem (from Latin: *qualis*: what is it like). Since the fundamental investigations of the logician and philosopher Clarence Irving Lewis in the early twentieth century [62], this problem has been linked to the question of how our subjective perceptions (or experience) of feeling, smell, color, beauty and the like are related to the material carriers of such properties. Today, however, the term *qualia* is used more broadly and is also applied to schemes of intersubjective perception. It is evident that the explanation of perceptions of properties places entirely different demands on the scientific concept of explanation than does the usual explanation of events and processes. Non-reductive physicalism is a possible answer to the "qualia" problem, although the solution is in effect only an explanation of inexplicability.

The model of a non-reductive explanation of mental phenomena initially resulted from studies on the logic of language. It is based quite fundamentally on the fact that the terms with which we describe mental phenomena cannot be traced back to the terms of the material level. In addition to this adverse finding, non-reductive physicalism surmises that there is no causal connection between the physical and psychological levels. However, this is a pure assumption ad hoc, since the non-existence of such a causal relationship cannot be proven in principle. This problem is comparable to the question of whether there is a causal relationship between the syntax and the semantics of information. Although it seems highly unlikely that there exists any hitherto unknown law that links syntax to semantics, such a possibility cannot be ruled out from the outset. However, lack of evidence does not constitute proof of non-existence.

To mark the scope of the model of supervenience let us consider, as a further example, the relation between the form and the content of a piece of meaningful information. As outlined in Sect. 1.5, information is always bound to a syntactic structure which is given by the form of the arrangements of signs, letters, sounds and the like. Above the syntactic level, the semantic–pragmatic dimension of information unfolds, which is related to the meaning content of a piece of information (see Sect. 6.8). Different syntactic states A have different semantic states B. However, even if the semantic state B is based on the syntactic state A of a piece of information, B is not caused by A.

Obviously, there is no law which would allow one to derive the content of a piece of information from its form, i.e., to reduce semantics to syntax. Instead, semantics seems to be a contingent property of its associated syntax. It is neither random nor determined by any laws anchored within the syntax. In other words: the syntax is a *sub*venient property of a piece of information, while its semantics is a *super*venient property. The arrangement of the signs determines the meaning of a piece of information, but it does not cause it. Instead, semantics represent a new quality that rises above the syntactic level of material signs.

The description of the relationship between syntax and semantics by appeal to the concept of supervenience seems entirely plausible because it characterizes the semantics of information as a new quality appearing above the mere structural level of information. However, the assumption that there is no direct causal relation between syntax and semantics does not necessarily imply that semantics have no causal origin at all. On the contrary, we have already outlined in Chap. 1 that the meaning of a piece of information is always "caused" by the recipient interpreting the information. It is the recipient who gives defined semantics to the initially undefined content of a piece of information. We have denoted this the general context-dependence of information.

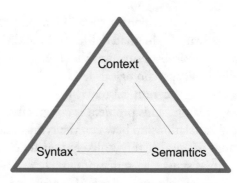

Fig. 3.7 Triadic relation of context, syntax and semantics of information. Although semantics are bound to syntax, there is no law-like relationship between semantics and syntax. Only the context, which functions as the recipient of a piece of information, defines the semantics of information. In a preceding step, however, the recipient of information must first recognize a syntactic structure as potential information. In this way, syntax and semantics are linked together by the context in which a piece of information is evaluated. The evaluation criteria may comprise quite different semantic aspects. Most studies have placed the pragmatic aspect of information in the foreground (Sect. 1.4)

We see that the concept of supervenience can only hold if one exclusively considers the logical dependencies between syntax and semantics. However, if one includes the genesis of this relationship, a new level of causal explanation opens up. For example, considering the semantics of genetic information, it is the creative process of evolution in an information-rich context—namely, the abiotic and biotic environment of the genome—which becomes the basis of a law-like explanation. Thus, regarding genetic information, evolutionary history is the link between syntax and semantics. Together, they establish a semiotic triangle of syntax, semantics and contextuality (Fig. 3.7). Similarly to a statue, which owes its beauty to the ingenuity of the artist, genetic information owes its admirable perfection to the creativity of the evolutionary process, based on the language of living matter (see Sect. 6.10).

3.9 Misunderstandings

The question of a mechanistic explanation of living matter has recurred perpetually in biological debate in the last two hundred years. Although it has become apparent in the meantime that organisms follow only physical and chemical laws, there are still hypotheses that postulate the autonomy of living matter. This is again a mistaken concept of wholeness that has contributed, and still contributes, to irritation and misinterpretations, as holistic concepts in biology often fulfil only a substitute function for actual or alleged gaps in explanation or understanding.

Nonetheless, in the nineteenth century, the elementary processes of life—such as self-preservation, development, shape formation and the like—appeared so enigmatic that many thinkers saw themselves forced to postulate the existence of a life-force (*vis vitalis*) to explain the apparent autonomy of living matter. Only toward the end of the nineteenth century did it gradually became clear that, apart from the known physical forces, there are no further natural forces. Nevertheless, the idea of a life-force has not disappeared. Instead, it was replaced by the concept of organic wholeness, which, from the viewpoint of physics, seemed to be less vulnerable to attack than the idea of a life-force.

The neo-vitalistic philosophy of Hans Driesch [63] is a clear example of the combination of vitalistic and holistic doctrines. Driesch, himself a biologist, had carried out a large number of experiments on developmental biology, in particular on the ability of sea-urchin germs to regenerate themselves. He demonstrated that the blastomeres of the two- and four-cell stages develop into normal (but somewhat smaller) larvae when the sea-urchin germ is halved or quartered. In this and other experimental discoveries, Driesch believed that he had found irrefutable proof of the holistic nature of the organism and the autonomy of life processes.

Embryonic cell systems whose elements, such as the sea-urchin germ in its earliest stage of division, have the same developmental potential, and for which it does not matter whether one takes away any part of these—or indeed whether their parts are shifted arbitrarily—were denoted by Driesch as "harmonic equipotential systems" [64]. To explain this phenomenon, Driesch introduced a teleological factor, which he referred to, following Aristotle, as "entelechy". Driesch regarded this as an immaterial and extra-spatial agent which, however, encroaches on space and regulates all organic events.

Today, the processes of biological development are very well understood without the need to resort to a life-specific factor or agent. Instead, they can be described as a process of material self-organization, one which depends exclusively upon the known laws of physics and chemistry. Even if we do not yet fully understand such developmental processes, because of their enormous complexity, no-one can seriously claim that they can only be explained holistically. Instead, the holistic doctrines of biology behave like the vitalism of which the molecular biologist Francis Crick once said: "When facts come in the door, vitalism flies out of the window." [65, p. 22]

At the beginning of the twentieth century, when the strange properties of quantum physics raised doubts about the reach of the analytical research method, even prominent physicists were prone to believe in the irreducibility of living matter. A decisive influence on this debate originated in a famous article by Niels Bohr in 1933, entitled *Light and Life,* in which he investigated the potential limitations of the physics of life (see also Sect. 5.8). To put it briefly: Bohr suspected an anomaly of explanation concerning living matter similar to that encountered in quantum physics. He pointed out that the complete physical description of a living organism would require rigorous dissection of the organism, down to the level of molecules and atoms—which however would inevitably lead to the loss of all life functions of the organism. "On this view", Bohr concluded, "the existence of life must be considered as an elementary fact that cannot be explained, but must be taken as a starting point in biology, in a similar way as the quantum of action, which appears as an irrational element from the point of view of classical mechanical physics, taken together with the existence of elementary particles, forms the foundation of atomic physics." [66, p. 458]

It may come as a surprise that such an eminent scientist as Bohr developed an almost vitalistic view of living matter. Looking back, however, this is entirely understandable. At that time, the ability of organisms to reproduce and maintain themselves appeared to be an inexplicable phenomenon from a physical point of view, which suggested that living matter could be subject to autonomous principles. Three decades later, however, Bohr corrected his opinion in a revised version of his article of 1933. He recognized that the discovery of the molecular structure of DNA in 1953 and the subsequent elucidation of the physical mechanisms of essential life functions left no space for holistic or even vitalistic speculations about the nature of living matter.

Nevertheless, some physicists still maintain that quantum physics questions biological reductionism. For example, Hans-Peter Dürr asserted: "An unbiased contemplation of living nature in its inextricable complexity seems to impose on us the idea that there must still be a formative force acting in the background, which coordinates all these processes in their interplay and provides for the essential initiations and necessary differentiations." [59, p. 14; author's transl.] Dürr ignores persistently the fact that the "formative force" is nothing more than the genetic information that is laid down in the heredity molecules and which regulates and controls all life processes. Contrary to his view, there is no need for advanced coding at the quantum-physical level to open up new dimensions of information storage for the living organism. The storage capacity of the genetic molecules is so immense that it cannot even be exhausted by evolution. Why should Nature make things more complicated than necessary?

There is another misunderstanding concerning living matter and quantum physics. Biological reductionism is sometimes overinterpreted in the sense that all life processes must, it is claimed, be actually reduced to the principles of quantum physics. Do we really have to trace life phenomena back to the quantum level in order to understand them? No one would seriously claim to have understood the function of a coffee machine just because they can describe it quantum-physically. This claim would be, to repeat our previous comparison, just as nonsensical as the statement that a fish can only be caught with a net whose mesh size is less than one-millionth of a millimeter.

Biological self-reproduction, for example, is already sufficiently explained by the structure and function of biological macromolecules. It does not require recourse to quantum physics. Likewise, with the help of Ohm's law or Kirchhoff's rules we can describe specific characteristics of an electrical circuit without having to refer to the fundamental laws of physics. At all levels of non-living and living matter, there are "special" (or "systemic") laws whose existence is due solely to the particular organization of the system in which they operate. In other words, those special laws only emerge when the effects of the general laws are channeled by the specific form of an artificial or natural organization. In the end, such differences in the organizational structures of matter also determine the differences between inorganic and organic matter.

Another gateway for conceptions of wholeness has been the psychology of the late nineteenth century. At that time, psychology was dominated by the "elementary psychology" of Wilhelm Wundt, according to which the phenomena of perception and mental experience are composed of basic psychological elements. In contrast, the so-called "Berlin School" of psychology, which included Max Wertheimer, Kurt Koffka and Wolfgang Köhler, took the view that the psychological elements were only parts of a superordinate whole, the meaning of which is only revealed by the perception of the whole, just as a melody only emerges from its single sounds when it is perceived as a whole. This idea led to the general thesis: The whole is more than the sum of its parts.

For the wholeness of a perception, Christian von Ehrenfels had introduced in 1890 the concept of "Gestalt" into psychology (compare also Sect. 1.8). By this term, he meant a meaningful whole in the broadest sense. In addition to concrete givenness such as melodies, sculptures, living beings and the like, he also included an abstract givenness such as thoughts or forms of government in the conception of Gestalt. The idea and the qualities of Gestalt were finally taken up by the Berlin School and developed into the so-called Gestalt psychology, which was a significant source for the development of holism in the first half of the twentieth century.

Gestalt psychology criticized not only psychological elementarism but also the analytical-reductionist orientation of psychology as represented at that time by the so-called "behaviorism". In a programmatic essay of 1913, the spokesman of this movement, John Broadus Watson, had argued that psychology must be founded as "a completely objective, experimental branch of natural science" [67, p. 158], the aim of which should be to predict and control mental behavior. In contrast to this, the Gestalt school advocated a holistic understanding of methods, in which phenomenology, i.e. the "unbiased" description of experienced events, should take center stage instead of the causal explanation of psychological processes.

The thesis that the whole is more than the sum of its parts was transferred from Gestalt psychology to biology. There, the thesis was intended to justify the otherness of living systems and demonstrate the limitations of biological reductionism. At the same time, new terms such as "emergence" and "fulguration" were introduced, which were designed to express the peculiarities of living organisms. Since then, the term emergence has been used to describe the appearance of qualitatively new properties that can only be observed in the whole system but not in its parts. Similarly, the term fulguration, derived from the Latin word *fulgur* (a stroke of lightning), was intended to express the fact that new features of a system may come about unforeseen and suddenly —in a flash.

How one is to imagine the sudden emergence of utterly new system properties that are not even rudimentarily present in the system's parts was outlined by the biologist Konrad Lorenz [68]. It is remarkable, however, that Lorenz, who intended to justify the peculiarities of biological systems, instead referred to an example from physics: the oscillating circuit. Such a circuit, Lorenz explains, can be thought of as the superposition of two circuits which, considered individually, do not have the properties of the overall system, namely to generate electrical oscillations (Fig. 3.8).

According to Lorenz, the oscillating circuit is a typical example of the phenomenon of fulguration, i.e. the abrupt appearance of new properties of a system that are not present in its subsystems. At the same time, however, the example of the oscillator demonstrates that the concept of "fulguration"

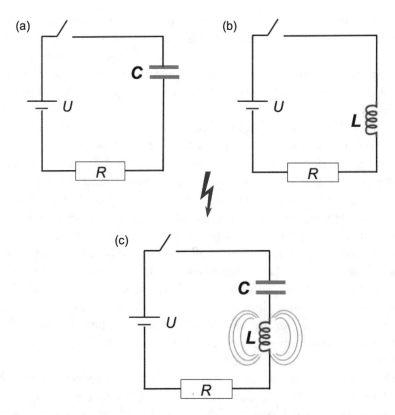

Fig. 3.8 A physical example, explaining the term "fulguration". The poles of a battery with voltage *U* are connected by a wire to a circuit with an Ohmic resistance summarized as *R*. **(a)** Circuit including a capacitor *C*. **(b)** Circuit including an inductor *L*. **(c)** By combining the upper circuits a series resonance circuit results which performs (damped) electrical oscillations. This example is due to Konrad Lorenz [68]. It was intended to illustrate the abrupt emergence ("fulguration") of a property that is present in **(c)** but is not present in either of its subsystems **(a)** and **(b)**

provides no justification of a different kind of science based on the statement "the whole is more than the sum of its parts". Instead, Lorenz demonstrates exactly the opposite, namely that the phenomenon of "fulguration" appears at all levels of reality, including in physical systems.

Lorenz himself called the statement "the whole is more than the sum of its parts" a "mystical-sounding sentence", which, he nevertheless believed to convey a profound truth. A sentence, however, that cannot even be meaningfully specified, and that also applies to everything in the world, actually expresses no truth at all. Since the whole always possesses qualities that differ from those of its parts, the truth content of the statement "the whole is more than the sum of its parts" is merely trite. As soon as one interprets this statement in the sense that the whole is more *abundant* than the sum of its parts, the thesis becomes false. Usually, the combination of parts to a whole restrains the multiplicity that is *a priori* present in the parts and the possibilities of combining them with each other. In this respect, the whole is even *less* than the sum of its parts.

Let us clarify this objection by taking an example from biology. We know today that the traditional view of genes as isolated and independent carriers of genetic information is no longer tenable in that simple form. Rather, the information content of a gene depends on its environmental conditions as well as on the information content of all the other genes in the genome. In that sense, one could well say that the genome is more than just the sum of its genes. However, a genome with a particular combination of genes represents a selection taken out of the variety of all possible combinations, and from this point of view a particular genome is less than the sum of its genes. This example illustrates how unclear the idea of the "supra-summative" whole is. It is no more than a metaphor which can be applied to all areas of reality. Therefore, the metaphor explains everything and nothing. In short: We are concerned here with an empty slogan, a flag under which all opponents of reductionism in science are sailing.

Let us now turn to another argument against strict reductionism in biology. This objection is associated with the idea of macrodeterminism and was propagated in particular by the biologist Paul Weiss [69]. He pointed out that in the organism, the incredibly complex interplay of life processes only could take place in a coordinated form because the system regulates the behavior of its parts at all levels of its organization. This kind of self-regulation, however, would require a variety of feedback loops through which the parts are connected to the whole and through which the whole controls the behavior of its components. Consequently, Weiss postulated two principles of order that he claimed are effective in living matter: "microdetermination" and "macrodetermination" (Fig. 3.9).

Microdetermination starts from the level of atoms and molecules and penetrates all levels of biological organization from bottom to top. This process is entirely determined by physical and chemical laws. Macrodetermination, in contrast, acts in the opposite direction, i.e., from top to bottom. In this case, the upper levels of the system's organization govern the lower ones. Macrodeterminism was thought by Weiss to be a systemic property of the living organism which ensures the stability and integral

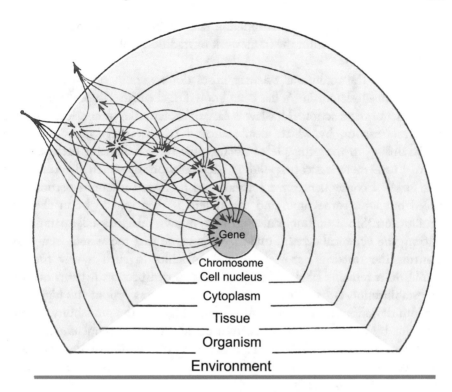

Fig. 3.9 Levels of microdetermination and macrodetermination in living matter. The biologist Paul Weiss believed that macrodeterminism and microdeterminism form a dual control system, which stabilizes and coordinates the action of all life processes. The figure shows the reciprocal relationship between the hierarchically ordered sub-systems of an organism. The arrows indicate possible interactions that determine the dynamics of the system as a whole. These interactions include microdetermination, starting from the gene, and macrodetermination, starting from the system in its entirety (adapted from [69])

activity of hierarchically organized systems. He was convinced that the scientific explanation of this property requires new concepts which go beyond those of reductionism and which are tailored to the effects that supra-molecular structures have on the dynamics of the lower levels of organization in living matter. Later, the philosopher Donald Campbell [70] interpreted macrodetermination as "downward causation".

The idea of macrodeterminism had come up to explain why living systems are stable against the fluctuations in their subsystems. Such fluctuations arise perpetually in consequence of internal and external perturbations. However, is self-regulation by macrodetermination really a property of living matter that eludes explanation within the framework of traditional physics and chemistry, as Weiss asserted?

A look at the traditional sciences raises doubts about this assertion. For example, macrodeterminism has been known in chemistry since the discovery of the law of mass action. This law is easy to explain. Let us consider a simple chemical reaction by which a substance A is converted into a substance B. According to the principle of microscopic reversibility, for each reaction from A to B there is a corresponding back reaction, by which B is converted into A. As a consequence, the forward and backward reactions become balanced out after some time, and the system is then said to be in chemical equilibrium. Since the molecules do not "know" that they take part in stabilizing the chemical system, one might imagine that the system as a whole controls the reaction behavior of the molecules in such a way that the equilibrium remains stable. However, on closer inspection it turns out that this stabilization is based on the mechanism of mass action: the higher the random deviations from equilibrium are, the higher is the probability of those reactions by which the deviations from equilibrium are compensated for. In this way, the system as a whole regulates the behavior of the molecules involved in equilibrium. The process of equilibration can be simulated by simple models (Fig. 3.10).

The law of mass action is an example demonstrating that macrodeterminism can be entirely understood within the framework of classical physics and chemistry. The stabilization of the equilibrium is based on a feedback between cause and effect that is already anchored in the reaction mechanism itself. Similarly, the cyclic causality of living matter is reducible to feedback loops that in turn can be explained in detail on a physical basis. The metabolic pathways of the living organism are paradigmatic for such feedback loops (Fig. 1.4).

Summarizing, one can state that macrodetermination in living systems is based on the mechanisms of feedback, mass action and cooperativity. All these mechanisms are fully accessible to physical analysis. Therefore, there is no need to regard macrodetermination as an irreducible property of living matter, involving a life-specific form of causality. However, in its holistic interpretation, the principle of macrodetermination becomes a phrase just as empty as the thesis of the "supra-summative property of the whole".

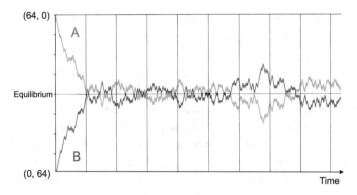

Fig. 3.10 Macrodeterminism as a chemical phenomenom. The chemical law of mass action can be simulated by a simple computer model (for details see [21, p. 130 ff.]). The figure shows the balance adjustment of a chemical reaction $A \leftrightarrows B$ where substance A is transformed into substance B and back (with equal rate constants for the forward and backward reaction). In the computer model, only a tiny population of 64 molecules is considered. The reaction starts from A (64, 0). First, the reaction will run more or less in one direction until the equilibrium (32, 32) between A and B is reached. In equilibrium, the concentrations fluctuate around the equilibrium value. The balance is stabilized by negative feedback: The larger the deviation from equilibrium, the higher is the probability of the reaction back toward equilibrium. Statistics tells us that the fluctuations around the equilibrium state have a bandwidth that is proportional to the square root of the total number of the particles involved. In the present case, the bandwidth of the fluctuations is $\sqrt{64} = 8$. Fluctuations of this size are clearly visible. However, the systems dealt with in chemistry contain vastly more particles. For example, a liter of a one-molar solution contains some 10^{24} molecules. The absolute size of the fluctuations is then immense, $\sqrt{10^{24}} = 10^{12}$ molecules; however, the *relative* deviation from the equilibrium position is very small: 10^{-12}. The fluctuations thus only make a difference in the twelfth decimal place, and could not be detected even with today's most sensitive instruments. That is the reason why chemists can treat the law of mass action as a deterministic law, although actually the concentrations of molecules in a chemical reaction system at equilibrium continuously fluctuate around their equilibrium values

3.10 Anything Goes?

As we now see, the objection that the analytical sciences proceed mechanistically and can only explain simple phenomena, i.e. those described by linear causal chains, belongs in the junk-box of a long-outdated understanding of science. In recent decades new disciplines have arisen among the analytical sciences, such as systems theory and cybernetics, which are focused on system properties, feedback mechanisms and control loops. Network theory, in turn, investigates the networking and integration of control loops into complex dynamic systems. Finally, synergetics examines the cooperative interactions by which the parts of a system are integrated into a coherent whole. Here too, micro- and macrodetermination play a central role, because the phenomenon of cooperativity, which is so fundamental to the understanding of living matter, can only be derived from the interaction of the two principles.

In the course of this development, numerous scientific disciplines have emerged and have nowadays become indispensable for the analysis of complex systems. Beside cybernetics, systems theory, network theory and synergetics, other disciplines such as information theory and game theory should be mentioned. All these sciences are aimed at investigating the overarching structures of reality, regardless of whether these structures occur in living or non-living, natural or artificial systems. By their very nature, these disciplines are all based on the methods of abstraction, idealization and generalization. Because of their high degree of abstraction, they are also referred to as "structural sciences" (Sect. 4.9). Physics, when dealing with complex systems, has incorporated many elements of these new disciplines and thereby changed its traditional character. Consequently, the concept of physicalism has also acquired an entirely new meaning. Thus, it is no wonder that criticism of reductionism increasingly comes away empty-handed.

The dispute over the adequacy of the analytical methods of science has also given rise to an ideological controversy about the scope of scientific explanations. From today's perspective, however, there are no signs that the reductionist research program of science is likely to come up against impassable limits. The holistic counterpositions to reductionism vanish into thin air as soon as they are confronted with the hard facts of science. The statement that the whole is more than the sum of its parts has turned out to be an empty slogan. In reality, the naïve reductionist, who desires to trace everything back to a heap of molecules, does not exist at all. In the words of the biologist Peter Medawar: The analytical-summative mechanist is "a sort of lay devil invented by the feebler nature-philosophers to give themselves an opportunity to enjoy the rites of exorcism" [71, p. 144].

When holistic doctrines are dissected with the scalpel of analysis, they usually burst like a soap bubble. Even the physicist Hans-Peter Dürr, who sympathized with holistic ideas, had finally to admit that holistic statements "can in principle no longer be put to the test in the sense that is considered necessary for a modern science" [22, p. 13; author's transl.]. His well-intentioned suggestion of searching for new criteria of truthfulness or coherence within the framework of holistic approaches is likewise meaningless, because it would result in abandoning the idea of the unity of science. In that case, the concept of critical and controllable science would give way to competing forms of science with their own claims to truth. The already deplorable separation of scientific disciplines would become even deeper and would divide the scientific landscape further.

The outstanding success of modern science has only been possible because it has steadfastly adhered to the idea of the unity of science—a unity that, not least, is reinforced by the reductionist research program. One would only consider supplementing or replacing this hugely successful program if fundamental anomalies arose that brought it seriously into question. However, as long as there are no indications of this, the reductionist method of research remains the sole basis for a self-critical science.

Of course, one may choose to interpret the methodological loyalty of science as arising out of a compulsive urge to constrain the free development of ideas. In the end, however, success and failure determine the choice of method. A colorful juxtaposition of research methods, which in themselves may even be mutually exclusive, offers absolutely no alternative to the analytical and reductionist method. This is all the truer for a counterposition to the methodological specifications of science as developed some years ago by the philosopher Paul Feyerabend [72]. Criticizing the methodological constraints which science imposes on the acquisition of knowledge, he propagated an "anarchistic theory of knowledge" instead. Under the slogan "Anything goes!" he initiated a frontal attack on the methodological self-image of science by questioning all the cornerstones of the scientific method. With his extreme counterposition to today's scientific rationalism, Feyerabend reached a public that was already biased and hostile to science. It is evident that from that point it is only a short step to pseudo-science and esotericism. In fact, "Anything goes" has become the *leitmotif* of a post-modern culture that propagates the free treatment of science. It is therefore all the more urgent that science maintain a critical awareness of its methods. Only the acquisition of methodologically sound knowledge makes it possible to keep science free of pseudo-scientific influences.

While the natural sciences are based on the analytical method, the humanities find it extraordinarily difficult to secure their alleged otherness with the development of an autonomous approach to research. This lack is undoubtedly a reason why the humanities are often on the verge of pseudo-science. Philosophical hermeneutics, to which the humanities frequently refer, attempts to circumvent the question of its method by renouncing from the outset any claim to be a science in the sense of the exact sciences. Instead, hermeneutics sees itself as looking at the world as a whole, and thus reviving the humanistic ideal of education. From the point of view of hermeneutics, the findings of the exact sciences always refer to methodologically worked-out excerpts of reality. These findings, it is however asserted, only appear as marginal forms of a world-understanding, access to which is only offered by hermeneutics.

Therefore, the problem of the unity of the sciences proves primarily to be a methodological problem, as the way to the unity of science obviously requires a standardized application of methods. On the other hand, there is only one cognition method that satisfies the conditions of critical objectivity at all. This is the analytical method, which Descartes already laid down at the very beginning of modern science. If one follows this idea, the scientific method of cognition has, in principle, to prove itself by solving all the as yet unsolved puzzles of science. This task poses an enormous challenge for the scientific method. Of course, it does not imply that we can expect a scientific answer to all these problems. It only means that a problem, as far as it is "scientifically" solvable at all, must be solved within the framework of the analytical method.

In this respect, modern science is indeed facing enormous challenges. The central research topics that are currently under the scrutiny of the sciences demonstrate this. The phenomena of living matter in general, and the appearances of mind and consciousness in particular, are in the foreground of scientific interest. Without prejudice to the fact that research into mental phenomena is still in its infancy, one can still be confident that here too the analytical and reductionist method will prove successful.

In any case, there have so far been no anomalies of explanation that would question this methodological approach as such. The same applies to the mysterious phenomena of quantum physics (Sect. 3.6). Since everything seems to indicate that quantum theory is complete, quantum phenomena, too, are in reality no longer a scientific problem, but rather a challenge for our traditional world-view. This is demonstrated not least by the fact that the scientific discussion in quantum physics revolves primarily around questions of its adequate interpretation.

However, what about the problems of history, for which the humanities traditionally feel responsible? Can the historical and cultural world of man, with its individual decisions and actions, its unique processes and its historically grown structures, ever become an object of exact scientific knowledge? Can historical events, in their immense complexity, ever be meaningfully dissected and dissolved into causal relationships? Are the special and unique features of historical processes also accessible to the exact sciences? In short, is the historical aspect of our world the point of divergence, at which the sciences inevitably separate into the natural sciences and the humanities? We shall address these questions more directly in Chap. 4.

References

1. Boulding M, Rotelle JE (eds) (1997) The Works of Saint Augustine: A Translation for the 21st Century, The Confessions, book 11. New City Press [Original: Confessiones, ca. 400 CE]
2. Polanyi M (2009) The Tacit Dimension. University of Chicago Press, Chicago
3. Hund F (1979) Geschichte der physikalischen Begriffe. Bibliographisches Institut. Mannheim
4. Jung CG (1968) Psychological Aspects of the Mother Archetype (transl: Adler G, Hull RFC) In: Adler G, Hull RFC (eds) The collected works of C. G. Jung, vol 9 (Part 1). Princeton University Press, Princeton [Original: Die psychologischen Aspekte des Mutter Archetypus, 1938]
5. Lacan J (2006) The Instance of the Letter in the Unconscious, or Reason Since Freud (transl: Fink B). In: Écrits: a selection. W. W. Norton, New York, pp 412–441 [Original: l'instance de la lettre dans l'inconscient ou la raison depuis Freud, 1957]
6. Jacob F (1977) Evolution and Tinkering. Science 196(4295):1161–1166
7. Koestler A (1964) The Act of Creation. Penguin Books, New York
8. Heisenberg W (1971) Die Bedeutung des Schönen in der exakten Naturwissenschaft. Physikalische Blätter 27(3):97–107
9 Weizsäcker CF von (1980) The Unity of Nature. Farrar, Straus & Giroux, New York [Original: Einheit der Natur, 1971]
10. Jacob F (1999) Of Flies, Mice, and Men (transl: Weiss G). Harvard University Press, Cambridge/Mass [Original: La Souris, la mouche et l'homme, 1997]
11. Kepler J (1997) The Harmony of the World (transl: Aiton EJ, Duncan AM, Field JV). Memoires of the American Philosophical Society, ccix. Philadelphia [Original: Harmonices Mundi, 1619]

12. Baumgarten AG (2007) Ästhetik, 2 Bd. Mirbach D (ed). Meiner, Hamburg [Original: Aesthetica, 1750–1758]
13. Boltzmann L (1979) Populäre Schriften. Vieweg, Braunschweig
14. Poincaré H (1913) Science and Method (transl: Halsted GB). In: The foundations of Science. The Science Press, New York/Garrison [Original: Science et méthode, 1908]
15. Hardy GH (1992) A Mathematician's Apology. Cambridge University Press, Cambridge
16. Dirac PA (1963) The evolution of the physicist's picture of nature. Sci Am 208 (5):45–53
17. Kepler J (1981) Mysterium Cosmographicum—The Secret of the Universe (transl: Ducan AM, intro Aitin EJ). Abaris Books, New York [Original: Mysterium Cosmographicum 1596]
18. Caspar M (1959) Kepler. Abelard-Schuman, London/New York
19. Weyl H (1952) Symmetry. Princeton University Press, Princeton
20. Müller A, Gouzerh P (2012) From linking of metal-oxide building blocks in a dynamic library to giant clusters with unique properties and towards adaptive chemistry. Chem Soc Rev 41:7431–7463
21 Küppers B-O (2018) The Computability of the World: How Far Can Science Take Us? Springer International, Cham
22. Ruff W, Rogers J (2003) The Harmony of the World—A Realization for the Ear of Johannes Kepler's Astronomical Data from Harmonices Mundi 1619. Audio CD
23. Robbin T (1992) Fourfield: Computers, Art & the 4th Dimension. Little, Brown, Boston
24. Peitgen H-O, Richter PH (1986) The Beauty of Fractals. Springer, Berlin/Heidelberg
25. Bense M (1969) Einführung in die informationstheoretische Ästhetik. Rowohlt, Hamburg
26. Brentano F (1968) Die Habilitationsthesen (1866). In: Kraus O (ed) Über die Zukunft der Philosophie. Meiner, Hamburg
27. Lange FA (1866/1974) Geschichte des Materialismus und Kritik seiner Bedeutung in der Gegenwart, 2 vols. Schmidt A (ed). Suhrkamp, Frankfurt am Main
28. Dilthey W (2002) The Formation of the Historical World in the Human Sciences (eds: Makkreel RA, Rodi F). Selected Works, vol III. Princeton University Press, Princeton [Original: Der Aufbau der geschichtlichen Welt in den Geisteswissenschaften, 1910]
29. Heidegger M (1962) Being and Time (transl: Macquarrie J, Robinson E). Basil Blackwell, Oxford [Original: Sein und Zeit, 1927]

30 Gadamer H-G (2004) Truth and Method. Continuum, New York [Original: Wahrheit und Methode, 1960]

31. Rorty R (2000) Der Vorlesungsgast. In: Figal G (ed) Begegnungen mit Hans-Georg Gadamer. Reclam, Stuttgart, pp 87–91

32. Gadamer HG (2004) From Word to Concept. The Task of Hermeneutics as Philosophy (transl: Palmer ER). In: Krajewski B (ed) Gadamer' Repercussions: Reconsidering Philosophical Hermeneutics. University of California Press, Berkeley [Original: Vom Wort zum Begriff, 1994]

33 Smuts JC (1926) Holism and Evolution. Macmillan, New York

34 Meyer-Abich A (1948) Naturphilosophie auf neuen Wegen. Hippokrates, Stuttgart

35. Harrington A (1996) Reenchanted Science. Princeton University Press, Princeton

36. Leśniewski S (1992) Collected Works, 2 vols. (Surma SJ, Srzednicki JT, Barnett DI, Rickey VF (eds). Kluwer, Amsterdam

37. Goethe JW von (1981) Werke, Bd 13. Beck, München

38. Mises R von (1968) Positivism. A Study in Human Understanding. Dover Publications, New York [Original: Kleines Lehrbuch des Positivismus, 1939]

39. Bohr N (1935) Can Quantum-Mechanical Description of Physical Reality be Considered Complete? Phys Rev 48:696–702

40. Born M (2005) The Born-Einstein Letters 1916–1955. Friendship, Politics and Physics in Uncertain Times. Macmillan, Basingstoke, Hampshire

41. Einstein A, Podolsky B, Rosen N (1935) Can Quantum-Mechanical Description of Physical Reality Be Considered Complete? Phys Rev 47:777–780

42. Zeilinger A (2010) Dance of the Photons: From Einstein to Quantum Teleportation. Farrar, Straus & Giroux, New York

43. Bell JS (1964) On the Einstein Podolsky Rosen paradox. Physics 1(3):195–200

44. Aspect A, Grangier P, Roger G (1981) Experimental tests of realistic local theories via Bell's theorem. Phys Rev Lett 47:460–463

45. Freedman SJ, Clauser JF (1972) Experimental test of local hidden-variable theories. Phys Rev Lett 28:938–941

46. Bohr N (1928) The Quantum Postulate and the Recent Development of Atomic Theory. Nature 121:580–590

47. Mittelstaedt P (2009) Quantum Logic. In: Greenberger D, Hentschel K, Weinert F (eds) Compendium of quantum Physics. Springer, Berlin/Heidelberg

48. Aguirre A, Foster B, Merali, Z (eds) (2015) It From Bit or Bit From It. On Physics and Information. Springer International, Cham

49. Bricmont J (2016) Making Sense of Quantum Mechanics. Springer International, Cham

50. Jaeger G (2009) Entanglement, Information, and the Interpretation of Quantum Mechanics. Springer, Berlin/Heidelberg

51. Küppers B-O (2013) Elements of a Semantic Code. In: Küppers B-O, Hahn U, Artmann S (eds) Evolution of Semantic Systems. Springer, Berlin/Heidelberg, pp 67–85

52. Heisenberg W (1954) Das Naturbild der heutigen Physik. Jahrbuch 1953 der Max-Planck-Gesellschaft zur Förderung der Wissenschaften, pp 52–54

53. Wheeler JA (1989) Information, Physics, Quantum: The Search for Links. In: Proc. 3rd Int. Symp. Foundations of Quantum Mechanics (Physical Society of Japan). Tokyo, pp 354–368

54. Küppers B-O (1990) Information and the Origin of Life (transl: Woolley P) MIT Press. Cambridge/Mass [Original: Der Ursprung biologischer Information, 1986]

55. Laplace PS de (1995) Philosophical Essay on Probabilities (transl: Dale AI). Springer, New York [Original: Essai philosophique sur les probabilités, 1814]

56. Humboldt A von (1797) Versuche über die gereizte Muskel- und Nervenfaser: nebst Vermuthungen über den chemischen Process des Lebens in der Thier-und Pflanzenwelt. Decker, Posen

57. Kuhn TS (1962) The Structure of Scientific Revolutions. University of Chicago Press, Chicago

58. Heisenberg WK (2008) Abstraction in Modern Science. In: Nishima Memorial Foundation (ed), Nishima memorial lectures. Lect Notes Phys 746:1–15

59. Dürr HP (1997) Ist Biologie nur Physik? Universitas 607:1–15

60. Davidson DH (1970) Mental Events. In: Foster L, Swanson JW (eds) Experience and theory. University of Massachusetts Press, Amherst, pp 79–101

61. Kim J (1993) Supervenience and Mind. Cambridge University Press, Cambridge

62. Lewis CI (1929) Mind and the World Order: Outline of a Theory of Knowledge. Scribners, New York

63. Driesch H (1929) The Science and Philosophy of the Organism. Gifford lectures delivered at Aberdeen University (1907–1908), 2 vols. A & C Black

64. Driesch H (1899) Die Lokalisation morphogenetischer Vorgänge, ein Beweis vitalistischen Geschehens. Engelmann, Leipzig

65. Crick F (1966) Of Molecules and Men. University of Washington Press, Seattle

66. Bohr N (1933) Light and Life. Nature 131(421–423):457–459

67. Watson JB (1913) Psychology as the behaviorist views it. Psychol Rev 20:158–177

68. Lorenz K (1977) Behind the Mirror: A Search for a Natural History of Human Knowledge (transl: Taylor R). Harcourt Brace Jovanovich, New York [Original: Die Rückseite des Spiegels, 1973]

69. Weiss PA (1968) Dynamics of Development: Experiments and Inferences. Academic Press, New York

70. Campbell DT (1974) "Downward causation" in Hierarchically Organised Biological Systems. In: Ayala FJ, Dobzhansky T (eds) Studies in the Philosophy of Biology. Macmillan, London, pp 179–186

71. Medawar PB, Medawar JS (1983) Aristotle to Zoos. Harvard University Press, Cambridge/Mass

72. Feyerabend P (2010) Against Method: Outline of an Anarchistic Theory of Knowledge. Verso Books, New York

4

Unity: The Deep Structure of Science

4.1 Toward the Unity of Science

In a world that is becoming increasingly complex and unmanageable, in which profound social, economic and political upheavals are taking place, aids to orientation and decision-making are expected, not least from the sciences.[1] However, modern science is more likely to offer a picture of inner strife. In the form of the natural sciences and the humanities, two intellectual currents are in collision, and they have quite different approaches to reality. Moreover, both currents are split up into numerous sub-disciplines and application areas. Beside, the importance of modern science for our world seems to be ambivalent: On the one side, science is significant factor in determining our conception of humankind and the world. On the other, owing to increasing specialization and compartmentation, science seems to be losing sight of reality as a whole and, thereby, to be bolstering a loss of reality.

The often-heard accusation that analysis reduces the wealth and abundance of our life-world to that which is scientifically experienceable and technically feasible is directed above all against the natural sciences. With this argument, the natural sciences are accused of being responsible for the progressive destruction of the natural basis of our existence. It is strange, in contrast, that the negative impact of the humanities on our life-world is only rarely

---

[1]In this chapter, the term "science" (*scientia*: knowledge), when used alone, is to be understood in its broadest sense, to include not only what we commonly term "science"—physics, biology and so forth— but also the social, political, economic and other sciences including the humanities. Where a narrower meaning is intended, this is made explicit. Common features and differences between the various types of science will emerge in the course of the discussion.

© Springer Nature Switzerland AG 2022
B.-O. Küppers, *The Language of Living Matter*, The Frontiers Collection,
https://doi.org/10.1007/978-3-030-80319-3_4

addressed in public debate. However, one can very well ask whether the humanities are not also involved in the dismantling of our life-world by, at least occasionally, bringing forward highly questionable and sometimes even dangerous ideologies.

This view of the intellectual landscape is further distorted by the fact that the humanities often present themselves as a counterbalance to the overpowering force of the natural sciences. This contention is usually justified by the argument that the humanities, unlike the natural sciences, have a true-to-life, unfalsified access to the world, because they focus their attention from the outset upon the realities of human life. Indeed, the historicity of human existence, man's inner experience and the world's artistic design are areas of reality that mostly elude the natural sciences. All in all, it appears as though the understanding of the historical and cultural world cannot be achieved by the analytical thinking of natural sciences, but only within the framework of the universal thought as practised by the humanities.

These are the reasons why the humanities claim sovereignty over the interpretation of all human knowledge. Not least, they seem to be the source of the orientational knowledge that is essential for shaping our scientific and technical world. Thus, in the public perception of science, an image has been consolidated over time, according to which it is the humanities that preserve man's historical and cultural heritage, while natural science and technology are leading us into a reality that is foreign or even hostile to mankind.

At this point, the question arises as to what status the humanities actually have in the scientific landscape. Do they provide any secure knowledge, comparable to that of the natural sciences, or is the notion of the humanities only a collective term for those disciplines which do not follow the mathematically based methods of analytical research? Do, in the end, the natural sciences and the humanities represent two different cultures whose representatives move in two completely different intellectual currents of society?

Already in the 1950s, the physicist and novelist Charles Percy Snow took up similar questions in a famous lecture entitled *The Two Cultures*. In this lecture, he criticized in sharp terms the mutual incomprehension between "literary intellectuals" on the one side and scientists, above all physical scientists, on the other [1]. The differences between the two cultures, Snow stated, had become so vast that an understanding between them was no longer possible. He went on to assert that the division of the intellectual world into two cultures was contributing to a narrowing of our mental horizons and to the creation of an intellectual climate that was characterized by an ever-deepening gulf of incomprehension, mistrust and dislike.

Snow's diagnosis that the findings of the natural sciences are not part of the educational canon of the general public is still correct today. However, in times in which science and technology are making rapid progress, the deficits in understanding scientific findings are becoming downright dangerous. After all, only a society that has an adequate level of scientific knowledge can make responsible and forward-looking decisions concerning the multilayered problems of technological progress. In short: only knowledge can control knowledge (Sect. 7.5).

This is not the place to follow in detail the educational and social conclusions that Snow drew from his thesis of the two cultures. Instead, we will ask, from the perspective of science, whether the gap between the natural sciences and the humanities can after all one day be bridged. This question is of general importance because it touches on the fundamentals of science.

To approach this, we must first take a brief look at the beginnings of scientific thinking (see also Sect. 2.2). In early antiquity, when the sciences were nothing other than natural philosophy, the world appeared to man primarily as a cosmos of things and facts. Consequently, the philosophers of antiquity concluded that the ultimate source (*archē*) of the world is to be found in the origin of all things. The first natural philosophers of the occident, Thales and Anaximenes, considered sensorily perceptible substances such as water and air to be the basis of all matter. Heraclitus, on the other hand, claimed fire to be the fundamental element of everything that exists. Empedocles, in turn, considered earth, air, fire and water to be the four elements from which everything that exists is built.

However, there were also amazingly abstract ideas of the primal ground of the world. Anaximander, for example, traced the origin of things back to an abstract substance that is neither qualitatively determined nor quantitatively limited. This primal substance, the basis of all being, with neither inner nor outer limits, was referred to as *apeiron* ("the unlimited").

The Pythagoreans, in turn, considered numbers and their ratios to be the essence of all that exists. Starting from this idea, they developed a "number atomism" which was thought to reflect the true nature of reality. Heraclitus probably had the most comprehensive and abstract view of the primal foundations of the world (Fig. 4.1). Going far beyond the idea of primal substances, he postulated the existence of a universal law, called *lógos*, that has a regulative character for the cosmic order as well as the social and political order. The *lógos* concept referred, for the first time, to the unity of reality mediated by a universal law.

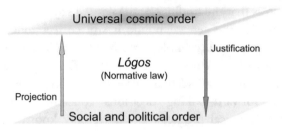

Fig. 4.1 The *lógos* concept of Heraclitus. In ancient times, Heraclitus was the first who tried to trace the unity of reality back to the action of a universal law. This law was not thought to be a law in the sense of the laws of Nature, expressing an unchangeable relation between things; rather, it was understood to be a normative law, i.e. a reason inherent to the world, one which material objects and humans were compelled to follow. The justification of the *lógos* doctrine serves, as it were, as a template for the later doctrines of natural law in which Nature (or the cosmos in the broadest sense) served to justify ethical and legal norms. Within this frame, social and political conditions are first projected to a metaphysical level of Nature (or the cosmos) to justify finally, on that basis, the socio-political conditions that existed, or those that should be established

All these "archaic" attempts to find a substance, principle or law as the primal cause of reality must be regarded as first steps toward a rational understanding of the external world. In modern times, the search for a unified world-understanding was pursued by philosophers such as Spinoza, Leibniz, Schelling, who developed their conceptions of a primal principle which was thought to overcome the divide between matter and mind (see Table 2.1). Last but not least, the search for unity is a decisive motif in contemporary thought, and above all in the natural sciences. This is demonstrated by, among other examples, the attempts of physicists to unify the four fundamental forces: gravitational force, electromagnetic force and the weak and strong nuclear forces.

To the extent that in the age of modern science detailed knowledge about the world's structure has increased, the problem of the unity of reality has shifted to that of the unity of scientific knowledge. In other words, the search for the ontological unit of reality has segued into the search for the epistemic unit of the scientific findings. There is a plausible reason for this change in perspective.

The profound insights that the natural sciences have attained into the fine structure of reality rest upon the analytical methods of dissection, abstraction, idealization and the like. A deeper understanding of scientific results, which are scattered over many disciplines, theories, models and ideas, presupposes the re-integration of findings into a cognitive whole. Only the unity of scientific knowledge leads back to a complete image of reality, the dissection of which necessarily precedes any analytical cognition.

The humanities approach reality differently. Their interests and efforts are directed toward the meaning-context of man's life-world, which they try to approach from quite different perspectives. Thus, the humanities investigate primarily historical events and the particular and individual phenomena of our cultural world, which mostly seem to elude any description by general laws. The lack of strict explanations in the humanities is the point where the paths of the natural sciences and the humanities begin to diverge (see Chap. 3).

In the following sections, we will consider the scientific landscape in more detail by examining the extent to which a rapprochement of the two cultures of knowledge is possible at all. We shall show that such a rapprochement is taking place, as is demonstrated by the present-day development of the scientific landscape.

In the humanities, for example, there is a new methodological approach which, starting from linguistics, has led to a structuralist branch of the humanities known as French structuralism (see Sect. 1.8). Employing mathematics and logic, structuralism searches for abstract models to explain the phenomena of historical, cultural, mental and social reality.

A similar development is taking place in the exact sciences. Starting from the investigation of complex systems, a new type of science has arisen, which deals with the abstract and overarching structures of reality, independently of whether these structures can be found in natural or artificial, in living or non-living systems. These cross-sectional disciplines, which I term collectively "structural sciences", have developed in parallel with digital technology. They can tap into segments of reality that previously seemed to be open only to the humanities. However, even in the natural sciences—in particular in physics and biology—more and more concepts from the structural sciences are being included. In consequence, the character of the natural sciences has also been transformed to a large extent (see Sect. 4.9). This development gives us reason to hope that the gap between the natural sciences and the humanities may one day, at least largely, be bridged.

Despite all the differences still existing between the natural sciences and the humanities, the search for their unity seems to reflect a central idea of scientific modernity. In contrast, however, some philosophers continue to espouse a pluralism of science. Above all, the philosophers of the so-called postmodernism believe that they have to reject the idea of the unity of science because, in their view, it is a relic of metaphysical fantasies. A prominent representative of this line of thought was, for instance, the philosopher Jean-François Lyotard [2].

One must be aware, however, that postmodern philosophy advocates the relativization of all structures that have become entrenched in the course of man's cultural history. For the philosophers of postmodernism, the sciences are *the* showcase example of the encrusted structures of modernity, which they believe have to be overcome. The postmodern critique of natural science, for example, has led to the absurd conclusion that its findings are only the result of random cultural development, and that they are as arbitrary as the historical and cultural circumstances under which they were made. As a result, the alleged objectivity of scientific knowledge is, it is claimed, nothing more than a powerful buzzword with which the sciences try to consolidate their social supremacy.

Some representatives of the exact sciences have described postmodern philosophy as mere "fashionable nonsense" [3]. Others even regard the postmodern trend with some concern. The physicist Steven Weinberg, for example, lamented a severe misunderstanding of the sciences, resulting from the postmodern world-view, which could have dangerous socio-political effects. He rightly warned: "If we think that scientific laws are flexible enough to be affected by the social setting of their discovery, then some may be tempted to press scientists to discover laws that are more proletarian or feminine or American or religious or Aryan or whatever else it is they want. This is a dangerous path, and more is at stake in the controversy over it than just the health of science." [4, p. 153] In fact, the history of the twentieth century teaches us what disastrous consequences can follow if our world-view is not oriented toward scientific knowledge, but, conversely, the sciences are oriented toward a given world-view or even serve it.

4.2 The General and the Particular

A closer look at the problem of the unity of science reveals that it has two different aspects. This becomes clear when one specifies the meaning of the term "unity" more precisely. The unity of science could mean that all differences between the scientific disciplines are eliminated, step by step. In the end, all boundaries between scientific disciplines would disappear, and all the sciences would merge into a unified science in its highest perfection.

It is hardly conceivable, however, that the development of science will actually take such a path. Rather, this kind of unity can only be considered to be a regulative idea, aimed at achieving the most compact possible representation of scientific knowledge. More realistic, in contrast, is an understanding of the unity of science according to which all scientific disciplines retain their autonomy, still retaining their different approaches to reality. In that case, the idea of unity would refer only to the search for the common abstract structures that underlie the various scientific disciplines. This path is taken by structuralism in the humanities, and by the structural sciences in the natural sciences. It will also the path taken in this book.

The search for the unity of science found its first climax in the philosophy of German idealism. Here, the question of the ultimate foundation of human knowledge was at the center of philosophical interest. However, the attempt to build up philosophy as the highest form of science, with a claim to absolute knowledge, unleashed a massive wave of contradiction by the majority of natural scientists (see Sect. 2.6). Mathematician and philosopher Auguste Comte argued around 1830 that physics was the science through which the unity of knowledge is realized. In his multi-volume work *Cours de philosophie positive* he set up a model according to which human knowledge necessarily passes through three stages: the theological-fictional, the metaphysical-abstract, and finally the scientific-positive stage [5]. Only at the final stage of development, Comte believed, does the previously fictional and abstract knowledge become real knowledge, as it emerges in physics as the purest and brightest form of all sciences.

Consistently with this, Comte also subordinated the science of social life to the cognitive ideal of physics. With the foundation of sociology as "social physics" Comte hoped that one day it would be possible to make precise predictions about the development of social and economic life that would in no way be inferior to the exact statements of physics. This idea soon found further supporters. In England, for example, John Stuart Mill—inspired by the work of Comte—attempted to apply the methods of the exact sciences to sociology and politics.

The three-stage law of the development of knowledge ultimately formed the basis for a "positivistic" world-view, which spread to large parts of science in the nineteenth century. This positivism only accepted knowledge that is based on "positive" (i.e. objective) and verifiable facts. He rejected all speculations that went beyond positive facts, writing them off as a relapse into pre-scientific or metaphysical forms of knowledge. In the wake of the aspiring natural sciences of the nineteenth century, whose supporting pillars were atomism and evolutionary theory, positivistic thinking gained still more ground. In the end, it guided the scientific world-view.

Against the increasingly positivistic Zeitgeist, the humanities finally tried to distance themselves by formulating their own methodology. This subsequently led to the dualistic conception of science, according to which there is a fundamental contrast between the outer and the inner world, between Nature and mind (see Sect. 3.3). However, the sharp separation of Nature and mind did not find favor with all scholars of the humanities. Criticism came above all from the philosopher Wilhelm Windelband, a leading representative of the German school of neo-Kantianism. In principle, Windelband shared the dualistic view of science, but he felt it necessary to blunten the sharp contrast between the natural sciences and the humanities. Nature and Mind seemed to him to be in far too close a relationship of mutual dependence for them to constitute the foundations of two fundamentally different sciences.

Furthermore, Windelband refused to accept that in the humanities the objects of cognition emerge only from inner perception and experience, as claimed by Dilthey (see Sect. 3.3). It also bothered him that such an essential discipline as psychology could not be included in the classification scheme of Nature/Mind at all. Judging by the its object, psychology belongs to the humanities, but according to its method, psychology seems to be one of the natural sciences. In any case, the experiments that Wilhelm Wundt carried out at the end of the nineteenth century on the human sensory and perceptive capacity left little doubt that psychological processes can be investigated appropriately by using the methods of the natural sciences.

Windelband believed that one could resolve the "incongruence" of the practical and formal classification of Dilthey's doctrine of science by replacing the contrasting pair "Nature/Mind" by that of "Nature/History". He was, of course, clear that the mere shift in perspective would not resolve the dualistic view of the sciences. He believed, however, that the underlying dichotomy would gain in conciseness because its conceptual and systematic positioning would now appear with far greater clarity.

A particularly illuminating statement of Windelband's view of science can be found in his lecture *History and Natural Science* [6]. In this lecture Windelband attempted to justify the division of the sciences into natural and historical sciences, as a manifestation of our understanding of the world that rests upon two basic elements: "law" and "event". According to Windelband, the natural sciences search for the "general", in the form of universally valid laws, while the goal of knowledge in the humanities is the unique event, the "individual" in its historically determined form. The one set of disciplines considers the form, which always remains the same, and the other the unique and intrinsically given content of the individual event. The one is concerned with what always *is*, the other with what once *was*.

To characterise this dichotomy, Windelband introduced the technical terms "nomothetic" and "idiographic". These terms are equivalent to "law-like generalisation" and "individual description". Later, the philosopher Heinrich Rickert sharpened the distinction between the nomothetic and idiographic sciences by re-interpreting the latter as a general cultural science, one that is primarily concerned with the determination and assignment of values and meaning in our cultural world [7].

According to Rickert, the difference between the systematic procedures of the different kinds of science lies in the fact that one set investigates objects, without reference to value or meaning, while the other set is concerned with objects that carry value and meaning. Even the first step toward an understanding of the historical and cultural world, Rickert claims, already requires a value decision, because one first must separate the meaningful from the meaningless in order to reduce the vast abundance of events to their essentials. This prior decision, Rickert emphasized, was indispensable to give historical and cultural studies any degree of generality. Without such value decisions, he declared, there would be no science of the historical, cultural world at all.

Rickert also pointed out that the general concepts of the natural sciences primarily emerged from abstractions, idealizations, simplifications and the like [8]. For this reason, according to Rickert, these concepts move away from empirical reality to the extent that their generality increases. The individualizing sciences, in contrast, required a terminological approach that allows a reproduction of the object of cognition as precisely as possible. In contrast to the concepts of the natural sciences, which, he asserted, become increasingly empty the more comprehensive they are, historical concepts become ever richer in substance and scope.

The contrast between nomothetic and ideographic sciences, as described by Windelband and Rickert, still shapes our understanding of the natural sciences and the humanities today. In the light of their considerations, the problem of the unity of science refers back to a fundamental problem of epistemology that is as old as philosophy itself: the relationship between the general and the particular. Can the particular, contrary to its terminological determination, ever become the object of a science, which has the general as the goal of its cognition? Windelband's answer was a definite "no". Indeed, this answer seems to be inevitable when one follows the tradition of thought, reaching back into antiquity, according to which the general and the particular are two incommensurable elements of our intellectual world.

Aristotle, for example, defined the general (universal), in contradistinction to the particular (individual), as that which by its very nature appertains to several things at the same time [9, book 7, 1038b]. In contrast to the universal, Aristotle claimed, the individual is something which is unique in itself and belongs to nothing else. Consequently, the individual (or particular) cannot be positively determined but only negatively, i.e., by the exclusion of all universal (or general) characteristics.

For this reason, Aristotle emphasized that cognition of the particular is not possible. Instead, he claimed, there could only be a sensory perception of the particular. The universal, on the other hand, was not an object of perception but rather a prerequisite of cognition. In other words, the manifoldness of individual phenomena only becomes understandable if one can trace it back to something universal. The latter finally manifest itself as a principle, rule or law.

However, what is the particular, in which we see something unique and individual? Since, by definition, the particular cannot be expressed in general terms, Aristotle had no choice but to understand the particular only in an exclusory sense, that is, as something that cannot be generalized. From this standpoint, general insights into the particular, ones that also involve the possibility of lawlike statements, seem unthinkable. This conclusion resulted from the fact that Aristotle looked above all at the general and the particular from a language-logical perspective. However, since the two terms are mutually exclusive, he saw himself forced to deny the possibility of defining the particular in general terms.

So is the idea of the unity of science, bringing together the nomothetic and ideographic sciences, only a fiction? Is this idea inevitably doomed to failure because of the irreconcilable antagonism of the general and the particular? Or is there, after all, a way to dissolve this *aporia*? Goethe offered a very enlightening *aperçu* that seems to suggest a possible solution. He wrote: "The general and the particular coincide; the particular is the general, appearing under different conditions." [10, p 433, no 491; author's transl.]

Goethe's thought was taken up by philosopher Ernst Cassirer in his treatise on *Substance and Function* [11]. Here, Cassirer emphasized that the particular is not something that can only be grasped by detailed description, as Windelband and others had claimed. The particularity, Cassirer explained, rather emerged as a consequence of the specific position which the particular takes within the network of the general relationships constituting reality. In other words, the particular does not acquire its unique character because it eludes all general relationship circles, but, on the contrary, because it enters into such relationships (Fig. 4.2).

With this interpretation, Cassirer dissociated himself from Windelband's concept of ideographic sciences, according to which the scientific approach to the particular is already exhausted in the individualizing description of its objects. Cassirer, in contrast, emphasized that it "is not evident that any concrete content must lose its particularity and intuitive character as soon as it

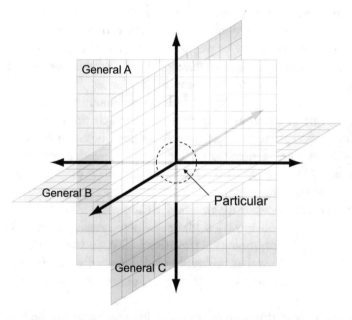

Fig. 4.2 Emergence of the particular by the interleaving of general properties. Schematic representation of the emergence of the particular within the network of its relationships to the general. The greater the number of planes intersecting within the circle that relates the particular to the general, the more the unique character of the particular is displayed. Accordingly, the particular is particular on account of its strong context-dependence

is placed with other similar contents in various serial connections, and is in so far 'conceptually' shaped. Rather the opposite is the case; the further this shaping proceeds, and the more systems of relations the particular enters into, the more clearly its peculiar character is revealed." [11, p. 224]

The unique character of the particular is thus to be seen as the ultimate case in an endless process of specification, through which the particular finally comes into a singular relationship to the reference field of the general. This "infinitely distant point" which lies "beyond the circle of scientific concepts and methods" is in the proper sense the particular [11, p. 232].

Cassirer rightly pointed out that one can only pursue scientific research if one abstracts from the particular and unique properties of things. Already the designation of a scientific object presupposes that properties with a certain degree of generality can be attributed to it. However, as soon as we reach such a level of intersubjective understanding, the individual and the unique are—at least in principle—accessible by sciences that employ general laws. Even Windelband had to admit this. At the end of his lecture about *History and Natural Science*, he pointed out that a science of the individual and the unique would probably not be possible without recourse to the methods of comparing, abstracting and generalizing. These methods, however, belong to the characteristic features of the nomothetic sciences.

At first glance, the philosophical discourse about the character of the particular seems to be far away from actual research practice. However, this impression is deceptive. In effect, one encounters here the principle of contextuality in its most general form (see Chap. 1) in that it relates the phenomena of individuality, their character and their origin to the general structure of reality.

In the following sections we are going to apply the idea of contextuality to various scientific problems. To these belong certain aspects of history (Sects. 4.4 and 4.5), psychological time (Sect. 4.6), and semantic information (Sect. 6.8). We shall see that this approach is a powerful tool to open up new ways to solve fundamental issues of science, ones that were not accessible to the exact sciences up to now.

Last but not least, this kind of contextuality also underlies the formation of individuality in social structures. An example of this is the social structure of Japan, which we examined briefly in Sect. 1.7. There is already a separate branch of sociology, denoted as relational sociology, which investigates, on the basis of modern network theory, the various concepts of the emergence of identities and individualities in social structures (see [12]).

4.3 The Structure of Scientific Explanations

The recognition that the gap between the general and the particular is not unbridgeable was an essential step toward the unity of science. Present-day development of the scientific landscape appears to confirm this. Thus, one can observe that the borders between disciplines are increasingly blurred, with consequent overlapping. In recent decades, this change has gathered enormous momentum, to the extent that it at the same time questions the dualistic understanding of science. The trend toward inter- and transdisciplinary research is characterised above all by the fact that the historically evolved, complex and unique structures of reality have increasingly been revealed as attributable to general laws, principles and rules.

In this respect, a development is continuing today that reached its first peak in the nineteenth century, when Charles Darwin succeeded in tracing the history of the development of life back to a general principle of Nature. Even Windelband had been aware that historical processes in Nature might one day find an explanation within the framework of the nomothetic sciences. However, he emphasized that, in this case, the actual target of cognition was not the unique manifestations of Nature, but rather only those aspects of Nature's history that show regular behavior. In contrast, the "historical" traits of living matter, which are manifest in the unique developmental characteristics of organisms, were closed off from any investigation within the nomothetic sciences. According to Windelband, these phenomena would always remain ordinary objects of the historical sciences, the goal of which is the accurate and detailed description of the objects of investigation. Consequently, Windelband regarded biology, which includes a major part of natural history, as a hybrid of nomothetic and ideographic sciences, which complement one another in their search for biological knowledge.

To understand Windelband's argument better, we must look at the deep structure of scientific explanations. Let us take, for example, physics. Physical explanations have a simple structure, because they rest upon only two elements: natural laws and the so-called initial conditions. The latter describe the physical circumstances under which the event to be explained has been occurred or will occur. These are either the conditions that the physicist can choose freely in an experiment or those that are already specified by natural circumstances.

The prototype of a physical explanation is that of planetary motion (Fig. 4.3). Here the general law that determines the movement of the planets

Fig. 4.3 Symbolic representation of sun's planetary system (not to scale). The movement of the planets around the sun is the perfect example of the dualistic structure of physical explanations. To determine the trajectory of a planet one needs two kinds of information: the general laws of motion and the initial conditions, i.e. the position and velocity of a planet at a given time. The explanation consists then in the derivation of the planet's state of motion from the general laws and the initial conditions. [Image: NASA]

is the law of gravity, while the initial conditions are the positions and velocities of the planets at a given time. The subsequent trajectories, along which the planets move about their central star, can then—by applying the law of gravity—be calculated precisely from the planets' initial positions and velocities. This applies in a broad sense. Thus, not only the future motion and relative positions of the planets can be determined, but also all of their past states. The planetary movement is, as it were, the paradigm for the universal predictability of the world, which Laplace had dreamed of at the height of the mechanical age (Sect. 3.7).

In this context one can distinguish two cases, depending on whether the explanation refers to the past or the future. If the event to be explained lies in the future, one speaks (in a narrow sense) of prediction. If the event has already happened, one speaks of a retrodiction. In an ideal case, such as that of planetary motion, any potential prediction is a potential retrodiction and vice versa.

Generally speaking, a law-like explanation consists in deriving the event to be explained from the initial conditions by employing natural laws. However, in a logically strict sense, this is only possible for cases in which the laws are deterministic. In the case of statistical laws, such as those that play an essential part in microphysics and sub-microphysics, the events to be explained can only be inferred with a certain probability. Accordingly, the philosophy of science distinguishes here between "deductive-nomological" and "inductive-statistical" explanations.

The planetary system that we have considered so far is comparatively simple. In biology, however, one is confronted with extremely complex systems. With a few exceptions, explanations that lead to precise predictions are no longer possible here. The explanations of evolutionary biology fall into this category. Such explanations are limited to finding plausible reasons why it makes sense that evolution has "invented" a given adaptation and not some other. One could denote such an explanatory attempt as an explanation *a posteriori*. Detailed predictions, in the sense of an *a priori* explanation, are hardly possible, especially since such adaptations are always the result of a long natural-historical process. Only to the extent that evolutionary phenomena could be reduced to physical mechanisms have exact predictions also become possible. The modern theory of the origin and evolution of life is paradigmatic of this. Already here, the natural sciences encounter the phenomenon of historicity, albeit initially only in the form of the history of Nature.

The nomological explanations described so far represent the prototype of explanations in the physical sciences. Their characteristic property is denoted as the "subsumption model" or the "covering-law model" of explanation [13]. Both terms make clear that the nomological sciences also deal with individual events, but, in contrast to ideographic sciences, they subsume these events under general laws.

It is worth taking a closer look at the deductive-nomological explanation, because its fine structure helps us greatly in understanding historicity. To avoid terminological ambiguities, we will use here the formal way of speaking, in which events, including facts in the narrower sense, as well as laws, are to be understood as statements expressed by sentences. In detail, a nomological explanation is based on two classes of statements which together constitute explanation, the so-called "explanans". They consist of the "antecedent" or "initial" conditions (statement A), and at least one general law (statement L). The antecedent conditions describe the circumstances which induce the event to be explained. The latter is the so-called "explanandum", described by the sentence E.

From a formal point of view, a deductive-nomological explanation consists in logically deducing the statement E from the statements A and L (Fig. 4.4). It is important to note that the explanation of natural events necessarily requires both sets of statements, A and L. The reason for this is immediately clear: statements about events cannot be deduced from general laws alone,

Explanans A L ⟶ E Explanandum

Fig. 4.4 General scheme of a nomological explanation. An event *E* is explained by two factors: the antecedent conditions *A* and (at least) one general law *L*. Two principal cases must be distinguished. (a) In the case of deterministic law(s), *E* can be deduced logically from *A*. This type of explanation is referred to as *deductive-nomological* explanation. (b) In the case of statistical laws, one can derive *E* from *A* only by statistical inference. This type of explanation is termed *inductive-statistical* explanation

because such laws—unlike events—are restricted neither in space nor in time. Therefore, laws cannot imprint any structure on reality. They can only establish connections between events.

Conversely, the same holds for events. For example, from the sentence: "Today the sun is shining" one cannot deduce any other substantial sentence. One can at best derive statements like "Today it is not raining" or "Today it is dry". These conclusions, however, are weak and tautological. In reality, they do not explain anything. In other words, nothing of interest can be deduced from the antecedent conditions alone.

To specify the concept of explanation, one has to consider the elements of an explanans carefully. Therefore, philosophers of science have formulated criteria that are intended to ensure the adequacy of scientific explanations. Thus, for example, the character of the laws must be specified. This leads, as mentioned above, to the distinction between deductive-nomological and inductive-statistical explanations. The adequacy of an inductive-nomological explanation requires that there be a criterion that guarantees the correctness of the statistical argument. Another criterion of adequacy, for example, is that the explanans must be verifiable in principle by experiment and observation, thus avoiding metaphysical explanations.

The deductive-nomological explanation can also be applied to explain special laws by means of general laws. A physical example is the derivation of Kepler's laws from the universal law of gravity. This is possible on the condition that the mass of the planets is negligible in comparison with that of the sun, and that the planets do not come too close to each other (see also Sect. 5.6). Metatheoretical explanations aim at reducing the number of special laws to general laws. This kind of intertheoretical reduction plays an essential part in unifying physics.

We can now ask the question that is central to the unity of the sciences: how far are historical explanations conceivable that fit into the explanatory schema of the exact sciences? In contrast to natural events, historical events are usually supposed to evade subsumption under general laws. They are rather thought to be contingent in the sense that they are neither random nor necessary. That is: they *can* occur because they are possible, but they need not *necessarily* occur. We can illustrate this with an example: The cup of coffee standing on my desk owes its place to the prehistory under which it found its way onto my desk. This prehistory is neither effected by a law nor by pure chance. Instead, it is the result of an intentional action by the person who placed it there.

For this reason, we first look for causal explanations of historical events, taking into account human interventions into reality. Here, however, we will approach the problem of historical events from another side. We will look at processes constituting history as such, i.e. processes independent of man's intervention into the course of history. The obvious model for this is Nature's history. In this case, only the natural laws have a primary effect on the historical happenings. Therefore, natural history is the real touchstone for the general question of how far a law-based understanding of historicity is possible. Strictly speaking, the question here is to what extent law-based explanations can nullify the contingent aspects of natural history.

Windelband had asked himself the same question. He, too, analysed this problem on the basis of the subsumption model of explanation. However, he focussed not upon the possible laws governing historical events, but upon the initial conditions from which historical processes start. By doing so, he anticipated an explanation scheme that is known today as "genetic explanation" (Fig. 4.5).

This scheme consists of a chain of deductive-nomological explanations in which each explanandum becomes the explanans for the next explanatory step. In other words, the historical process to be explained is first broken down into its individual events, which reflect the order in time of their occurrence. However, even if one supposes that there are laws determining the transition from one event to the next event, the problem remains that the antecedent conditions of the first step do not have an explanation.

Every genetic explanation starts from a premise that cannot be explained by the chain of explanations that follows it. Even if we knew all the laws that underlie a historical process, and even if all these laws were deterministic, the primary initial conditions of the process would always remain unexplained. All attempts of physics to derive the beginning of the world from general laws must inevitably fail. The "beginning of the beginning" cannot be explained within the framework of a relativistic understanding of the world, underlying

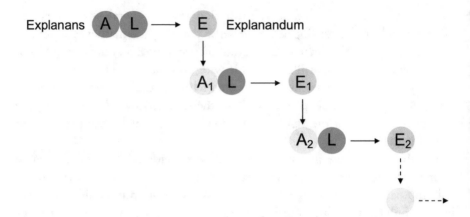

Fig. 4.5 Basic pattern of genetic explanation. If one can split up a historical process into a sequence of events, the explanation of the overall process is constituted by the sum of all consecutive steps of nomological explanation according to the schema of Fig. 4.4. The linear chain of explanatory steps, shown here, is the simplest case one can think of. In reality, the relationships between historical events are much more complex: chains of events occur that are cyclic, cross-linked, split by bifurcation, overlapping and much more

the empirical sciences. Even so, physicists such as Stephen Hawkins have tried to develop cosmological models in which universal laws may themselves create the initial conditions of the universe [16]. This is, however, where metaphysics comes in.

Windelband had already, and rightly, pointed out that interleaving explanations in which the initial conditions themselves become the item to be explained cannot entirely explain away the initial conditions. Instead, he was convinced that "law" and "event" are two basic aspects of our scientific world-understanding that cannot be abrogated through any act of explanation. He concluded that idiographic sciences could never merge into nomothetic sciences. The ideographic sciences would, therefore, always remain an independent form of science.

At first glance it seems that the diversity of explanatory concepts, as developed in science, would verify to the fullest extent Windeband's dualistic understanding of the sciences. Thus, beside the deductive-nomological and the inductive-statistical explanations of the natural sciences, there are dispositional, causal and historical explanations, which are applied in the humanities in general and in the historical sciences in particular (Table 4.1).

Table 4.1 The diversity of scientific explanations. (1) *Deductive-nomological explanation*. According to this model, an explanation consists in deriving the event to be explained from its initial conditions, on the basis of general law [13]. This explanation represents the prototype of a scientific explanation. There are different variants of this explanation schema. If the event to be explained lies in the future, one speaks of a *prediction*; if it lies in the past, of a *retrodiction*. If the subject of explanation is a special law derived from a universal law, one denotes this type of explanation as a *metatheoretical explanation*. The so-called *genetic explanations* occupy a special position because they open up a nomological approach to historical processes (for details see text). (2) *Inductive-statistical explanation*. When the law underlying an event is statistical, an explanation cannot be offered by logical deduction, but only by statistical inference. (3) *Dispositional explanations* are a familiar method to explain human actions in terms of purposes, beliefs, character traits and the like [14]. (4) *Causal explanation*. Of particular interest in the humanities are nomological and non-nomological causal explanations. They play an essential role, for example, in the so-called "interventionist" theory of human actions (for further details see Sect. 4.5). (5) *Historical explanations* cover a wide range of models that explain, or help to understand, historical phenomena (see [15]). The empirical approach in the historical sciences, for example, distinguishes between *a priori explanations* and *a posteriori explanations*. This distinction corresponds to the difference between prediction and retrodiction in nomological explanations. (6) *Teleological explanation*. In philosophy and science, there have at all times been attempts at explaining natural and historical events by postulating an inherent directedness of the happening. A well-known example of this is the Aristotelian doctrine of movement (see Sect. 1.1)

Classification	Subcategories
(1) Deductive-nomological explanation	Prediction
	Retrodiction
	Genetic explanation
	Metatheoretical explanation
(2) Inductive-statistical explanation	
(3) Dispositional explanation	
(4) Causal explanation	
(5) Historical explanation	Explanation *a priori*
	Explanation *a posteriori*
(6) Teleological explanation	

However, a closer inspection shows that even explanations in the humanities include nomological elements. For this reason, Windelband had drawn the boundaries between nomological and ideographical sciences not only along the lines of their different explanatory concepts, but also of their methodological peculiarities in defining their objects of research. From the perspective of today's scientific knowledge, however, there is no borderline that could justify making a sharp distinction, as suggested by the dualistic understanding of science, between the objects of research.

Even those aspects of the course of history that are bound to values have nowadays become the object of nomothetic explanation in the natural sciences. For example, the self-organization and evolution of life can only be explained by reference to a physical "value principle" which governs the transition from non-living to living matter [17]. The models show that already in the history of evolution—albeit at a level of complexity far lower than that of human society—"event" and "value" are inseparably linked one to the other. The naturalistic concept of value, however, has a fundamentally relative character. This is because the values inherent to Nature depend upon the perpetually changing conditions of the natural order.

On the problem of the unity of science, we can now draw a first conclusion: The dualistic conception of science that Windelband, Rickert and others set out seems today to be undergoing a process of gradual dissolution. The central issues of the ideographic sciences—such as individuality, uniqueness, value and the like—no longer lie outside the realm of the nomothetic sciences. On the contrary, they are increasingly becoming the object of strict explanations.

Historical explanations in the nomothetic sciences do not transform the differences between natural and human history into a contourless picture. Nonetheless, it is conceivable that the two forms of historical development might show unified features, ones that could be described within the framework of a general theory of structural changes in history. In precisely this sense, the philosopher Hermann Lübbe has spoken of a possible unity of natural history and cultural history—a unity that could consist of common development structures [18]. Likewise, the social-anthropological studies of Claude Lévi-Strauss, for example, suggest that the transition from Nature to human culture must be seen as gradual, allowing both aspects of historical events to be described by unified structures [19].

In view of the arguments presented above, the sharp distinction between natural history and human history cannot be sustained. Instead, we are compelled to assent to the verdict of Richard von Mises: "The natural sciences on the one hand, and history or humanities on the other, form the two hemispheres of the 'globus intellectualis' which we distinguish easily as long as we rely on naïve intuition. Much more difficult is it to draw some kind of exact boundary lines, and on closer inspection the contrast seems to vanish altogether." [20, p. 213]

4.4 Can History Become the Subject of Exact Science?

Given the enormous progress that contemporary science has made on all fronts, one cannot duck the fundamental question of whether and to what extent one might conceivably develop a theory of historical processes based closely upon the methods of nomothetic sciences. Will historical processes one day become predictable? Such questions are already likely to give rise to an uneasy feeling among most scientists; to them it should seem so absurd that to occupy oneself with it would raise the suspicion of frivolous speculation.

Is history not the sheer epitome of unpredictability? In his essay *"Objectivity" in Social Science and Social Policy,* Max Weber even speaks of the "chaos" of history and historical reality as a "vast chaotic stream of events, which flows away through time" [21, p. 111]. Given the unprecedented complexity of the course of history, one is forced to ask how one could possibly entertain the idea that history could be predictable in any form.

The fact that the general course of world events is indeed wholly irregular, and even chaotic, seems to have a large number of reasons. Here we mention only the three most important of these. First of all, history does not seem to be subject to general laws that—like the laws of Nature—lead to a reproducible and thus predictable event. Secondly, history is determined widely by man's autonomous decisions of the will, which gives world events their unique and unpredictable course. Finally, historical events usually have so many different causes that it seems impossible to isolate those that are relevant to the facts and make them the basis for a causal explanation. No wonder, then, that—given the complexity of the interrelationships—world events present themselves as an opaque network of decision processes, which must appear to any external observer, with an overall view, as unique chaos.

Of course, this cannot just be a matter of realizing an old human dream by searching for the general laws of world affairs—laws of which we do not even know whether they exist at all. Just as little do we intend to make a case for a teleologically argued ideology of history, one which believes that it must impose universal laws on human history. Rather, we just want to show a conceivable way of approaching the phenomenon of historicity as such, within the framework of the exact sciences. In other words, we are going to develop arguments demonstrating that the question in the title of this section is not as absurd as it first appears.

Strictly speaking, we want to inquire about the methodological and conceptual prerequisites that could lead to the establishment of a structural theory of historical processes. Such fundamental questions are by no means foreign to historians. Already at the beginning of the last century, the economist Franz Eulenburg asked, in an essay in honor of Max Weber, the question: Are historical laws possible? [22]. The goal Eulenburg had in mind was the one just outlined, that is, to clarify the methodological conditions under which a nomothetical justification of past events might seem conceivable.

Even if historical processes are hopelessly complex, this does not necessarily contradict the idea of using analytical methods to investigate such processes and to interpret them theoretically. However, this idea is in opposition to any understanding of history according to which the essence of history manifests itself precisely in the uniqueness of historical events. This view of history is obvious because history bears the clear stamp of man's free-will decisions.

Furthermore, research into history aiming at dissection, simplification, and abstraction of historical events seems to reach its limits very fast. One may argue, for example, that a historian investigating history is at the same time also a participant in history and thereby contributes to the historical happening. This argument is central to the hermeneutical philosophy of history, referring to the "rootedness" of human understanding in history (Sect. 3.4). Following this view, one must interpret history as a context of experience in which man and his historical existence form a unity. However, a holistic approach to historicity that is based on the unity of the researching subject and the researched object contradicts the imperative of objectivity of the scientific method and therefore cannot make any claim to provide an explanation in a scientific sense.

On the other hand, if one wishes to adhere to the analytical science model with its ideal of objective cognition, one must approach the problem of historical processes with a research strategy that is free of any ontological references to history. In particular, one will not attempt to grasp the phenomenon of historicity from the outset in all its wholeness. Instead, one will first try to reduce the complexity of the problem by using appropriate abstractions and idealizations. In other words, one will first look at the minimum requirements that are characteristic of a historical process and study the general characteristics of such processes under precisely these conditions. Only after completing that step will one try to reconstruct the original complexity of the problem by adding further properties to the concept of historicity, to arrive finally at a detailed picture that depicts with sufficient precision the manifold features of the historical event or the course of history.

The procedure just outlined is nothing more than the reductionist research strategy that René Descartes set out in his methodology and which has proved to be of such outstanding value in the natural sciences. One might think that the causal-analytical method has only been so successful in the natural sciences because, in the past, the natural sciences have always dealt with relatively simple natural phenomena. However, this assumption is not correct: the natural sciences have long been confronted with the complex issue of historicity.

In physics, this problem is known as the problem of irreversibility. Every spontaneous process in Nature is in principle irreversible. This irreversibility, in turn, is an expression of the historicity of natural events. It is therefore of great importance to see how physics has mastered this problem. The most famous example of this is the so-called Carnot engine, which represented the first successful approach to the problem of irreversibility.

When steam engines were introduced in the eighteenth century, physicists and engineers were faced with a seemingly impossible task. All the basic laws of mechanical physics known at that time were reversible. Accordingly, they did not contain any indication of a preferred direction of time. In contrast, technical processes are (and were known to be) irreversible, i.e., they can only take place in one direction, as specified by the engineer. To be able to apply the physics of reversible processes to mechanical structures such as steam engines, the physicist and engineer Sadi Carnot developed at the beginning of the nineteenth century the model of an idealized steam engine in which all the processes taking place are reversible. In this way, he was able to calculate essential features such as the efficiency of a machine, even though all processes in a steam engine are in reality irreversible (Fig. 4.6).

Carnot treated a steam engine, in his idealized model presentation, simply as an equilibrium system, even though no steam machine can ever operate under equilibrium conditions. This example of simplification is not an isolated case. Nearly all theoretical models in physics are based on simplification, idealization and abstraction (see, in particular, Sect. 3.8).

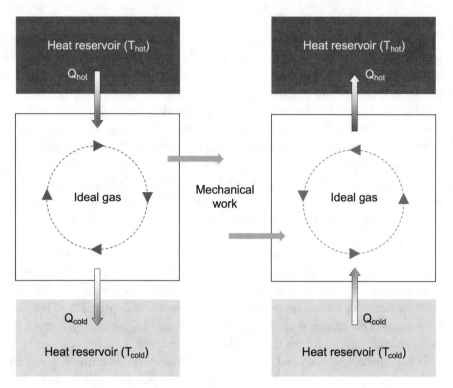

Fig. 4.6 Mastering the problem of irreversibility in physics. The Carnot engine is a thought experiment describing a heat engine which in its duty cycle converts heat into mechanical work. In a simplified representation, the Carnot engine consists of an ideal gas interacting with a hot thermal reservoir (red) and a cold thermal reservoir (blue). Both reservoirs are held at constant temperatures (T_{hot} and T_{cold}). In the "engine" shown in the left-hand box, a gas is present which absorbs a quantity of heat Q_{hot} from the hot reservoir, transforms it into mechanical work W and gives off a residual quantity of heat Q_{cold} to the cold reservoir. In the idealized case, all these processes are reversible, i.e. they also can run in the opposite direction: a reverse (counterclockwise) cycle becomes a heat pump (right-hand box). In natural systems, however, reversible processes do not exist. Spontaneous processes in Nature are irreversible, a property that reflects the essential historicity of all natural processes. Nevertheless, Carnot's idealized model allows one to calculate precisely properties such as the efficiency of a steam engine. The model demonstrates the power of abstraction in physics and is one of many that together illustrate how the methods of abstraction and idealization can be applied to historical processes

Why, one will justifiably ask, should these procedures not also apply to historical processes? Why should world events not be explainable by laws? Is it because there are no general laws governing world affairs? Does causal-analytical reasoning break down because of the complexity of the problem, because the solid network of historical events cannot be penetrated to reveal the all-decisive causes? Or is it because man's influence on history—his deliberate actions, purposeful and purpose-determined, but occasionally also random and irrational, is fundamentally unpredictable, so that any search for a law-like explanation of world events is doomed to failure from the outset? These are some of the difficult questions that are inevitably raised by any attempt to find a law-like description of historical processes and their possible mathematical modeling.

4.5 Mass Action and Cooperativity

We will now take a closer look at the objections to a nomothetic explanation of history as discussed above. First of all, are there really no general laws governing world events? This question cannot be answered offhand, as the concept of world events is an extraordinarily far-reaching one, encompassing as it does not only the history of humankind but also the history of Nature. For natural history, however, we know that there are general laws governing development. Thus, a causal and analytical explanation of history by general laws is very well possible, at least for that part of world events that is not determined by the purposeful behavior of higher, organized beings. Secondly, it cannot be ruled out in principle that the history of humankind is also subject to specific structural laws that are still hidden from us. Even though there may not be any evidence for this, we cannot necessarily conclude from this that law-like connections do not exist. Generally speaking, it is not possible to prove that there are no laws. Instead, it is only possible to go the other way: first to assume that rules or laws do affect history, and then to verify this.

Let us first ask how much weight we can attach to the belief that historical events are incalculable because they have numerous interconnected causes. This is tantamount to saying that one cannot penetrate to a first, objectively given cause. One can distinguish between a primary and a secondary cause only by value judgments, and even then one cannot be sure that one has grasped all the possible causes in the first place. However, this applies to everything and to all happenings in the world. Strictly speaking, the entire world is always the cause of every single event, since no subsystem can ever be completely self-contained: ultimately, everything is connected to everything else. Thus, the difficult problem of model-like isolation of events, systems, processes and causes is fundamental in nature. It is by no means limited to historical events, although there it appears in an intensified form.

Far more difficult is the question of how human activity is manifested in history. How significant is the influence of man's "free" will on the course of history? Of course, there can be no doubt that decisions by individuals have decisively determined the history of humankind at all times. However, individual decisions require appropriate mechanisms of reinforcement. In the age of mass media, the alignment and the synchronization of opinions are mainly due to mechanisms of mass action. Without this mechanism, individual decisions that have proved to be of historical significance would most probably not have had the chance to become effective.

Anyone who has ever been exposed to the suggestive power of a mass event knows how strong an effect the associated "suction" into a crowd of people can have on the will, opinion and action of an individual. Defamatory rumors, collective panic or even the waves of enthusiasm that roll through sports arenas are examples of the self-reinforcing power of mass action. Such phenomena of collective action, which appear as typical expressions of complex and unpredictable behavior, are quite accessible to the exact sciences. They do not represent anything mysterious. On the contrary: not only can they be precisely classified; they may even lend themselves to description in a mathematically detailed form.

From such investigations, we know, for example, that the phenomena of mass action can be based on different mechanisms. Let us first take chemical equilibrium and its description by the chemical law of mass action. This is the classic case for a mass action based on a mechanism of self-control. As described in Sect. 3.9, a chemical equilibrium is controlled by the system as a whole: The higher the random deviations from the equilibrium value, the higher the probability—and thus the rate—of the opposite chemical reaction; this tendency necessarily leads to a reduction in any deviations. This principle is "negative" feedback; it regulates reaction behavior in such a way that the reaction partners remain in equilibrium—that is, their quantities (or concentrations) remain constant over time.

The chemical law of mass action is based on a purely statistical effect. Coherent individual behavior, however, can also result from "cooperative" interactions. For example, cooperative behavior patterns are seen wherever the atoms or molecules of a system tend to align with their immediate neighbors. Cooperativity can be positive or negative, i.e. neighbors can tend to be similarly or oppositely orientated, and cooperativity can thus have either a strengthening or a balancing effect. In cases of positive cooperation, small random deviations are amplified, as in an avalanche; in cases of negative cooperation, they will be dampened, like ripples on a lake.

Cooperative interactions play an essential part in both inanimate and animate Nature. The evaporation and condensation of a liquid can be explained by cooperative models, as can the aggregation of virus particles or the coherent behavior of nerve cells. At the level of biological macromolecules, cooperative interaction is indeed the most important principle, as it allows the rapid formation and degradation of large molecular structures without sacrificing their stability. Likewise, it is the basis of the regulation of elementary life functions (Fig. 4.7).

This becomes particularly evident in the replication of the hereditary material. Genetic information is usually present in double-stranded DNA molecules, whose strands must separate for replication. For this purpose, however, the bonds between the individual base pairs must not be too strong, as otherwise replication would be impeded. On the other hand, the integrity of the DNA must be maintained over generations, to ensure the overall stability of the organism. These two opposing requirements (weak bonds within individual base pairs and robust stability of the structure as a whole) can only both be met at the same time by means of cooperative interactions. Proteins, the second important class of biological macromolecules, likewise use the principle of cooperation to perform regulatory functions. A well-known example of this is the protein complex hemoglobin, which regulates the strength of its binding to oxygen through cooperative conformational changes (Fig. 4.7).

Of particular importance are the cooperative processes that control the individual behavior of living beings. In animal populations, numerous examples of this are known. It is seen above all in the fascinating images of the abruptly changing, but always coordinated, movements of a swarm of birds, fishes or insects (Fig. 4.8). Today, scientific theories are already dealing with the phenomena of mass action and cooperativity in entirely different areas of research. Of particular interest in connection with the questions raised above is the concept of phase transitions, which describes in general terms transitions between physical aggregation states, or between biological population states. It is well conceivable that this kind of theory will also be able to represent historical changes as a sequence of phase transitions induced by mass action and cooperativity.

Of course, we must be aware that, in contrast to systems in physics, phase transitions in social systems also depend on free decisions by individuals. For this reason, one cannot expect historical changes to appear as simple as physical phase transitions do. This leads on to further complex questions of historical changes. How can we calculate the deliberate and purposeful behavior of humans? Can the instruments of statistics be helpful here? Is it

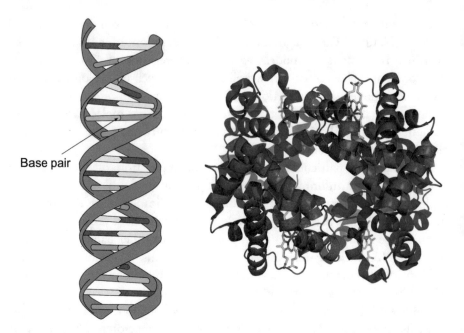

Fig. 4.7 Examples of cooperativity in biological macromolecules. Left: The molecules of heredity usually have the structure of a double helix, in which a DNA strand and its complementary copy wrap around each other like a twisted rope ladder. Cooperative stacking interactions between the base pairs imparts a high degree of stability to the DNA structure as a whole. Nonetheless, the individual bonds that hold the double helix together, are weak. In consequence, the double helix can open and close like a zipper. Right: The protein complex hemoglobin has the task of binding oxygen and transporting it to the metabolic centers of the cell. The hemoglobin molecule consists of four closely similar subunits (shown here as two red and two blue); each has an active binding site (green) for oxygen. Each subunit, however, can adopt a second conformation with a drastically altered affinity for oxygen. Which of the two possible conformations is present at a given moment depends on the concentration of oxygen in the haemoglobin molecule's local environment: at higher oxygen concentrations, the high-affinity state is found and at low oxygen concentration the low-affinity state. The transition between the two states is a result of cooperative interaction among the four subunits. [Left image: Sponk, Wikimedia Commons. Right image: Richard Wheeler, Wikimedia Commons]

possible to record comprehensively the actions of individuals and to predict them on the basis of probability statements?

At first glance, intentional actions seem to be largely inaccessible to statistical analysis, because spontaneity and the freedom of human decisions entail a fundamental element of randomness, which leads to the well-known unpredictability of human behavior. However, a closer look at this problem reveals a peculiar situation. If "free-will" decisions were free in an absolute sense, then their effects would presumably be relatively well calculable by

Fig. 4.8 Swarm behavior of birds. A flock of starlings that, by chance, has taken on the form of a giant bird. Apparently, such swarm behavior offers a degree of protection against larger predators, which often lurk where starlings meet. The swarming of birds, fishes and insects is due to cooperative interactions. [Photo: courtesy of Daniel Biber]

statistical methods, like the statistical behavior of molecules. However, human decisions and actions are not entirely free. They are always subject to complex boundary conditions that can have biological, psychological, biographical or social causes. It is here that the real difficulties in making human behavior predictable seem to lie. In other words, human behavior appears to us to be unpredictable not because it is based on free decisions of the will, but because an immense number of boundary conditions ("constraints") restrict or otherwise influence the freedom of human choice.

Regardless of these difficulties, decision-making processes have already become the subject of research. The theories that deal with the problems of decision-making processes are *decision theory* and *game theory* (Sect. 4.9). Decision theory concerns processes in which the decision of an individual is independent of that of other individuals and vice versa, while game theory concerns decision-making in which the choices of different individuals do affect each other. Both theories provide access to the complex decisions of individuals that shape the course of history.

Last but not least, models have been developed in the philosophy of science that should make the intentional actions of man comprehensible within the framework of general explanatory schemes. The theory of action, developed by philosopher von Wright [23], deserves special mention in this context. His theory is an *interventionist* concept of action, because it especially considers the fact that man, through his actions, always intervenes in the natural cause-and-effect relationships of his world.

Von Wright's theory makes it clear that causal explanations of intentional processes have a character different from that of causal explanations of natural processes. In the latter, the connection between cause and effect is supposed to be a necessary one laid down in the natural process itself. In the explanation of an intentional process, however, the connection between cause and effect is given solely by the expediency of human action. Accordingly, von Wright saw himself compelled to outline a theory of action that was based upon a logic of reasons.

From such considerations, the modern *deontic* logic has emerged. Deontic logic represents on the one hand a branch of formal logic. On the other, it can be regarded as a special case of modal logic, where the modal-logical terms "possible, impossible and necessary" are replaced by the terms "allowed, forbidden and compulsory". Consequently, deontic logic is a logic of normative statements which make regulations such as "it is allowed that …", or "it is obligatory that …" and the like.

With his interventionist model of explanation, von Wright offered the humanities a concept of understanding based on linguistic logic, one that bears a formal analogy to the causal explanations of the natural sciences. Von Wright went so far as to interpret even the concept of empirical causality from the perspective of the acting human being, and thus to subordinate it to interventionalist causality. According to this view, we experience and understand empirical causality only in the context of the possibilities of our actions, or more precisely: by our abilities to intervene in reality.

Here, we cannot go further into the diverse arguments for or against law-like explanations of historical processes. Nevertheless, it is already possible to draw a first conclusion: It is not a lack of theory that makes the predictability of historical processes seem impossible, but rather the tremendous complexity of the historical process that stands in the way of doing so. Science, however, has generally found ways to solve allegedly "unsolvable" problems (see Sect. 6.2). Therefore, there is no reason why science should not try to search for the general features of the course of history and to describe them by using the instruments of analytical research.

4.6 Temporal Depth

The understanding of history begins with the study of the sources that tell us facts about past events. History's task is then to process the sources methodically and to identify the causes and reasons that make it possible to explain, i.e. to understand, history. This is undoubtedly an arduous task. On the one hand, historical events are of immense complexity. On the other, historical documents are mostly incomplete and thus are only selective testimonies from the past. Therefore, the proper and objective development of sources is always problematic. Nonetheless, it is evident that the natural sciences can be of great help to historians in this respect; compare, for example, the findings of brain research on perception, memory and forgetting, which can undoubtedly make an essential contribution to the objectification of the historical sources (see [24]).

In this, the step that involves the most significant problem is the processing of the sources for the reconstruction of past events. This is a matter not only of structuring the facts, but also—in particular—of researching the motives, intentions, causes and reasons that are hidden behind historical processes and which make it possible to classify past events in their overall historical context. Therefore, the reconstruction of historical facts is always accompanied by an evaluation of the sources by the historian.

One can even aggravate this argument by claiming that, through the reconstruction, the historical object is constructed in the first place. This is the Achilles' heel of the science of history, because subjective, non-scientific and ideological elements flow into the science of history as into no other. Moreover, this is the place where the hermeneutic approach to history has a disastrous effect, because it undermines the scientific requirement of an objectifying method and replaces it by an existential understanding of history, one that places the participant and co-creator of history at the center of the investigation (see also Sect. 3.4).

Notwithstanding the philosophical discussion about the essence of history, the analytical sciences have already taken first steps on the way to a theory of historical processes. It is already very well understood which laws determine the development of living matter, and how certain levels of values are thereby built up automatically—values that in turn exercise a selective function on the developmental history of life. For their part, the models again enable conclusions to be drawn on the general characteristics underlying the processes of the generation of information in Nature and society.

An essential step on the way to a general theory of historical processes is also provided by the natural sciences in the context of the mechanisms of mass action and cooperative behavior. Another is the successful application of game theory and decision theory to human decision-making processes, showing that the individual and collective actions of human beings are also predictable to a certain degree. In all of these models, man now appears only as a "virtual subject", one that has become utterly objectified in the analysis of historical events.

At first glance, a strange, even paradoxical situation seems to emerge. The further one penetrates historical reality by the causal-analytical methods of the exact sciences, and the more the veil of its unpredictability is lifted, the more contradictory historicity itself seems to become. This is because historicity as such appears to dissolve to the same extent as it becomes possible to narrow down the historical happening by general concepts. However, this impression is deceptive. As we saw in the preceding section, the historicity of the world will by no means disappear under the scalpel of causal analysis. Instead, its details will become better understandable, because the analysis will shift from the understanding of single historical events to a detailed knowledge of the dynamic relationships between the events. These relationships are the proper core of the structural changes in human history.

Searching for the causes of the structural changes in history, we again encounter the idea, outlined in Sect. 4.2, according to which the particular crystallizes out of the network of its various determinants. In the present case, the peculiarity of historical events proves to be a consequence of the particular constellation of causes in which a specific historical happening is, or was, embedded (Fig. 4.9).

A structural change in history usually has numerous causes which together generate the tremendous complexity that characterizes the course of history. Even just the task of isolating and identifying the relevant causes is already extremely difficult. No wonder that most historical explanations are only explanations *a posteriori*. At best, these provide plausible reasons for the selection of the causes that are supposed to have been decisive for a specific structural change in history. Because of historical events' temporal depth, mono-causal explanations of history and the direct comparison of historical events are hugely problematic.

Of the many problems that come into play here, we highlight just one, namely, the temporality of the causes themselves. The causes underlying historical processes are usually themselves processes of the past, which have

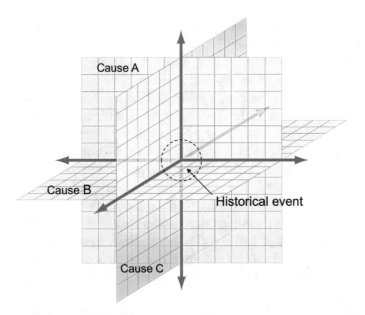

Fig. 4.9 Schematic drawing of the entanglement of historical causes. A historical happening does not have only one cause. Instead, it usually has many causes, which contribute with different weights to the induction of a specific event. Moreover, the causes belong to different overlapping layers of time, determining the temporal depth of a historical event. Despite the complexity resulting from this, it is conceivable that one day the properties of such causal relationships will be describable within the framework of modern *network theory* (Sect. 4.9)

their own historicity. This means that a proper understanding of a specific historical happening must take account of the time structure underlying its network of causes. This time structure does not exhaust itself in the one-dimensional, steadily passing time of physics, where each cause is associated with a point somewhere along time's arrow. Rather, the causes themselves belong to different time layers that overlap. This gives a historical event its temporal depth, which ultimately determines the scope of the event in the overall context of history.

We can go even one step further and claim that temporal depth is a property of our time consciousness as such. To illustrate this idea, let us consider the relationship between experience and time. Experience is, without any doubt, a primary category of human existence. However, experience is always within time. The idea of a timeless experience makes no sense at all. Thus, any experience necessarily presupposes a "before" and an "after", separated by the "present". The different time modes ultimately define the temporality of all happening.

However, the application of this thought to the experience of temporality itself inevitably leads to an entanglement of time modes (Fig. 4.10). For example, experiencing the past, present and future in the flow of time, one can meaningfully speak of the "present of the past (Pr of Pa)", the "present of the future (Pr of Fu)", but also of the "past of the present (Pa of Pr)" or the "past of the future (Pa of Fu)" and the like. In total, there are nine possible entanglements of the first order. If we now repeat this process by exposing the time entanglements of the first order to reflected experience, then twenty-seven second-order entanglements emerge which interlace past, present and future to give three-digit time modes such as the "past of the (present of the past) [Pa of (Pr of Pa)]" or the "future of the (present of the past) [(Fu of (Pr of Pa)]" and so on. By continued iteration, a detailed and extremely complex time structure builds up that reflects all temporal forms of human experience.

The entanglement of time modes can be represented schematically in the same way as the entanglement of historical causalities (Fig. 4.11). This shows once more that the particular and individual can be traced back to the superposition of general categories, as shown paradigmatically in Fig. 4.2. We have derived the entangled modes of time from a general reflection upon the prerequisite of experience. These time modes are not to be confused with aspects of psychological time based on our subjective feeling for time. Nevertheless, the concept of entangled time modes may bridge the gap between a rational and scientifically justified understanding of our time awareness and the idea of psychological time as it underlies the so-called life philosophy developed by Henri Bergson and others at the beginning of the twentieth century.

Bergson, for example, believed that "experienced time", which—in contradistinction to physical time (*temps*)—he associated with duration (*durée*), can only be grasped by intuition. In experienced time, as Bergson writes, the freedom, the spontaneity and the uniqueness of the processes of human consciousness take shape and permeate ever deeper layers of consciousness until they finally merge with one another, condensing into a non-repeatable experience. Starting from a dynamic understanding of time, Bergson outlined a comprehensive metaphysics of time, going far beyond the physical understanding of time, which he regarded as a mere counting measure [25].

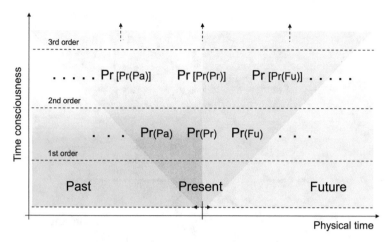

Fig. 4.10 Entanglements of time modes constituting human time-consciousness. Physical time is structured by the time modes past, present and future. It is the time of clocks. Above physical time, the time of consciousness unfolds. It emerges from the entanglement of physical time modes with the iterative experience of temporality. The diagram shows some cases of the hierarchy of entanglements of the present. In contrast to physical time, the "present" in human time-consciousness consists of a small window of time in which the simple acts of consciousness become integrated into a perception unit. In each step of the iteration, a new level of entanglement is reached. There are nine entanglements of the first order, 27 entanglements of the second order etc. In this way, the deep structure of temporality emerges

With Martin Heidegger's interpretation of temporality as a trait of human existence in general, the philosophy of time once again took a radical turn. Heidegger even saw in the shift from the philosophy of being toward a philosophy of time a fundamental break with classical occidental ontology. In his treatise on *Being and Time*, he speaks of the "destruction of the history of ontology, with the problematic of temporality as our clue" [26, p. 63].

According to Heidegger's understanding of time, being can no longer be interpreted as a substrate upon which time acts. Instead, Heidegger claims that all existence must itself be regarded as something temporal, i.e. as being in time. This understanding of time has been expressed by the philosopher of religion Georg Picht in the following words: "Everything that is at all, the planets, the fixed stars, the solar systems and the Milky Way, the atmospheric phenomena, the mountains, streams, seas, plants and animals, the molecules, atoms and atomic nuclei, and everything that happens and can happen at all

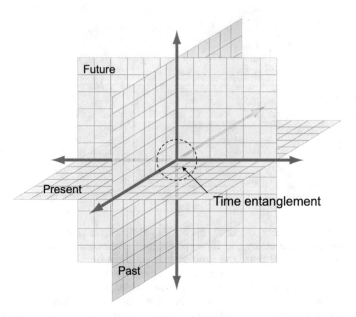

Fig. 4.11 Entanglements of time modes constituting temporal depth. Any process takes place in time and thus belongs to a specific time layer. Accordingly, the superposition of processes leads to superposition of time layers and thus to entanglement of time modes. This entanglement finally determines the temporal depth of a historical happening and, therefore, its "impact history"

is in time. Even the mathematical structures, the numbers and the laws are in time, for their truth is shown to be always valid, and 'always' is a mode of time. Time is, therefore, the universal horizon of everything we can say that it is." [27, p. 143; author's transl.]

Bergson already believed that in his metaphysics of time he had grasped the essence of being as such, and thus also the essence of historical being. However, the understanding of history that can be derived from his philosophy is aimed primarily at the intuitive experience of the individual. Since outer and inner time, i.e. physical time and the time of inner experience, differ in their essence, Bergson's idea of historical being is beyond the sphere of experience given by the material world.

Let us leave metaphysics and come back down to earth, and consider the concept of temporal depth that results from the entanglement of time modes. There are good reasons to conclude that time entanglements are not just an epiphenomenon of human time consciousness, having only an individual, inward-looking character. The mere fact that man's memory records temporal happenings means that the past radiates into the time consciousness of man, inducing the entanglement of the past with the present. On the other hand, man's goal-directed actions are nothing less than manifestations of the future. The most interesting question, however, is that of the entanglement of the present with the present itself. This immediately leads to the question of what the present is at all. Or, asked the other way round: Is the "present of the present" just a mathematical abstraction, or does the present actually exist as a mode of time?

To approach this problem, let us first consider the lowest level of our time consciousness as manifested in physical time. For the physicist, time is merely a one-dimensional "time arrow". With the help of periodic processes, it is possible to define a measurement unit on the time axis and thereby standardize our clocks. This allows past events to be timed accurately, which indeed is the first prerequisite of any theory of history.

With the concept of time's arrow, the physical idea of time is already exhausted. The fundamental equations of mechanical physics, which describe the movements of bodies, do not include any preferential direction of time. Instead, time appears in the equations of mechanics only as a directionless mathematical parameter. Such processes, in which time occurs only as a time index, are in principle reversible. This means that an avalanche rolling down a mountain can change direction at any time and roll back up to its starting position. The laws of mechanics actually allow for such a bizarre world.

The idea associated with the concept of reversibility is that of a time-symmetric, and thus ultimately timeless, reality in which all historical processes are blanked out. Even though such a world is in blatant contradiction to our experience, reversible physics has proved to be hugely successful, as the example of the Carnot engine shows (Sect. 4.4). It was only with the discovery of the second law of thermodynamics that a natural principle came to the fore which considers the irreversibility, and thus the temporality, of natural events. It states that in an isolated system (i.e., in a system that exchanges neither energy nor matter with its surroundings) only those processes occur spontaneously for which a certain physical measure, known as entropy, increases.

Entropy provides information about the degree of order in a physical system: The higher the entropy, the lower is the system's order. An increase in entropy is thus equivalent to an increase in disorder. The fact that ordered states do not build up spontaneously, but tend to dissipate, is an experience that we also have in our daily lives in many ways. Everyone knows the growing disorder that spreads in a children's room if one leaves the room and the child to themselves. As with the physical entropy principle, this is because the number of possible disorderly states in a nursery is far higher than the number of ordered ones. Therefore, as long as there is no external intervention, ordered states will change over time into more probable—that is, more disordered—states.

The entropy principle determines the direction of natural processes, whereby the magnitude of the increase of entropy is a direct measure of the irreversibility of such processes. Assuming that the universe can be regarded as an isolated system, the second law of thermodynamics has the rank of a cosmic developmental law (see also Sect. 5.5). Irreversible processes are, therefore, processes that occur spontaneously in only one direction. Radioactive decay, the propagation of a light wave or a wave on a water surface, and also the evolution of life, are typical irreversible processes that run only in one direction. Only such directed processes document the temporal asymmetry of reality, in which past and future are distinct from one another.

Where is the present? In physics, the present shrinks to a mere point on the time arrow dividing the past from the future. However, this abstract image is in contrast to our real time consciousness. Here, the perception of the present extends to a small time-window. One can even measure the duration of the present. Neuropsychological investigations have revealed that, in the human brain, simple acts of consciousness, which usually have a duration of 30 ms, are automatically integrated to perception units of about 3 s' duration [28]. In other words, the present as a physical point and the present of which we are aware become separated. Which of the two ideas should we give priority to?

The answer is given by physics itself. Two findings of relativity theory and quantum theory are instructive in this respect. Let us first look at relativity theory. The principle of "time dilation" states that in a physical system which is in motion relative to the observer, time runs more slowly, and this time dilation is the more significant, the larger the relative velocity. Because of time dilation, it is not possible to make a statement about the simultaneous occurrence of two events that is valid for all observers. This means that the passage of time is context-dependent. It depends, in each case, upon the velocity of the observer relative to his reference system. Accordingly, absolute time is only thinkable in a world without any movement. In such a world, however, clocks would not even run. It would be a timeless world, in which the present would extend to eternity.

Quantum physics reveals a further aspect of time which seems to correspond more to the idea of the present as a small interval of time rather than as a mathematical point. Thus, a corollary of the uncertainty principle, known as energy-time uncertainty, states that the measurement of energy always takes a finite time, for fundamental reasons. This means that the term "instantaneously" loses its meaning in the measurement of quantum objects, but this also opens the door for a physical understanding of the human time consciousness. Conversely, is it perfectly conceivable that the concept of temporal depth, which at first appears to be peculiarity of our time consciousness, may have repercussions for our understanding the strange phenomena of quantum physics such as those of causality and non-locality (see Sect. 3.6).

4.7 The Double Nature of Causality

So far, we have examined several features of historicity: the nature of historical explanations (Sect. 4.3), the possible mechanisms of historical change (Sect. 4.5) and the temporal depth of historical causes (Sect. 4.6). Now, we will take up the question of whether there are general laws that could lead to the unique characteristics of a historical happening. At first, the law-like and the unique seem to be irreconcilable opposites. We circumvented this *aporia* by a figure of thought, according to which the historically unique crystallizes in the network of its general relations (Sect. 4.2). In this approach, however, the individual is not the consequence of any law. The individual only emerges from the particular position that it occupies within the network of its relation to general determinants. This raises the all-decisive questions: Can the unique forms of historical processes emerge directly from general laws too? What is the essence of law-like behavior?

The epitome of a nomothetic science is physics. The laws of Nature on which physical events are based leave no doubt whatever about the idea that the notion of law is synonymous with that of repeatability. This seems to answer the question of whether general laws can be a source of individuality and uniqueness. However, what exactly does "repeatability" mean? In order to delve deeper into this question, we must first of all bear in mind that physics is an experimental science in which nomothetic relationships are examined with the aid of controlled experiments. In the end, a test must show whether, and to what extent, a physical process is repeatable. Therefore, the technique of experimentation should first and foremost be able to provide more information about the phenomenon of repeatability.

So let us take a closer look at the basics of scientific experimentation. In the natural sciences, we distinguish between two types of experiment. One type consists in disturbing natural happenings in a targeted way, to infer the underlying natural mechanisms from Nature's response to the disturbance. In genetics, for example, the question of which biological function is encoded in a particular gene can usually be answered by examining the effect of focused mutations in the genetic material of the organisms. From the functional changes that the mutations they bring about, one can infer the details of the actual process.

A completely different form of experimentation is preferred in physics. Here, the system whose law-like behavior is to be investigated is first put under defined initial conditions. By varying these conditions, the experimenter then tests whether there is a nomothetic relation between the processes triggered by the experiment. Only the method of the controlled test has made it possible to achieve the enormous rise of the natural sciences in modern times. It was none other than Galileo Galilei who (both figuratively and literally) set the ball rolling here. He let an object roll down an inclined plane, from different starting positions, to examine the laws of falling which he had previously inferred with the aid of thought experiments. His experiments on the free fall of bodies are famous—not least because they have become an exemplar of the experimental method of physics.

Since that time, the controlled experiment has played a crucial part in scientific research. On the one hand, it makes possible the discovery of natural laws. On the other, the experiment is also the authority that validates the claim of scientific findings by excluding false conjectures. Here we will focus our attention on the first aspect: the experimental discovery of hitherto unknown laws. The mere fact that one can observe, in an experiment, a correlation between the results of a measurement—a correlation that can be represented by a curve or a mathematical formula—is not yet sufficient to justify the existence of a law-like correlation: the demonstration of such a relationship must also be reproducible. However, every experimenter knows that in reality, no experiment can be repeated down to every detail. Even the initial conditions of a test, such as the numerical values that are set on a measuring instrument, can never be reproduced with a hundred percent accuracy. Beside, every reading of the control and measurement values has its errors, no matter how small those may be.

This consideration makes it clear that every experiment, like every other event in this world, is always a historical event. However, if it is said that a particular experimental result is reproducible, this can only be meant in the sense that the result can be reproduced within certain error limits. If—and only if—similar experimental conditions always lead to similar results, then small deviations in the experimental execution do not have a significant impact on the reproducibility. Only on that condition are there stable error limits, so that the "true" values can finally be filtered out by appropriate error statistics.

Reproducibility represents a further idealization of physics, because it presupposes a particular form of causal determinacy. It must guarantee that the same causes not only have the same effects, but also that similar causes have similar effects. Only if the latter condition holds can one speak of the "regularity" of a natural occurrence, although in reality, all events differ from each other. This, in turn, means that one has to distinguish between a mere "causal principle" and a "causal law".

The causal principle is merely a physical interpretation of the philosophical principle of sufficient reason. This says that nothing happens without reason, so that one can assign a cause to each effect. The causal law, on the other hand, makes a detailed statement about how cause and effect are related. The precise correlations are then expressed in the form of causal judgments, which in the natural sciences are called "laws of Nature." However, for such laws to appear as a regular natural phenomenon at all, the causal relation must also apply, according to which similar causes have similar effects. Only this particular form of causality leads to an (approximately) identical, and thus reproducible, behavior of matter.

The causality associated with "reproducibility" is described as "strong" causality, because under this form of causality small changes in the causes bring about only minor changes in the effects. Mathematically speaking, this means that the connection between cause and effect is based on "linear" coupling (Fig. 4.12, left). However, there are also "nonlinear" couplings. As a result, similar causes can have quite different effects. In this case, even the smallest changes in the causes can lead to major changes in the effects (Fig. 4.12, right). This form of causality is called "weak" causality. It can be observed in physical systems governed by nonlinear laws.

Nonlinear systems have aroused great interest in science because they bring us a good deal closer to an understanding of complex phenomena. Particularly noteworthy are systems that react sensitively to changes in their initial conditions. In these systems even the smallest deviations in the initial conditions can be amplified in an avalanche-like manner. This is also referred to as exponential error propagation, as it restricts drastically the computability of the systems.

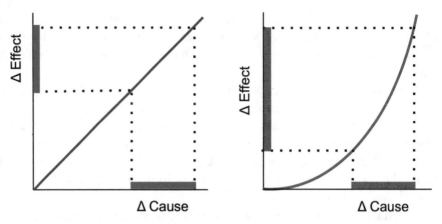

Fig. 4.12 Different types of coupling between cause and effect. Left, linear coupling: Similar causes have similar effects. This causality is generally termed "strong" causality. Right, nonlinear coupling: Similar causes have different effects. For this type of causality, the term "weak" causality is used. The Δ sign represents "difference", and shows the change in effect resulting from a given change in cause

Sometimes the dynamics of such systems appear to be so irregular that the impression is evoked that only chance and chaos are operating (Fig. 4.13). Systems with these characteristics are, in principle, still calculable because they obey strict laws. In fact, however, the behavior of such systems cannot be calculated, since the long-term prediction of their dynamics would require an absolutely exact determination of their initial conditions. Yet the less precisely the initial conditions can be specified, the more drastically the possibilities of prognosis are reduced. Physicists have coined the term "deterministic chaos" for this phenomenon. Behind this term, which seems to be self-contradictory, lies the fascinating realization that even in natural processes that appear chaotic a law-like order can still be present.

The prime example of such nonlinear and unpredictable natural phenomena is the set of complex physical processes that determine our weather. These processes are also subject to strict laws. Nevertheless, they are to a large extent unpredictable, because even the slightest atmospheric disturbances can cause significant fluctuations, owing to nonlinear amplification mechanisms.

The fact that even the smallest causes can have a significant impact on the development of the weather is occasionally referred to as the butterfly effect: even the beat of a butterfly's wing can trigger a hurricane elsewhere. The "butterfly effect" is undoubtedly a very suggestive picture. However, it may give the wrong impression, as though the slightest local fluctuations in the atmosphere could have a global impact on the weather, making weather

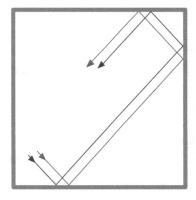 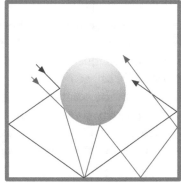

Fig. 4.13 Regular and irregular behavior of deterministic systems. The phenomenon of behavior that is irregular but is nevertheless governed by deterministic law can be illustrated by a billiard ball that moves on a billiards table in the middle of which an obstacle is placed (right). This arrangement is called "Sinai billiards" after the mathematician Yakov Sinai. In physics, one uses the mathematical model of "phase space" to describe the dynamics of such a mechanical system. Phase space consists of all the possible values of the position and momentum of a body. Each dynamic state can be represented by a point in phase space, so that changes in the state describe a trajectory in phase space. Left: If there is linear coupling between cause and effect (similar causes have similar results), then the path in phase space is stable. This means that small changes in initial conditions have only correspondingly small effects upon the final state in the phase space; the trajectories remain close together, i.e., the process is reproducible. Right: If the coupling of cause and effect is nonlinear, the trajectory in phase space is unstable, and minute changes in the initial conditions have strong impacts on the system's subsequent development; the trajectories in phase space diverge. Consequently, even initial states that are close to each other lead to completely different final states; the process is irreproducible

forecasts ultimately impossible. Yet the butterfly raises critical questions concerning the reliability of computer simulations when nonlinear equations are involved.

Although the phenomenon of deterministic chaos has been known since the end of the nineteenth century—thanks to the work of the mathematician Henri Poincaré—it was rediscovered only some decades ago. Remarkably, it was the meteorologist Edward Norton Lorenz who came across the butterfly effect when he encountered inexplicable anomalies in computer simulations. This underlines the problems resulting from computer simulations when the complexity of a system makes it impossible to carry out real experiments. However, the far-reaching predictions that computer simulations are often expected to deliver about natural events—such as the climatic change on Earth, for example—demand the most critical examination (see also Sect. 7.9).

Let us return to the general properties of chaotic processes. What makes these processes so instructive for the questions that we raised at the outset is the fact that they have the essential characteristics of a historical process. They appear random, unique and unrepeatable. Surprisingly, however, the seemingly irregular behavior of these processes is not an expression of a lack of general laws. On the contrary, it is a direct the consequence of deterministic laws. This apparent contradiction is resolved as soon as one distinguishes between determinism on the one hand and predictability on the other. As already mentioned, the behavior of chaotic systems is by no means an expression of a lack of determinacy, but is merely a sign of the limited predictability of such systems.

The essence of a causal law now appears in an entirely new light. It becomes clear that the ideal of law-like reproducibility, exemplified in the laws of Nature, is bound to a particular form of causality, namely the strong causality described above. In chaotic systems, on the other hand, weak causality still leads to law-like behavior, which permits reproducibility only in the limiting case of infinitely exact repeatability of the initial conditions. However, since no state can ever be repeated exactly, the nonlinear laws of chaotic systems are an immediate source of individuality and uniqueness. This unexpected finding steers in an entirely new direction our discussion of the issue of whether (and if so, then in what form) laws of history are conceivable.

4.8 The Abstract World of Boundary Conditions

Let us deepen our discussion by focusing on the initial conditions from which a physical process starts. The initial conditions are indispensable for any nomothetic explanation in physics (Sect. 4.3). The Newtonian laws of motion are the best example of this. We can only get information from the laws of motion when we have information about the initial conditions of a body, i.e. when we know the body's position and velocity at a specified time.

Since, on the other hand, the end of a physical process can itself become the starting point of a new process, the initial conditions themselves must have a history. Thus, if one could trace the development of the initial conditions right back to their origin and even explain the development nomothetically, one would already have an abstract theory describing the historical changes of Nature. At the same time, this theory could also serve as a model for the understanding of structural changes in human history.

For a long time, such considerations played no part in physics. The laws of mechanics, examples of the physical understanding of reality, are reversible and do not distinguish direction in time. In other words, in the mechanistic view of the world, reality has no historical dimension, since all processes are symmetrical in time and are in that sense without history. This is the logic behind the idea of the Laplacian demon, who can look into both the past and the future.

However, when the limits of reversible physics became visible, and the principles of irreversible physics gained importance, initial conditions also became a focus of interest. Ludwig Boltzmann was the first to recognize this, when he tried to explain the direction of time by an increase in the entropy of the universe. For this purpose, he saw himself forced to suppose that the development of the world started from exceptional initial conditions (for details see Sect. 5.5). Today, the physics of chaotic systems demonstrate that the initial conditions can become a generally critical factor, since under the influence of nonlinear laws even the smallest random disturbances and fluctuations, so-called instabilities, can intensify and give a unique course to physical events.

The critical behavior of the initial conditions is particularly important in cases where feedback is included. Here, the initial conditions can develop a life of their own, as it were because they are themselves modified by the physical processes that they trigger. Feedback mechanisms of this kind are characteristic of biological amplification processes. They fall under the term "autocatalysis", which basically means self-reinforcement. In autocatalytic systems, however, it makes little sense to discuss "initial" conditions, since the beginning of the system becomes lost in the history of its development. Here, the initial conditions have rather the character of permanent boundary conditions, ones that channel the development of the system.

The term "boundary conditions" is borrowed from mathematical physics, where it denotes the specific constraints associated with the differential equations that express a natural law. These constraints determine, together with the natural laws, the natural course of events. To be quite precise: constraints are selection conditions that narrow the immense number of processes allowed in principle by a given natural law down to the processes that actually take place. Under certain prerequisites, which are specified in Sects. 6.9 and 6.10, physical boundary conditions can optimize themselves, in that they adopt increasingly special configurations and thereby raise themselves step by step to a higher organizational level.

The self-supporting development of increasingly specific boundary conditions in physical systems is also called material "self-organization". In principle, this process can lead from the non-living states of inanimate matter up to the complex organizational forms of life. In this case, material self-organization merges into a process of molecular evolution. This makes boundary conditions the key to understanding complex systems [29]. At the same time, however, the mathematical concept of boundary conditions leads to an abstract world that is usually inaccessible to the non-specialist. For this reason, let us take a look at boundary conditions from another perspective—one that is more familiar to us.

We have said that boundary conditions "channel" the development of a system in that they determine selectively the system's dynamics. Something similar can be observed on a machine (Fig. 4.14). A machine is also characterized by special boundary conditions that determine how it functions. These boundary conditions are prescribed by the machine's construction plan. This determines, among other things, which components the machine consists of, and what material properties, shape and arrangement they have. With the specification of these boundary conditions, all physical processes that run in the machine and give the machine its function are automatically defined. In principle, a machine is thus nothing other than a big physics experiment that forces the laws of Nature to work under constraints that are fixed by the machine's boundary conditions.

The image of a machine working under specific boundary conditions can be applied directly to the living organism. The organism, too, contains a hierarchy of boundary conditions that regulate all its essential processes such as to preserve the system. They steer the function of the organism in a manner similar to that in which boundary conditions determine the operating principles of a machine. From this point of view, the so-called "machine theory" of the organism seems to have become relevant again. Descartes had put this idea into circulation when he described animals as soulless automatons. An even more radical interpretation of it came from the physician and philosopher Julien Offray de La Mettrie, who concluded in his *L'homme machine* of 1748 that man too, as the title asserts, is a machine [30].

The machine theory of the organism, which later became the guiding principle of mechanistic biology, has repeatedly been rejected because it contradicts the organismic view of man and Nature. Moreover, the biological

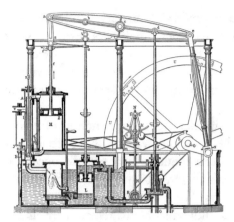

Fig. 4.14 Boundary conditions in a machine. In a machine—just as in a physics experiment—natural laws are forced to operate under defined boundary conditions. These constraints, in turn, result from the machine's construction plan, which lays down the shapes of, and the boundaries between, its parts. The pictures show an industrial steam engine from 1788, built by Boulton & Watt, that turns heat into mechanical energy. [Left: Meyers Großes Konversationslexikon, 1905. Right: German Museum, Munich]

experiments of the eighteenth and nineteenth centuries, which proved the reproductive and regenerative capacity of organisms, seemed to contradict the machine theory of the organism, since a machine that can be taken apart cannot regenerate itself. Thus it seemed unthinkable that a machine might be able to reproduce itself. However, in the middle of the twentieth century the mathematician John von Neumann was able to prove that self-reproductive machines, for which he used the term "automata", are possible in principle.

At first glance, one might get the impression that the boundary conditions governing a machine, or an organism, are different from the physical boundary conditions emerging by feedback in autocatalytic systems (described above). However, closer examination shows that all boundary conditions in machines and organisms can be traced back to physical boundary conditions. Thus, even if only in a thought experiment, the boundary conditions of a machine can be given a precise physical description—the shapes and arrangements of its parts and, ultimately, the properties and mutual positions of all the atoms involved.

The same applies to the boundary conditions of the living organism. Here, the circumstances are even more straightforward than in the case of a machine, since all biological boundary conditions can be traced back to the molecules of heredity as the primary physical boundary condition. This applies, notwithstanding, to the fact that these molecules are themselves embedded into particular environmental conditions, which in turn have the character of general boundary conditions. These two classes of boundary conditions together determine biological events, in the sense of genetic determinism (see [31]).

We have already emphasized the point that boundary conditions function as selection conditions, restricting the unlimited variety of physically conceivable processes to the particular process sequence that actually occurs. In short: They represent constraints to which physical processes are subjected. These constraints can vary enormously. The examples range from a rigid body, where the distances between the individual atoms remain constant, or the walls of a gas container that limit the movement of the gas molecules within it, or to the restricted movement of a bead on a wire (Fig. 4.15). In all three examples, the special features of the system are the result of a constraint. This demonstrates the universal character of boundary conditions.

Let us now take a look at the constraints in macrophysical structures such as the Boulton & Watt steam engine (Fig. 4.14). From a physical and technical point of view, the parts of the machine are a number of rigid bodies. Their movement is subject to particular constraints given by the arrangement of the machine's rods, surfaces, walls and so forth, constituting the machine's macroscopic structure. The constraints determine the material organization of the engine and thereby its function. Moreover, the fact that the device is designed to serve a purpose indicates that, at the same time, the constraints are carriers of information—in this case, as laid down in the machine's construction plan.

This general description of the principles of a machine applies to technical as well as biological systems. Let us consider the boundary conditions in a living organism. These are laid down in the system's genetic blueprint, which in turn is encoded in a biological macromolecule, in most organisms familiar to us a DNA molecule (Fig. 4.16). The genetic information carrier operates like a physical constraint: it aligns all physical processes of the living organism in such a way that its essential functions are maintained.

Furthermore, the boundary conditions in a living organism are organized hierarchically. The hierarchy emerges from the DNA as the primary physical boundary condition during the development of the living organism from a fertilized egg cell. In this process, the information in the genome is expressed step by step into the structure of the organism. Since at each level of expression the physicochemical environment of the genetic information alters, the information content changes too. In other words, the overall

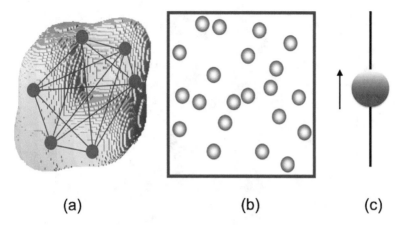

(a) (b) (c)

Fig. 4.15 Three examples of physical boundary conditions. All three sets of conditions have in common the property that they restrict freedom of movement. (a) A rigid body (all distances between the atoms are unchangeable). (b) A gas in a container (the movements of molecules are restricted by the container's walls). (c) A bead restricted to move on a wire. Physical constraints can be classified according to various criteria. Usually, they are distinguished according to their mathematical representation. If they can be described by equations they are called "holonomic". If they cannot, then they are termed "non-holonomic". In addition, they can be classified by whether they are independent of time ("scleronomic") or whether they depend explicitly on time ("rheonomic"). The boundary conditions in (a) are holonomic, while those in (b) are non-holonomic. (c) shows a rheonomic boundary condition. Most of the constraints that one encounters in physics are non-holonomic (for details see [32])

process of development is subject to internal feedback of an extremely complex nature. As the process passes through its various stages of organization and degrees of complexity, ever newer feedback loops are formed, leading finally to a stringent hierarchy of boundary conditions constraining the organism and thus to its organizational structures.

All the boundary conditions present in the differentiated organism can be traced back to its genetic information. More precisely, the genome represents by its particular structure a primary physical constraint, which, together with its physicochemical environment, is a necessary and sufficient condition for the reproductive maintenance of the organism. The fact that the information content of the sophisticated structure of the organism is more complex than that of its genome is due solely to the fact that the physicochemical environment of the genome also contributes to its information content (see Sect. 6.9). Thus, the thesis of Chap. 1, according to which information only exists relative to its context, is impressively verified in modern genetics.

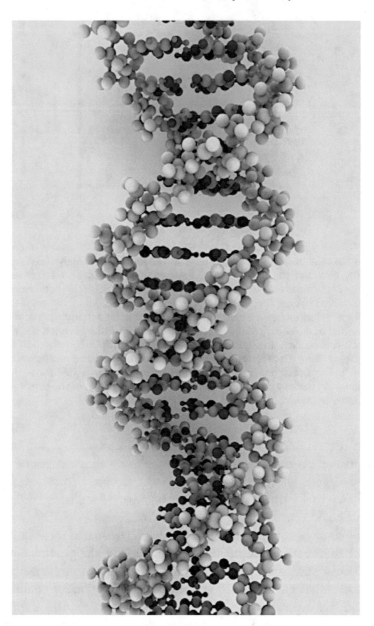

Fig. 4.16 The molecular structure of DNA, shown here as a ball-and-stick model, represents a biological boundary condition which, in principle, could be described as a physical constraint. For this purpose, however, one would have to specify the exact spatial position of all atoms of the molecule, including all the environmental conditions to which the DNA is exposed (temperature, ionic strength, etc.). [Image: ynse/Flickr]

Michael Polanyi was the first to point out the importance of boundary conditions for the understanding of life phenomena [33]. However, instead of pursuing the forward-looking aspect of his reflections, Polanyi took an anti-reductionist position concerning the origin of the boundary conditions. He believed that boundary conditions transcend the explanatory power of physics. In a machine and in a living organism, Polanyi argued, the boundary conditions are carriers of meaningful information. However, neither the laws of physics nor the principles of statistics, he believed, could ever make comprehensible the selection of information-carrying boundary conditions from the unlimited variety of physically equivalent alternatives.

At the time when Polanyi published his ideas about the characteristics of living matter, the physics of self-organization was still in its infancy, and the theoretical insights on which we can currently rely were not yet available. Today we know that Polanyi's conclusion, according to which biological boundary conditions are irreducible to physics, is wrong. The modern theory of the origin and evolution of life has shown that molecular self-organization and evolution can very well allow information-carrying boundary conditions to emerge from unspecific initial conditions [31].

To put it concisely: the real character of living matter lies in the evolutionary self-optimization of its physical boundary conditions. Since developmental biology and neurobiology also rely on models of self-organization, the boundary conditions gain universal significance for the understanding of life phenomena. In fact, the central concepts of biology—such as complexity, organization, functionality, self-organization and information—are interlinked by the idea of boundary conditions.

Since the concept of boundary conditions is closely related to the concepts of information and organization, there may be a justified hope that one day the idea of boundary conditions may also be expanded beyond physics and applied to human forms of organization. However, this requires a new type of science, which deals with the abstract, overarching structures of reality and is called "structural science" (Sect. 4.9).

It is an obvious step to transfer the idea of boundary conditions to human action, because all human activities are fundamentally integrated into a complex hierarchy of boundary conditions. These include the social environment of the actor, his cultural horizon, his biography, his personality structure, his living conditions and so forth. All these boundary conditions channel human action. They weigh on man's acts like a constraint, and thus they determine his social behavior to a very great extent.

The economist Herbert Alexander Simon, for example, has investigated these constraints upon human actions, in particular in connection with individuals who have to make decisions in economic and political contexts [34]. His model has become known as "bounded rationality". It considers that acting rationally is limited by many factors such as lack of information, a limited amount of time, the complexity of the problem and many others. Here, the enormous significance that boundary conditions have for understanding reality becomes clear. A theory describing the development of boundary conditions might not only form the basis for a unified understanding of the processes of self-organization in Nature and society; it could also open up an entirely new perspective on the phenomena of history. The latter aspect brings us back to the problem of the unity of the sciences.

4.9 The Ascent of Structural Sciences

Examination of today's scientific landscape shows that the borderlines of traditional scientific disciplines are beginning to dissolve. As we have seen in previous sections, this development is mostly because the complex, historically evolved structures of reality are increasingly coming into the focus of the nomothetic sciences. An example we have discussed in detail is the physical concept of boundary conditions. However, the extension of this idea to human actions and finally also to human history requires an exceedingly abstract theory of boundary conditions—one that, in the first instance, is free of all concrete references to reality.

The theory of boundary conditions that we can envisage, dealing with the general structures of reality without first asking where these structures occur and whether they are found in natural or artificial, in living or in non-living systems, falls into the broad spectrum of "structural sciences". These sciences belong to the category of exact sciences because their aim is to find the laws to which the abstract structures of reality are subjected. However, in contrast to the traditional sciences, their intended field of application is, *a priori*, the totality of reality. The structural sciences include, among other fields, cybernetics, game theory, information theory and systems theory. In recent times, they have been supplemented by synergetics, network theory and the theory of self-organization, to mention but a few (Table 4.2).

Today, the structural sciences have become powerful instruments for exploring the complex structures of reality. They are classified according to the overarching aspects of reality, described by generic terms such as "system", "organization", "self-regulation", "information", "cooperation", "network", "decision" and the like. The development of the structural sciences began just over half a century ago. Among their founding fathers were Norbert Wiener

Table 4.2 Structural sciences. The current development of science indicates that there is an interconnection between the structural sciences. There is much evidence that this connection is mediated by the concept of initial and boundary conditions expressing the essence of reality. This concept is an essential element of scientific explanations because it gives laws a meaning in the first place

The "universe" of structural sciences
Cybernetics
Information theory
Game theory
Systems theory
Network theory
Synergetics
Complexity theory
Catastrophe theory
Theory of self-organization
Chaos theory
Decision theory
Semiotics
Structural linguistics

(cybernetics), John von Neumann and Oskar Morgenstern (game theory), Claude Shannon (information theory) and Ludwig von Bertalanffy (systems theory).

Mathematics is the archetype of structural science. Therefore, one could be inclined to see in the structural sciences that form of science which is referred to as "formal science". However, the concept of formal science has a long-standing meaning that points in a different direction from the idea of structural science introduced here.

The formal sciences, which are traditionally considered to be the counterpart of the "real sciences" (such as the natural, economic and social sciences) are based on experience-independent insights, as expressed *inter alia* in the formally correct statements of logic. In contrast to the formal sciences, however, the structural sciences cannot be demarcated from the real sciences. On the contrary: the structural sciences are the abstract framework of the empirical sciences. This is precisely why the structural sciences possess a bridging function between the natural sciences on the one side and the humanities on the other [35]. This is also reflected by their methodology, which abstracts from the various manifestations of reality and replaces them with abstract terms and symbols. For this reason, all things, states and processes of reality are first treated as abstract structures without reference to their respective real forms of expression.

Nevertheless—and this is the real difference between structural and formal science—the laws of the structural sciences are derived from experienceable reality. Thus, for example, the laws of the structural sciences are gained from real laws by eliminating all constants that refer to reality and replacing them with logical and mathematical constants. The laws obtained by this method are "structural" laws, i.e. they have the same syntactical structure or logical form as the real laws from which they were derived. However, they can now be applied to entirely different areas of reality. Taking up a phrase of the mathematician Hermann Weyl, one could consider the structural law as an "empty logical form of possible science" [36].

Concerning the question of the unity of the sciences, the structural similarities that apply to laws in different fields of science are of particular importance. A simple example is the structural equivalence between the physical law that describes the volume–pressure behavior of an ideal gas and the economic "law" of supply and demand. According to the rules of the market, a fall in the price of particular goods leads to an increase in demand for them, and a rise in their price leads to a corresponding drop in demand. In this case, the product of demand and price is just as constant as the product of volume and pressure of an ideal gas described by Boyle's (or Mariotte's) law.

However, one encounters structural equalities not only at the level of empirical regularities, but also at the theoretical level of science. An example is the so-called Fourier equation, which can be applied to all physical transport processes, regardless of whether the process in question is the transport of heat, electricity or a liquid. Another example is the theory of Brownian molecular movement, which describes the thermally induced motion of molecules. From this physical theory, Robert Merton, Fischer Black and Myron Scholes have worked out a general theory of random processes, which has been used successfully in economics for the valuation of derivative financial markets [37, 38]. The interaction between physics, economics and sociology mediated by the methods of structural science has already led to an independent research field, which is called "physics of socio-economic systems". The main focus here is on the structural theories of phase transitions and mass effects derived from the physics of multiparticle systems.

The structural sciences already interconnect large areas of the natural sciences, economics and the humanities. For this reason, the structural sciences could also be called "cross-sectional sciences". Without their integrative effect, research into complex systems would not only stagnate, but in many areas it would not even be possible. This is particularly evident in modern biology, the theoretical rationale of which is based mainly on the structural sciences. Last but not least, the structural sciences have also become an indispensable theoretical basis for economics. In addition to game theory, which owes its genesis mainly to economic problems, it is above all decision theory, the theory of self-organization, chaos theory and the theory of fractal geometry that are currently attracting attention in the economic sciences [39]. Furthermore, with semiotics and linguistic structuralism, structural thinking has long found its way into the humanities and has spread to other areas such as anthropology, sociology and psychoanalysis. Sociology, however, has gone its own way. An example of this is the systems theory of society propounded by Luhmann [40].

In mathematics too, the archetype of all structural sciences, there are structural approaches to reunite and standardize the multitude of mathematical sub-disciplines by way of abstract hyperstructures. For a long time, the trend-setter here was a group of mathematicians founded in France under the pseudonym "Bourbaki" [41], which pursued the goal of building up all essential areas of mathematics from three fundamental hyperstructures, which they termed mother structures (*structures-méres*). These and other attempts to justify mathematics more deeply demonstrate that, even within the most abstract branch of science, the idea of unity is a driving force behind progress.

The computer is the weightiest instrument of the structural sciences. Computer-aided processing of vast amounts of data, accompanied by the possibilities of computer simulation, computer graphics and computer animation, forms an indispensable basis for research into complex phenomena. Thus, alongside the experimental method of the natural sciences, computer-aided analysis of reality has become established as a new and powerful method of gaining knowledge.

The fact that computer simulation is increasingly replacing conventional laboratory experiments in the natural sciences constitutes clear evidence that the structural sciences are already having a profound impact on traditional science. The question of whether virtual experience employing computer simulations leads to a world-view that is only partially consistent with the reality that can be experienced, or whether such simulations are just an extremely efficient application of conventional experimental methods, cannot be conclusively answered at present. It is probably best to compare the use of the computer with the use of the microscope. Just as this opened up a new realm of reality to man, one beyond his natural perception, computer simulation enables us to gain insights into complex interrelationships that are not accessible to us by conventional analytical methods. The critical question of how far computer simulations are reliable, and whether they really do what we think that they do, we shall take up in Sects. 7.9 and 7.10.

4.10 What Task Remains for the Human Sciences?

Given the overwhelming progress of the exact sciences and their associated impact on society, the humanities are seeing themselves ever more strongly forced to prove that they too have some significance for the modern "knowledge society". In any case, the tranquility of the ivory tower of the humanities seems to be disturbed. However, the call to take part in designing the future scientifically very often meets with resistance in the humanities, because this task is not consistent with their traditional self-image. Consequently, the statements in which the humanities attempt to assert their social relevance and necessity often reflect a rather defensive attitude.

In the course of their self-discovery, scholars of the humanities have developed a number of bizarre ideas about the task of their disciplines, often hand in hand with polemic attacks against the exact sciences. Natural scientists, for example, are often accused of uncritically adhering to a naïve naturalism that in no way does justice to the human mind. Only a knowledge of the humanities, as runs the argument already mentioned several times, represents an appropriate level at which to reflect on and understand our world, allowing a critical assessment and classification of scientific knowledge. With this and similar reasoning, the humanities try to secure their claim to the sovereignty and superiority of their knowledge, which they fear may be refuted by the progress of the exact sciences.

However, there is no uniform opinion among the scholars of the humanities as to precisely what task the humanities have within society. For example, some years ago, the philosopher Odo Marquard adopted a course that was opposed to the traditional self-understanding of the humanities [42]. Marquard fully acknowledged the supremacy of the natural sciences, and assigned a role to the humanities in which they merely appear as a necessary complement to the natural sciences. He argued that the humanities had to compensate for the "modernization damage" caused by natural science and technology. In that they counterbalance such life-world losses, Marquard claimed, the human sciences have a function that makes modernization possible in the first place.

Marquard based his "compensation theory" on the argument that the exact sciences, oriented to experimenting and measuring, were forced to hide or to "neutralize" all historical aspects of their research objects in order to make them suitable for the laboratory. This, he argued, is the deeper reason why the humanities are indispensable for society. The humanities were primarily concerned with the historicity of the world, which the exact sciences inevitably exclude. According to Marquard, it has always been the task of the humanities to compensate for the ahistorical approach of the natural sciences to the world. This, he argued, indeed made the humanities the younger form of science and not, as wrongly assumed by most, the natural sciences.

The idea of placing the historical world at the center of the humanities is not new. On the contrary, it forms an essential part in the line of thought in the humanities, rooted in the historicism of the nineteenth century (see Sect. 3.4). New, however, is the way in which Marquard downgrades the task and thus the significance of the humanities. Following this, it is no longer up to the humanities to contribute to scientific progress with new findings; rather, the humanities are thought to provide a contemplative and constructive counterbalance to the natural sciences with their inherent drive toward modernization. Consequently, Marquard's compensation theory demands that the humanities should be concerned with the grand narratives of the world, which are supposed to fill the gap left by the exact sciences—with themes such as preservation, sensitization and orientation. This compensation function justifies, as Marquard puts it with a touch of irony, the "unavoidability" of the humanities.

The compensation theory was brought into play to restore significance to the humanities, which were ailing in respect of their relevance. However, this idea does so at the price of dismantling the humanities. It may therefore seem understandable that many scholars in the humanities are reluctant to accept the role of a mere storyteller who fills the features pages of the newspapers—for, with the compensation idea, the aspiration of the humanities to be a science is finally abandoned. The role of the humanities would then be reduced to that of a stopgap, with the sole task of explaining what the hard sciences leave to them as a provisionally inexplicable residue or *quantité négligeable*, as the literary scholar Mattenklott [43] laments; Mattenklott warned the humanities against resigning or even "capitulating to the new ideology of naturalism" [43, p. 581; author's transl.].

Here, we encounter the language of the culture war again. It makes clear that the mutual misunderstanding between natural scientists and scholars of the humanities which Snow bemoaned is still present. However, the arguments that are brought to bear stem from yesterday's battle. The development of the modern structural sciences has moved the goalposts and given the explanatory concept of the natural sciences a new face. In particular, the natural sciences have moved away from the mere naturalistic description of the world in favor of a structural world-understanding. As a consequence, the rugged contrasts between mind and matter, animate and inanimate, history and Nature, which have shaped the dualistic understanding of modern science, are dissolving.

In the face of the upheavals in our life-world brought about by science and technology, it is the distinct task of the humanities to participate in the development of a future-orientated rationality. Using the definition of rationality given by Nicholas Rescher, we can say: "Rationality consists in the intelligent pursuit of appropriate ends. It pivots on the use of intelligence or reason, the crucial survival instrument of the human race, in the management of our affairs." [44, p. 1] As the key link between thinking and acting, rationality enforces the unity of theoretical and practical knowledge. We cannot act rationally without having reasonable findings and objectives, and we cannot have reasonable objectives and findings without their being geared to rationally substantiated action.

Future-orientated rationality must give scientific knowledge an unrestricted priority in shaping society. Only rationality that is free of moral indoctrination, and free of ideologies that are hostile toward science, will lead us into a worthwhile future. Such rationality will see its goal in openly shaping our living world, instead of following a creation myth that obliges us to preserve the world in its given form. The task to contribute to a forward-looking rationality will challenge the humanities to a large extent. It requires, in particular, that the humanities do not abandon their self-understanding as a mere narrative from the past. Instead, the humanities are called on to contribute their historical knowledge, their sharply focused view of cultural happenings, social developments and perspectives, options for action and the like. If the humanities take this path, they can undoubtedly play an essential role within the modern knowledge society.

References

1. Snow CP (1959) The Two Cultures. Cambridge University Press, Cambridge
2. Lyotard J-F (1984) The Postmodern Condition: A Report on Knowledge (transl: Bennington G., Massumi B). University of Minnesota Press, Minneapolis [Original: La condition postmoderne: Rapport sur le savoir, 1979]
3. Sokal A, Bricmont J (1998) Fashionable Nonsense. Postmodern Intellectuals' Abuse of Science. Picador, New York
4. Weinberg S (2001) Facing Up. Science and Its Cultural Adversaries. Harvard University Press, Cambridge/Mass

5. Comte A (1988) Introduction to Positive Philosophy (transl: Ferré F). Hacket Publishing, Indianapolis [Original: Cours de philosophie positive, 1830]

6. Windelband W (1998) History and Natural Science (transl: Lamiell JT). Theory Psychol 8:6–22 [Original: Geschichte und Naturwissenschaft, 1894]

7. Rickert H (1962) Science and History. Van Nostrand, New York [Original: Kulturwissenschaft und Naturwissenschaft, 1926]

8. Rickert H (1986) The Limits of Concept Formation in Natural Science: A Logical Introduction to the Historical Sciences (abridged edition, transl: Oakes G). Cambridge University Press, Cambridge [Original: Die Grenzen der naturwissenschaftlichen Begriffsbildung, 1896]

9. Aristotle (1933–35) Metaphysics. In: Tredennick H (ed) Aristotle in 23 volumes, vols 17, 18. Harvard University Press, Cambridge/Mass [Original: Tà metà tà physiká, 350 B.C.]

10. Goethe JW von (1981) Werke, Bd 12. Beck, München

11. Cassirer E (1923) Substance and Function & Einstein's Theory of Relativity. Open Court Publishing Company, Chicago [Original: Substanzbegriff und Funktionsbegriff, 1910]

12. White H (2008) Identity and Control. Princeton University Press, Princeton

13. Hempel CG, Oppenheim P (1948) Studies in the Logic of Explanation. Philos Sci 15:135–175

14. Hempel CG (1978) Dispositional Explanation. In: Tuomela R (ed) Dispositions. Synthese Library, vol 113. D. Reidel Publishing, Dordrecht, pp 137–146

15. Munslow A (2006) The Routledge Companion to Historical Studies. Routledge, New York

16. Hawking S (1982) The boundary conditions of the universe. Pontif Acad Sci Scr Varia 48:563–574

17. Eigen M (1971) Selforganisation of matter and the evolution of biological macromolecules. Naturwissenschaften 58:465–523

18. Lübbe H (1981) Die Einheit von Naturgeschichte und Kulturgeschichte. Akademie der Wissenschaften und Literatur Mainz, Abhandlungen der Geistes- und Sozialwissenschaftlichen Klasse 10:1–19

19. Lévi-Strauss C (1963) Structural Anthropology. Basic Books, New York [Original: Anthropologie structurale, 1958]

20. Mises R von (1968) Positivism. A Study in Human Understanding. Dover Publications, New York [Original: Kleines Lehrbuch des Positivismus, 1939]

21. Weber M (1949) "Objectivity" in Social Science and Social Policy. In: Weber M. On The Methodology of the Social Sciences (transl: Shils EA, Finch HA). Free Press, Glencoe, Illinois, pp 49–112 [Original: Die "Objektivität" sozialwissenschaftlicher und sozialpolitischer Erkenntnis, 1904]

22. Eulenburg F (1923) Sind "Historische Gesetze" möglich? In: Palyi M (ed) Hauptprobleme der Soziologie. Erinnerungsband für Max Weber, Bd 1, Duncker & Humblodt, München und Leipzig, pp 23–71

23. Wright GH von (1971) Explanation and Understanding. Cornell University Press, Ithaca
24. Singer W (2015) Neural synchrony as a binding mechanism. In: Wright JD (ed) International encyclopedia of the Social and Behavioral Sciences, vol 16. Elsevier, pp 634–638
25. Bergson H (1913) Time and Free Will. George Allen & Company, London [Original: Essai sur les données immédiates de la conscience, 1889]
26. Heidegger M (1962) Being and Time (transl: Macquarrie J, Robinson E). Basil Blackwell, Oxford [Original: Sein und Zeit, 1927]
27. Picht G (1970) Zukunft und Utopie. Die großen Zukunftsaufgaben. Piper, München
28. Pöppel E, Bao Y (2014) Temporal Windows as a Bridge from Objective to Subjective Time. In: Arstila V, Lloyd D (eds) Subjective Time: The Philosophy, Psychology, and Neuroscience of Temporality. The MIT Press, Cambridge/Mass
29. Küppers B-O (1992) Understanding Complexity. In: Beckermann A, Flohr H, Kim J (eds) Emergence or Reduction?: Essays on the Prospects of Nonreductive Physicalism. De Gruyter, Berlin, pp 241–256
30. La Mettrie, JO (1996) Machine Man and Other Writings (transl: Thomson A). Cambridge University Press, Cambridge [Original: L'homme machine, 1747]
31. Küppers B-O (1990) Information and the Origin of Life (transl: Woolley P) MIT Press. Cambridge/Mass [Original: Der Ursprung biologischer Information, 1986]
32. Goldstein H, Poole CP, Safko, JL (2011) Classical Mechanics. Pearson Education, Harlow, Essex
33. Polanyi M (1968) Life's irreducible structure. Science 160(3834):1308–1312
34. Simon HA (1957) Models of Man: Social and Rational. Wiley, New York
35. Küppers B-O (2018) The Computability of the World: How Far Can Science Take Us? Springer International, Cham
36. Weyl H (2009) Philosophy of Mathematics and Natural Science (based on transl: Helmer O) Princeton University Press, Princeton [Original: Philosophie der Mathematik und Naturwissenschaft, 1927]
37. Black F, Scholes MS (1973) The Pricing of Options and Corporate Liabilities. J Polit Econ 81(3):637–654
38. Merton RC (1973) Theory of rational option pricing. Bell J Econ Manage Sci 4:141–183
39. Mandelbrot B (1997) Fractals and Scaling in Finance. Springer, New York
40. Luhmann N (2012) Introduction to Systems Theory. Polity Press, Cambridge [Original: Einführung in die Systemtheorie, 2002]
41. Corry L (2009) Writing the ultimate mathematical textbook: Nicolas Bourbaki's Éléments de mathématique. In: Robson E, Stedall J (eds) Oxford Handbook of the history of mathematics. Oxford University Press, Oxford, pp 565–587

42. Marquard O (1991) On the Unavoidability of the Human Sciences (transl: Wallace RM). In: Defence of the Accidental: Philosophical Studies. Oxford University Press. Oxford [Original: Über die Unvermeidbarkeit der Geisteswissenschaften, 1989]

43. Mattenklott G (2001) Wider die Resignation der Geisteswissenschaften. Forschung & Lehre 11:581–582

44. Rescher N (1988) Rationality: A Philosophical Inquiry into the Nature and the Rationale of Reason. Oxford University Press, Oxford

5

Limits: Insights into the Reach of Science

5.1 The Enigmatic Reference to the "Self"

Occasionally, politicians are said to have an ambivalent attitude to truth. Imagine a politician going public and confessing "I always lie." How would the public react? Most people would probably be highly irritated. Others, however, might muse about this statement, because the self-recrimination of our politician is as self-contradictory as it could possibly be. In fact, his confession is only valid if it is false! Does this mean that we can't trust a politician even when he is telling the truth?

With a similar assertion, "All Cretans always lie", the Crete-born philosopher Epimenides confused his contemporaries in antiquity. Epimenides, who was highly esteemed for his wisdom, had revealed a boundary of human reflection related to the tricky problem of self-reference. The liar-antinomy, of which there are various versions, probably goes back to the philosopher Eubulides of Miletus.

However, not all versions of this antinomy prove to be equal under the keen scrutiny of the logician. The original Cretan version, "All Cretans always lie", is one of the weaker variants, because it does not rule out the possibility that some Cretans occasionally tell the truth: there remains a possibility that the statement "All Cretans always lie" is wrong, which means that not all Cretans always lie. Epimenides could have been one of these exceptions, so his statement "All Cretans always lie" would just have been an ordinary wrong statement.

© Springer Nature Switzerland AG 2022
B.-O. Küppers, *The Language of Living Matter*, The Frontiers Collection,
https://doi.org/10.1007/978-3-030-80319-3_5

In contrast, the slightly modified statement "I always lie" represents a real contradiction. The truth value of this sentence is fundamentally undecidable. In this case, the sentence is, because of its direct self-reference, at the same time, both true and false. If the person in question is telling the truth, his assertion: "I always lie" is false. If, on the other hand, the person is telling an untruth, the statement is true.

The antinomy of the liar is probably the most famous example of a self-referential statement that leads to a real contradiction. Such contradictions used to be called antinomies and were distinguished from paradoxes, which are merely apparent contradictions. In today's usage, however, this subtle distinction is frequently not observed. Thus, in the present case, one often speaks of the "liar paradox", although strictly speaking it is an antinomy.

Paradoxes have been known, in numerous variants, since ancient times. Zenon of Elea, a disciple of Parmenides, was a master in the formulation of paradoxes. Plato reports that Zenon is said to have described up to forty paradoxes (*logoi*). Only a few, however, have been handed down to us, but those we know are highly sophisticated. An example is the paradox of Achilles and the tortoise, which was set up by Zenon to demonstrate a fundamental contradiction between experienceable reality and logical reasoning (Fig. 5.1). The paradox seems to prove that in a race between Achilles and a tortoise, Achilles, although running faster, will never be able to catch up with the tortoise if he gives his opponent a head start.

For Zenon, the truth of logical reasoning took precedence over observation and experience. Therefore, he held fast to the assertion that the appearances of reality deceive us into believing things that in fact are not possible, in this case that Achilles catches up with the tortoise. This and other paradoxes that Zenon invented are based on a sophisticated thought-game with the idea of infinity. Only in modern mathematics has it became clear that the summation of an infinite number of summands (which also underlies the Achilles paradox) can still result in a finite answer, thereby resolving the paradox. This was not known to the mathematicians of antiquity, and Zenon was thus led to regard the logical conclusions of his paradoxes as unshakable truths.

The interpretation that Zenon gave to the Achilles paradox was entirely in keeping with the thinking of his teacher Parmenides, who had claimed that reality as it appears to us has only the character of an illusion (Sect. 2.2). So Zenon concluded that the paradoxes are based on a false premise, one that is the opposite of the actual truth: in fact, he argued, true reality undergoes no movement or change.

Fig. 5.1 Achilles paradox set up by philosopher Zenon in the fifth century B.C. Achilles and a tortoise take part in a seemingly unequal race. Achilles, the faster of the two, is fair and gives the slower tortoise a head start. Let us assume that Achilles is ten times faster than the tortoise, and the latter enters the race with a lead of hundred meters. When Achilles has run the hundred meters, the tortoise has completed ten meters. Once Achilles has caught up with these ten meters, the tortoise has advanced by one more meter. This argument can be continued endlessly and seems to lead to the conclusion that Achilles comes ever closer to the tortoise, but never catches it—a conclusion in complete contradiction with experience. However, the paradox represents only an apparent contradiction, one that can be resolved within the framework of modern mathematics by employing the concepts of limit value and of the convergence of infinite geometric series

In the case of the liar antinomy, however, the problem is a different one. Here it is the self-reference of linguistic expression that creates confusion. Nevertheless, there have been repeated attempts to resolve such "semantic" antinomies. For example, as Alfred Tarski [1] has shown, clean separation of object language from metalanguage is a proven means of avoiding semantic antinomies. "Object language" is the language with which we talk about things in the world. However, if we talk about statements that are formulated in object language, then we already have moved up to the level of "metalanguage" (a specialized form of language or set of symbols describing the structure of a language). For example, the sentence:

This object has property A

is a statement within object language. If, in contrast, one says:

The statement 'This object has the property A' is true

then this is a metalinguistic statement. Certain logical and philosophical problems can be attributed to the simple fact that they mix object language and metalanguage together (Sect. 2.1).

The paradoxical characteristics of self-reference always turn up when thinking itself is made the object of thinking, so that thinking is "reflected" at its own limits. In the borderland of human reflection, we realize clearly that we are always prisoners of our thought because we cannot distance ourselves from the structures of thought that are given to us. On the other hand, the capacity for self-reflection is undoubtedly the most striking characteristic of our consciousness. It represents the highest possible level of consciousness, where consciousness returns to itself and the world-reference of the Ego arises in its delimitation from the world of the non-Ego.

The separation of all that exists into subject and object, which goes hand in hand with self-reflection, forms the basis of human judgment. It seems imperative that a subject which is to reflect on its Ego must for this purpose step out of the identity with itself and view itself, as it were, from the outside. Fichte's doctrine of science was the attempt to reconstruct philosophically this process of externalization of the Ego.

The reference to one's being and the resulting self-consciousness is the beginning of all human knowledge. Accordingly, Fichte placed the self-assurance of the Ego ("I am I"), as the most elementary act of human knowledge, at the beginning of his doctrine of science (Sect. 2.6). Nevertheless, the idealistic philosophers viewed the division into an ideal world and a real world, accompanied by self-reflection and self-analysis, as an unnatural disruption of the primary unity of self and world, of man and Nature, of subject and object. Schelling, for example, believed that it was through reflective thinking, with its dissecting approach, that the intellectual world had been filled with "chimeras" [2]. He went so far as to see in mere reflection a "mental illness of man; moreover, the most dangerous of all, that kills the germ of his existence and exterminates the roots of his being" [2, p. 338 note 1; author's transl.]. However, the idealists were not able to explain convincingly how to rid themselves of reflection, which they regarded as an annoying nuisance. They probably felt that all attempts to overcome the separation of subject and object are tantamount to squaring the circle. How, they argued, can a prerequisite for thinking be abolished by thinking itself?

The separation of the existing into subject and object appears indispensable because the subject can only become aware of itself through reflection on an object that is distinct from the reflecting subject. Descartes, on the other hand, believed that the Ego alone attained self-assurance through the mere fact of thinking. This is certainly the core of his existence statement *Cogito ergo sum* (Sect. 2.4). However, although this sentence may give the Ego certainty, it provides no ultimate justification of the Ego's existence for a simple reason: in the "thinking Ego", which according to Descartes justifies the existence of the Ego, this existence is already presupposed.

No thinking can begin without preconditions. Likewise, we can never remove the self-reference of our thought by thinking. We have no external reference point outside our thinking from which we can look at our thinking. All forms of our thinking can only be illuminated from the inside perspective of the thinking Ego. Self-reference does not provide a reference point outside the "self". Only the famous Baron Münchhausen was once able to overcome the limits of the self, when—if one is inclined to believe his story—he pulled himself out of the swamp by his own pigtail.

The fact that a "self" is constituted by thinking at all could be considered a unique feature of human thinking. However, the distinction between "self" and "other" as a preliminary stage of self-knowledge seems to exist at all levels of living systems. Even cells can recognize kinship and distinguish between self and non-self. This ability forms, among other things, the basis of the immune system. Furthermore, behavioral biology studies have shown that the distinction between "self" and "other" is an essential element in the social behavior of animals. Thus, in populations of animal species that show particularly pronounced social behavior toward close relatives, an equally strong rejection of foreign individuals is generally observed (see for example [3]).

Thus, self-reference is not a quality that can only be ascribed to the human individual, serving his or her self-image. Instead, this property seems to be an indispensable element in the development of higher forms of life. In fact, there could be no information and communication structures between living beings if the individuals involved could not distinguish between the "sender" and the "receiver" of information, i.e. between "self" and "other". Without information and communication, the development of higher life forms is inconceivable [4].

The phenomenon of self-reference seems to provide a general key to understanding the world. Only when we know how the appearances of the world act upon themselves can we understand the world's internal connectivity. This was already recognized by the natural philosophers of antiquity when they sought to elucidate the "self-movement" of the world. They understood the term "movement" to mean all quantitative and qualitative changes in the world. Their reflections finally led to the idea of a "world-soul" as the universal principle that on the one hand moves the world and on the other hand also holds it together.

The philosophical idea of the world-soul originated among the Pythagoreans, and is later found in the philosophy of Plato. Rediscovered by the philosopher Giordano Bruno in the Renaissance, the concept of the world-soul occupied a central position above all in the philosophy of German idealism. One finds this idea in the writings of Johann Gottfried Herder and of Goethe, as well as in the philosophical work of Schelling and, last but not least, in Hegel's conception of the "world-spirit".

The question of the self-motion of matter also plays a central part in modern science. Today, however, one no longer speaks of "self-movement", but of "self-organization". In the broadest sense, this term denotes the process of an overarching, material evolution that spanned billions of years, from the beginning of the universe up to the evolution of living matter. In a narrower sense, the term "self-organization" denotes the manifold forms of self-causing of living beings, such as their self-reproduction, self-preservation, self-regulation, self-renewal and the like. Only its self-causation makes an organism appear as an autonomous whole.

In the organism, self-causation becomes apparent in the cyclic causality of life processes. Here, the chain of causes and effects flow back into itself, so that each effect becomes a cause and each cause at the same time an effect. Therefore, the causal-analytical sciences, which are based on the separation of causes and effects, long regarded organic causality as a strange and incomprehensible phenomenon. No wonder, then, that a fierce dispute has arisen over the reductionism inherent in modern science. Its opponents have attempted to plug the explanatory gaps—real or apparent—in the causal-analytical understanding of the organic world by means of vitalistic or holistic doctrines.

5.2 Incomplete and Undecidable

Self-reference, which at first appears to be an annoying limitation of the human cognitive faculty, has led to profound insights into the foundations of rational thought. The self-reference of thinking is the deeper reason why a final, and thus absolute, justification of human knowledge is not possible. Not even the findings of mathematics are exempt from this caveat. Therefore, it may seem almost paradoxical that this insight itself can be proved with mathematical rigor.

To understand this, we must first take a look at the discussion of the basics of mathematics at the end of the nineteenth century. At that time, mathematics, which in comparison with all other sciences seemed to be on the safest ground, had fallen into a deep crisis. The writer Robert Musil described this situation vividly: "And suddenly, after everything had been brought into the most beautiful kind of existence, the mathematicians—those ones who brood entirely within themselves—came upon something wrong in the fundamentals of the whole thing that absolutely could not be put right. They actually looked all the way to the bottom and found that the whole building was standing in mid-air." [5, p. 42]

Among those who tried at that time to substantiate the edifice of mathematics were, above all, Gottlob Frege, Bertrand Russell and David Hilbert. However, in their efforts to advance to the fundamentals of mathematics, they took different paths. Frege and Russell aimed to secure the claim of mathematical statements to truth with the help of logic. They were guided by the idea that mathematical truths are ultimately nothing more than logical truths —a concept that founded the so-called "logicism".

Hilbert, on the other hand, proposed to formalize mathematics completely, in order to keep the mathematical methods of proof free from all subjective influences, such as inevitably intrude during the intuitive comprehension of mathematical correlations. The program he developed for the foundation of mathematics was the so-called "formalism". According to this, the claims of mathematical statements to be true should be guaranteed by certain characteristics of formal systems, to which belonged, *inter alia*, consistency and completeness.

Frege's and Russell's first concern was to replace the clearly and intuitively acquired terms of mathematics with precise, logically founded terms—as Leibniz, in the seventeenth century, had already had in mind when he developed his symbolic artificial language, the so-called "*characteristica universalis*". The same goal was pursued in particular by Frege in his *Begriffsschrift* of 1879, in which he attempted to develop a "formula language, modelled upon that of arithmetic, for pure thought" which is free of all non-logical and descriptive references [6].

With the help of his new formula language, Frege believed that he could verify his view that all essential parts of mathematics (except for geometry) can be rigorously justified and that mathematics is only a branch of logic. In the preface of the *Begriffsschrift*, Frege wrote: "The most reliable way of carrying out a proof, obviously, is to follow the pure logic, a way that, disregarding the particular characteristics of objects, depends solely on those laws upon which all knowledge rests." [6, p. 5] As a touchstone for the viability of the logicist approach, he chose number theory: "Arithmetic", Frege stated, "thus becomes simply a development of logic, and every proposition of arithmetic a law of logic, albeit a derivative one." [7, p. 99]

Indeed, in his major work *Basic Laws of Arithmetic*, Frege finally succeeded in deriving the tenets of number theory from a finite system of logical axioms [8].

At the same time, however, when Frege published the second volume of his *Basic Laws of Arithmetic*, Russell noticed that an antinomy could be inferred from Frege's axiomatic system. Now, at one fell swoop, fundamental doubts arose about the logistical rationale of mathematics. While Frege immediately stopped work on his program, Russell tried to save it by making various modifications. He finally published the modified program, together with his mentor Alfred North Whitehead, in the years from 1910 to 1913 under the title *Principia Mathematica* [9].

A popular version of this antinomy, which Russell ascribed to an unknown source, illustrates the logical problem that Frege's logicism program questioned. It is the story of a village barber who shaves all the men of the village who don't shave themselves. It is evident that this statement is contradictory: while it does not allow the conclusion that the barber shaves himself, neither does it allow the opposite conclusion that the barber does not shave himself. This form of logical antinomy, according to which a statement is both true and not true at the same time, Russell had discovered in the foundations of set theory. Here, it is the definition of the set S of all sets that are not members of themselves. This definition leads to a glaring contradiction in the sense mentioned above, because the set S both includes itself and does *not* include itself.

Russell had encountered this antinomy during his studies of the set theory founded by Georg Cantor. Ernst Zermelo, a mathematician from Hilbert's circle, had also come across the antinomy, but in contrast to Russell he did not publish it. At first, the antinomy was considered a curiosity of set theory. Then, however, it became clear "that something was rotten in the foundations of this discipline. But not only was the basis of set theory shaken by Russell's antinomy: logic itself was endangered." [10, p. 2]

Indeed, the antinomy found by Russell and Zermelo "came as a veritable shock to those few thinkers who occupied themselves with foundational problems at the turn of the century" [10, p. 2]. Russell, however, believed that the contradiction in set theory could be avoided by terminological clarification. For this purpose, he developed a so-called type theory, which treats the elements of a set and the set itself as different types.

For example, the set of planets is not a planet itself. It is of a different type from its elements. Russell continued along the same lines and placed the set, whose elements are sets themselves, above their elements, and so on. With type theory, Russell believed he could avoid the antinomy that had called

Frege's logicism program into question. Only later was it to emerge that further problems arise in the logistical reasoning program, so that from today's point of view the program can be considered to have failed. Looking back, the mathematician Hermann Weyl observed in the 1940's that: "From this history one thing should be clear: we are less certain than ever about the ultimate foundations of (logic and) mathematics. Like everybody and everything in the world today, we have our 'crisis'. We have had it for nearly fifty years." [11, p. 13]

Hilbert also stated that Russell's antinomy had "a downright catastrophic effect in the world of mathematics" [12, p. 375]. Given this antinomy, Hilbert raised the fundamental question of the extent to which the mathematical procedure of justification and proof can still be trusted at all. One thing seemed certain: as long as mathematical proofs depend on human intuition and insight, they cannot be free from mistakes of rational thought, just as they cannot be free from paradoxes and antinomies. To eliminate these uncertainties once and for all, Hilbert suggested that the mathematical method be formalized such that proofs can be carried out mechanically, i.e., without recourse to the human ability to judge.

At the heart of Hilbert's work on the foundations of mathematics was the construction of so-called "formal" systems that are based on an artificial language. This language consists of a finite alphabet of symbols, by which all possible statements of the system can be formulated, and a grammar, determining which of the possible statements are to be considered as meaningful. With the introduction of an artificial language, Hilbert believed that the inaccuracies and ambiguities inherent in human language could be excluded from the outset. Furthermore, a formal system contains a finite number of basic statements ("axioms") as well as fixed rules of logical deduction, by which all mathematical theorems can be derived from the axioms. In such a formal system, a proof is provided by applying the same basic logical and arithmetic operations again and again according to a given scheme, known as an "algorithm."

Hilbert was convinced that in this way all mathematical theorems could be found and proved automatically—provided that the system is designed consistently. A formal proof would then proceed in the following way: First of all, one would formulate all symbol sequences and thus all conceivable statements that are possible according to the alphabet and the grammar of the language used. In the next step, one would list all the symbol sequences by their length and their position in the alphabet. Finally, one would have to check each sequence by applying the proof-adducing algorithm to determine whether the sequence is a correct statement or a correct proof, i.e., whether it is consistent with the rules of derivation.

It is evident that the program of formal systems follows the ideal of the axiomatic-deductive method, which Descartes, Spinoza and others had already hoped would show an absolutely safe path to true knowledge (Sect. 2.4). Hilbert's program, however, did not only see itself as an alternative to logicism in mathematics, but it was also a reaction to a sometimes polemical dispute with the so-called intuitionists, whose most important representatives were Luitzen Brouwer and Henri Poincaré. Both rejected Hilbert's formalistic-axiomatic program and regarded mathematics as a constructivist form of cognition based on intuition. Following their own understanding of mathematics, the intuitionists recognized only those mathematical objects that could be explicitly constructed.

Hilbert, on the other hand, insisted that a mathematician "must be able to say at all times—instead of points, straight lines and planes—tables, chairs and beer mugs" (after [13, p. 57]). He argued that it was not the intuitive content of the mathematical objects that ultimately matters. Instead, he declared, mathematical objects and their relationships existed exclusively as abstract variables, which were to be judged solely by whether they satisfied the axioms that were considered to be valid. For this reason, Hilbert rejected intuitionism with the same vehemence as that with which the intuitionists opposed Hilbert's formalization program.

Within the framework of Hilbert's program, mathematics became a general theory of formalisms, the task of which is to demonstrate the absence of contradiction in content-free mathematical theories by "finite" means, i.e. in a finite number of operations. In a formal system, everything that can be deduced from the system's axioms in a consistent form can claim validity, regardless of whether it is understandable or not. Moreover, Hilbert combined the formal systems with the requirement for completeness. According to this, it must be possible to check each statement that can be formulated in a system by the means offered by the system, and thus to find out whether the statement is true or false.

Hilbert was regarded as the leading mathematician and thinker of his time. In 1900, at the 2nd Congress of Mathematicians in Paris, Hilbert presented a collection of 23 unresolved problems that he considered to be fundamental, and he challenged the mathematical experts to solve them [14]. Among other things, this included the task of proving that arithmetic is free of contradictions. Hilbert's problem collection had an enormous influence on the development of mathematics in the twentieth century. In 1928, at a Congress of Mathematicians in Bologna, Hilbert expanded the list of problems and challenged his colleagues to prove the completeness of logic and arithmetic as well.

Only a short time later, however, the logician Kurt Gödel destroyed Hilbert's dream of a perfect procedure for adducing mathematical proof [15]. In a brilliant paper, he showed that even in a consistent formalization of arithmetic there can be a statement "A" with the property that neither A nor non-A can be proven within the framework of this system. This means that formal systems, no matter how carefully they are constructed, always have only a limited capacity for proof and are therefore in principle incomplete. In any sufficiently complex and consistent formal system, as Gödel found, there exists at least one statement whose truth cannot be proved within the system. The validity of unproven statements can only be determined if one extends the system. However, the extension of the system is only possible at the price of new statements, the truth of which cannot be demonstrated within the system—an argument that can be continued ad infinitum.

Already in the nineteenth century, Bernard Bolzano had recognized that the set of sentences and truths is infinite in itself. This is easily verified: Let us consider a statement A. The statement "A is true", is a new statement B that differs from A. This thought can be repeated indefinitely. Just as one has derived from A the different sentence "A is true", one can derive from B the different sentence C "'A is true' is true" and so on. The paradox of the endless interweaving of true propositions makes it clear that every truth is embedded in the context of an overarching truth, whereby the chain of truths has no end. This is in accordance with Gödel's findings.

Interestingly, the proof of the "incompleteness theorem" is based on a modified form of the liar paradox, namely the phrase: "This statement is false." As demonstrated before, this sentence can be neither true nor false, as it is only true if it is false. According to Hilbert, however, it should be possible in a formal system to prove each statement clearly to be true or false, which in the above case is not possible. It was an ingenious thought of Gödel to replace the concept of truth with that of provability. Thus, replacing "This statement is false" by "This statement is unprovable", Gödel could demonstrate that this substitution leads in formal systems to an unsolvable problem: either a false statement can be proved correct, which would lead to a contradiction, or a valid statement is unprovable, which would mean that the system is incomplete.

A mechanical procedure of proof cannot replace mathematical intuition and insight. The idea of truth is more fundamental and comprehensive than that of provability. This is the lesson to be learned from Gödel's incompleteness theorem. Gödel regarded this as a confirmation of his philosophical attitude, which was close to Platonism. He believed that, in addition to the real world of physical objects, there also exists the ideal world of mathematics, to which man only has access through intuition. Furthermore, Gödel considered that all statements have a truth value, i.e., that they are either true or false, regardless of whether their truth can be determined empirically or not. With this view, he took a clear counter-position to the philosophers of the Vienna Circle, who only wanted to accept such statements as meaningful that can be confirmed or refuted empirically (Sect. 2.7).

In the crisis of the foundations of mathematics, different interpretations of mathematical reality became apparent. The Platonists took the point of view that there is a world of mathematical objects that exists independently of the recognizing subject and which is to be discovered. Accordingly, mathematical truth is regarded as an absolute property of mathematical objects. Others, such as the formalists, viewed mathematics exclusively as a sequence of logical deductions within a formal system that is created by the recognizing subject itself. Here, mathematical truth appears in the form of mathematical provability. The incompleteness theorem reveals the essential difference between the two positions, by demonstrating that mathematical truth and mathematical provability are in no way congruent. As a result, non-trivial logical and mathematical deductions cannot be automatized.

Because of Gödel's theorem, mathematicians and logicians—such as John Lucas, Roger Penrose and others—have concluded that the human brain, which can unearth such profound truths, must function in a manner fundamentally different from the mechanical working of a computer [16, 17]. If they were right, then it would never be possible for the intuitive and creative power of the human mind to be simulated by a computer. All attempts to build an "intelligent" computer worthy of the name would then be doomed to failure from the outset. Evidently, such a conclusion depends on a detailed knowledge of the structure and function of the human brain—knowledge from which we are still miles away. Moreover, we cannot anticipate the future development of computer architectures. Today, advances in this are so rapid that almost anything seems possible (Sect. 6.7).

5.3 How Complicated is the Complex?

Gödel's incompleteness theorem states that there are no effective decision-making procedures for formal systems. Instead, each formal system contains at least one undecidable statement, the truth of which cannot be determined within the system given. Although Hilbert's formalistic program ultimately failed because of Gödel's finding, the idea as such, to turn mathematical decision problems into problems of computability, had significance for the future: the approach that Hilbert had proposed for formal systems can in principle be carried out by a machine working purely mechanically. This became clear in the 1930s, when the mathematician Alan Turing developed the model of an idealized machine that routinely executes mathematical operations prescribed by a fixed system of rules. Initially, Turing's machine was a pure thought construction intended to formalize

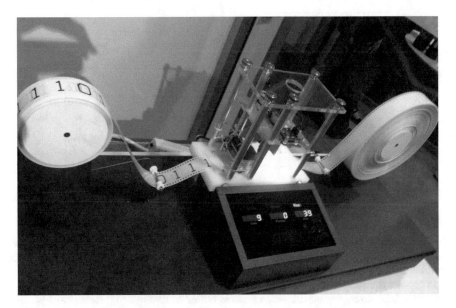

Fig. 5.2 A "real" Turing machine. In its simplest form, the Turing machine consists of three components: an indefinitely long "tape" divided into a series of adjacent squares, or cells, each of which contains a symbol; a "head", or monitor, that can read symbols on the tape or write symbols onto it; and a "state register" that stores the machine's "state" (of various possible ones). The machine is supplied with a "table" of instructions, i.e., what we today would call a computer program. Beside the one-tape device, shown here, there are also multi-tape machines. [Photo: Mike Davey, Wikipedia Commons]

mathematical reasoning. Only a few years ago was a "real" Turing machine constructed for illustrative purposes. It is shown in Fig. 5.2.

A Turing machine works like a modern digital computer, performing a computation based on a program (software) and some given initial information ("input"). Input and output are represented in "binary" coding, i.e. as a sequence of binary symbols (0 and 1). The operation principle of a Turing machine is explained in Fig. 5.3, which shows a short calculation of a binary sequence B (output) from a sequence A (input). After a certain number of calculation steps, which follow the instructions laid down in the program, the machine stops, and the result is printed out.

As the machine shown in Fig. 5.3 can only adopt two states, it can only carry out trivial computations. More complex Turing machines, in which the machine can adopt more than two states and that therefore have a greater

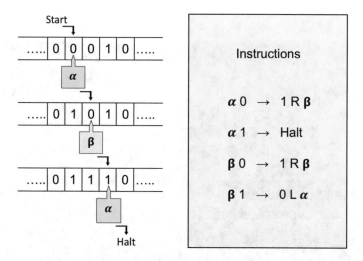

Fig. 5.3 Operational scheme of a Turing machine. The diagram shows a one-tape machine operating on an indefinitely long tape divided up into squares, each containing a binary symbol (0 or 1). The tape is read by a monitor. In each step, the monitor can read a symbol, or print one out, and it then moves one square to the left or the right. Moreover, the machine can possess either of two internal states (α or β). The "program" for movement of the monitor consists of a set of instructions; these determine, according to the four possible combinations of the monitor's position and the machine's state, in which direction (to the left, L, or right, R) the monitor is to move, which symbol (0 or 1) it is to print out and which internal state (α or β) the machine is to adopt. The illustration (based on [19]) shows a short calculation, in which the input "00010" is turned into the output "01110", and the program stops after two steps of calculation

spectrum of transitions rules, are *universal* in the sense that they can simulate any computer, and in particular those that are much larger and more complex than Turing machines are. The complete logical state of the larger machine is stored on the in/out tape, and every computational step of the larger machine is broken down into such small steps that it can be simulated by the smaller machine.

The Turing machine was initially designed as an abstract mathematical model of a calculating machine designed to work automatically. Later on, the mathematician John von Neumann developed a real computer architecture for the Turing machine, specifying the structure of the computer as well as the organization of the work processes running in it. The von Neumann computer, in turn, formed the basis for the development of the modern digital computer. However, while the von Neumann computer processes information sequentially, according to the model of the Turing machine, current developments in computer technology are focusing increasingly on parallel information processing. At present, the most important application of this is the technology of "neural networks". This computer architecture is designed to model cognitive functions of the brain (Sect. 6.7). On the basis of this technique, information can be processed much more efficiently than with the conventional von Neumann computer.

The importance of the Turing machine for basic research lies in the fact that it has led to an entirely new view on significant mathematical problems. For example, for a Turing machine, the mathematical problem of decision-making presents itself as a so-called "halting problem": if a mathematical problem can be solved, then a Turing machine must stop after a finite number of calculation steps and display the result of its calculation. If, on the other hand, the problem is unsolvable, the computation never comes to an end, because the Turing machine gets caught in an infinite calculation loop.

Problems of computability, as simulated on a Turing machine, are many and various. They are not only of interest for basic mathematical research; they also provide information on fundamental questions of the empirical sciences. In biology, for example, it was precisely the concept of computability that led to a precise definition of the idea of complexity and, subsequently, to profound insights into the possibilities and limits of objective knowledge ([18] and Sect. 5.4).

Since the problem of complexity is of central importance for understanding reality, let us take a closer look at it. First of all, we notice that everyday language uses the concept of complexity in a remarkably vague way, which tells us little about the actual essence of complexity. It is usually said that something is "complex" or "complicated" when it is opaque and obscure.

However, such intuitively used terms are of little use in the sciences. Instead, we here need objective and precise definitions. This requirement presupposes a long developmental process of scientific terms and theories, in the course of which the precise terms of scientific language gradually emerge. Thus, an expression in everyday language that at first has a broad meaning can disintegrate within the sciences into a spectrum of variants, each with its own precise meaning. In the case of "complexity", this process has not only given the idea of complexity an entirely different meaning in physics from in biology, but even within physics the term "complexity" has various meanings.

In physics, a well-known example of the concept of complexity is that of thermodynamic complexity, referring to the structural disorder of a physical system. Its measurand (that which is to be measured) is entropy (Sects. 5.5 and 5.6). In biology, on the other hand, the notion of complexity usually refers to the functional order of living matter. Here, at first glance, an appropriate measurand appears to be the information content of the genetic program. However, just as little as the sheer quantity of letters tells us anything about the actual complexity of a text, the sheer length of the genome tells us nothing about the real complexity of the life processes that it serves. Therefore, the functional complexity of the organism requires a measurand different from the number of the symbols that constitute its genetic information. However, as we shall see in Sect. 5.6, there is a primary connection between entropy and information, which reveals that information is also a fundamental concept in physics.

The search for an adequate measurand of complexity has led to a mathematical concept, which is referred to as the "algorithmic" theory of complexity (or of information) and which was developed around 1965 by Ray Solomonoff [20], Andrej Kolmogorov [21] and Gregory Chaitin [22], independently of each other. The idea underlying the algorithmic theory of complexity can be illustrated with a simple example. For this purpose, let us consider two sequences, each consisting only of the letters A and B. The sequences have the same length n (here $n = 40$).

The first is a more or less irregular sequence of A's and B's, as might be produced by repeatedly tossing a coin (heads = A, tails = B):

ABABBABBAABAABBBBBBBAABBAAAABAAABBBBBBBAB.

The other sequence is completely regular:

ABABABABABABABABABABABABABABABABABABABAB.

From an information-theoretical point of view, both sequences have the same decision content, i.e., at each of the forty positions one has to decide whether it is occupied by an A or a B. The decision content refers exclusively to the information that is necessary to determine the precise succession of the letters in an *a priori* unknown sequence. The decision content may equally well be termed the *structural information* of the sequence in question (Sect. 1.4).

The decision content is the same for all sequences of a given (same) length, regardless of whether the sequence carries a meaningful message or not. Nevertheless, there is a difference between the sequences shown here. The difference becomes even more obvious if we need to transmit them in the most efficient way (the shortest form) possible, for example to an astronaut who is taking a walk on the moon.

In the case of the irregular sequence, the succession of the letters A and B appears so disordered that it is hardly possible to find any significantly shortened representation of the sequence. Consequently, to communicate the series of letters to our astronaut, we have to transmit it in its full length.

The situation is entirely different in the case of the highly ordered sequence in which the letters A and B are strictly alternating. In that case, there is a simple algorithm by which the sequence can be generated. It is as follows: "x times AB", where x is the number of AB pairs contained in the sequence. The algorithm by which the totally ordered sequence can be generated keeps its simple form, irrespective of how many AB pairs there are. In other words: Ordered sequences are distinguished from unordered sequences by the fact that they can be compressed.

In the "language" of the Turing machine, we describe algorithms by a sequence of binary digits (0 or 1). Therefore, taking up the above consideration, we will call a sequence "complex" if it can only be generated by an algorithm that is (nearly) as complex as the sequence itself (Fig. 5.4). Thus, the complexity of a sequence depends on the degree of its incompressibility. In other words: In a complex sequence, the succession of the individual characters is so irregular that it appears to be random to a high degree.

At this point, however, we must avert a possible misunderstanding. The term "random" here refers solely to the pattern of a sequence. This kind of

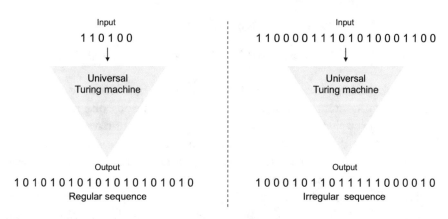

Fig. 5.4 The algorithmic definition of complexity. Left: A regular sequence can be generated by means of a Turing machine with an algorithm that is smaller than the sequence itself. Right: A random sequence cannot be compressed. It can only be generated with an algorithm that is as complex as the sequence itself. However, the transition from irregular (complex) to regular (non-complex) sequences is a gradual one. The algorithmic definition of complexity is an important tool to specify randomness, irregularity and disorder

randomness says nothing about the genesis of the sequence in question. This also applies in a converse sense to ordered sequences. Thus, as a result of repeatedly tossing a coin, even a highly ordered sequence could arise. This highly ordered sequence would have the same expectation probability as any one of the unordered sequences generated in the same way. Since, however, among all possible sequences of a given length the number of possible unordered sequences is many times greater than the number of possible ordered ones, unordered sequences will occur more frequently when sequences are produced by coin-tossing.

Furthermore, it is remarkable that according to the definition of "randomness" given above the transition from random (unordered) to non-random (ordered) sequences is smooth: the degree of incompressibility is a direct measure of the degree of randomness. The same holds for the terms "complex" and "not complex". Within the framework of algorithmic complexity theory, a phenomenon only appears to us inherently complex as long as we lack an algorithm to reduce its complexity. However, as soon as we discover a regularity in a complex phenomenon, the "complex" only proves to be "complicated", although it may be, at least in principle, transparent and thus understandable. In short, "regularity" is synonymous with "law-like": that which allows reduction of a complex phenomenon to something less complex.

The degree to which one can be mistaken in the search for a complexity-reducing algorithm is shown by the following example. The number series

$$3.141592653589793238462643383832\ldots$$

appears at first to be wholly disordered and does not seem to offer any rule for its (self-consistent) continuation or extension. Yet the mathematically trained reader will immediately see that this number series is the beginning of the irrational number π. For this number, however, there is indeed an algorithm by which the above sequence of numbers can be calculated and continued (with the help of computers) up to millions upon millions of digits. Thus, the number π does not make up a digit string of maximum complexity. At the same time, this example also shows that one cannot conclude from the apparent lack of order in a given pattern that in reality is completely—or even partly—disordered.

Whether an algorithm can be found for a given pattern of digits depends on the available resources, which include both human intelligence and time. No wonder, then, that almost every intelligence test includes a task in which the examinee has to continue correctly a given pattern of symbols within a given time. Incidentally, the "intelligence" of machines also depends on resources such as program size, running time and storage space.

It is challenging to discover a hidden pattern of order in a series of binary digits, but even more so—if one fails to find any pattern—to prove that there *is* no such pattern. In the first case, one has only to find a corresponding algorithm that is significantly shorter than the sequence in question. This may be difficult, but it is solvable at least in principle. In the second case, one has to prove that the sequence is random and thus of maximum complexity. However, according to a theorem formulated by Chaitin, any such proof of randomness is subject to the same limits that Gödel's incompleteness theorem imposes on mechanical proof in formal systems [22].

Chaitin's theorem is nothing other than the information-theoretical variant of Gödel's theorem. The starting points are also comparable. Both theorems are based on the Cretan paradox. While Gödel transformed this paradox into a statement about numbers, Chaitin converted it into a decision problem for a Turing machine. The proof then follows a standard mathematical procedure: The assertion to be proved (in this case, the theorem of randomness) is first transformed into its opposite assertion, and subsequently it is shown that the latter leads to a contradiction. The original statement is then considered proven.

This line of evidence is based on the logical principle of the excluded middle (*tertium non datur*). It says that either the statement A or its negation ("non-A") is valid, and there is no third possibility. Therefore, if one can demonstrate that the statement "non-A" leads to a contradiction, then statement A must be valid. It should be noted, however, that the above-mentioned controversy between the intuitionists and the formalists concerned, among other things, the question of whether the application of "tertium non datur"—which was doubted by the intuitionists—is a generally valid principle of mathematical reasoning.

The algorithmic concept of randomness is of great importance, because it reveals a fundamental limit of objective cognition in science. It demonstrates that it is impossible to classify something as inherently complex. This is because one can never exclude the possibility that there may be a law, a rule or a general pattern that can reduce the complexity in question. Consequently, there is no definitive answer to the question of whether anything of irreducible complexity exists at all. In other words, any complexity is to be regarded as complex only for as long as no algorithm is known that can reduce the complexity. Against this background, it makes sense to differentiate between the terms "complex" and "complicated" in an attempt to describe the difference between objective and subjective complexity.

5.4 Limits of Objective Knowledge

Complexity theory belongs to the branch of science that we have described as "structural science". A brief reminder: The structural sciences examine the structures of reality at its highest level of abstraction, i.e. detached from the concrete manifestations of these structures (Sect. 4.9). However, abstract structures must correspond to concrete, known objects, if the structural sciences are to have any relation to reality. In connection with complexity theory, for example, we must ask which structures of reality can be represented in the abstract form of binary strings.

Let us look at two examples from the realm of the natural sciences: A pure salt crystal is made up of two kinds of atoms, the negatively charged chloride ions (Cl^-) and the positively charged sodium ions (Na^+). Since carriers with the same electrical charge repel each other and attract those with the opposite charge, sodium ions and chloride ions build an alternating arrangement of the following type:

$$\ldots Na^+ Cl^- Na^+ Cl^- Na^+ Cl^- Na^+ Cl^- Na^+ Cl^- Na^+ Cl^- Na^+ Cl^- \ldots$$

which extends in three dimensions of space as shown in Fig. 1.3. In binary coding the lattice structure is represented by a periodic sequence of binary digits:

...1010101010101010101010101010101010101010101010101010101010...

Structures that are built up from a few basic elements according to a kind of crystal lattice can also be found in living Nature. The best examples are nucleic acid molecules. Like the salt crystal, these molecules consist of only a few basic building blocks, the nucleotides (see Fig. 1.7). However, since the components of nucleic acid molecules form only a one-dimensional lattice, they can fold in various ways and thereby take on a more complex and unique three-dimensional structure than a salt crystal does. Some examples are shown in Sects. 6.5 and 6.9. Moreover, the final structures of such molecules depend on the detailed sequence of their building blocks.

To recapitulate: The nucleic acids are built from four classes of nucleotides (Sect. 1.5) and in higher organisms can comprise several billion building blocks. In the case of ribonucleic acid (RNA), these are the nucleotides A, G, C and U. In this notation, the nucleotide sequence of an RNA molecule takes the form of a long sequence of letters, for example:

...G U U C U C C A A C G G U G C U A U U A U U C C G A A A C...

or in binary coding (with A = 00, U = 11, G = 01 und C = 10):

...0111111011101000001001011101101100111100111110100100000010...

The advantage of the binary coding is that it allows to establish a direct connection between genetic scripts and the theory of Turing machines.

In contrast to the salt crystal, the lattice of which is strictly periodic, genetic sequences seem to be more or less aperiodic, i.e. highly irregular. It is remarkable that the physicist Erwin Schrödinger [23] already spoke of the aperiodicity of chromosomes before the molecular structure of DNA was discovered. It should be noted that at that time the idea was still prevalent that the growth of the living organism took place in analogy to the growth of a crystal. Thus, it was supposed that the chromosomes instructed the formation of the living organism, just as a microscopically small crystal nucleates the growth of a crystal. According to this idea, the living organism appeared as the expression of the complex microstructure of the chromosomes, just as the crystal is the visible image of a tiny crystallization nucleus.

Schrödinger had realized that the conventional image of the regularly structured crystal does not correctly describe the complexity, diversity and individuality encountered in living matter. In contrast to a simple crystal, as Schrödinger argued, living matter cannot originate by the "dull" repetition of the same microstructure [23, p. 4 f.]; rather, it is the aperiodicity of the material composition on which the characteristic properties of living matter are based. Therefore, Schrödinger concluded, the chromosomes, as carriers of hereditary traits, cannot have the simple order of a "periodic" crystal. Instead, he suggested that the chromosomes must be regarded as "aperiodic" crystals. Only in this way, he believed, could the ordered diversity of living matter be conceivable.

Chromosomes, Schrödinger argued, "contain in some kind of code script the entire pattern of the individual's future development and of its functioning in the mature state" [23, p. 21]. Today we know that the "miniature code" anchored in the chromosomes is nothing other than the genetic program of the organism, written down in molecular language (Sect. 1.5). Moreover, Schrödinger's guess that the miniature code must have an aperiodic structure matches exactly the prerequisite which any language must fulfil, namely aperiodicity of its syntactic structure.

Indeed, any language—including the language of genes—presupposes this kind of aperiodicity. There is also a plausible reason for this. Language, as the fundamental instrument of communication, must be able to map the immense diversity and richness of reality appropriately. This, in turn, requires that language should have the potential to generate an almost unlimited variety of linguistic expressions from a finite stock of symbols. The more such expressions can be formed, the more effective language will be as a tool of communication.

Be that as it may, mathematical analysis already reveals that the possible number of aperiodic sequences is in any case many times greater than that of periodic sequences. A numerical example demonstrates this: For simplicity, we consider a series of binary digits and determine the proportion of aperiodic sequences within all conceivable sequences of a given length n. To do so, however, we must bear in mind that the transition from periodic to aperiodic sequences is a gradual one. Thus, it is first necessary to define the degree of complexity above which a series of binary digits is to be classified as aperiodic. So let us denote all binary sequences that cannot be compressed by more than ten binary numbers as aperiodic. A simple counting procedure then shows that of all binary sequences of length n only every thousandth can be compressed by more than ten binary digits (for details see [24, p. 46 ff.]). In other words, almost all binary sequences are aperiodic. This means that for sequences of given length the number of periodic (or ordered) sequences is vanishingly small compared with the number of aperiodic ones. Consequently, nearly all binary sequences are *de facto* aperiodic.

Against this background, it becomes evident that periodic sequences are unsuitable for encoding any meaningful information. Just as little as a poem by Goethe can be written down as a regular series of letters, the richness and and complexity of a living system can be never encoded in a regular nucleotide sequence of genetic script. In general, the higher the degree of the periodicity of an information-carrying sequence of signs is, the less complex and extensive is its information content.

Furthermore, the fact that any meaningful sentence is subject to the constraint of aperiodicity of its syntactic structure implies that such sentences cannot be compressed without changing their content, even if only slightly. This becomes clear, for example, when we receive a message in "telegram style". This restriction is an essential and at the same time momentous aspect of semantic information. Although the rules of grammar, syntax and the like impart a structure to linguistic expressions, their aperiodicity as such is not abolished by this. Instead, the rules of language merely narrow down the set of aperiodic sequences that can be built up from a given alphabet.

At this point, it should again be pointed out that the classification of a sequence as aperiodic always represents a preliminary state of cognition, which cannot be considered as valid for all time. Thus, it could well be that behind a given sequence of letters, for example, a poem by Goethe, there is a hidden algorithm that nobody has yet discovered. Although this possibility can be shown to have a probability bordering on zero, the non-existence of such an algorithm can never be proven.

Comparable considerations apply to the language of genes. Here too, one may ask whether some law-like regularity might be detected in the genetic programs of organisms. This question was at the center of the book *Chance and Necessity* by the molecular biologist Jacques Monod [25]. However, Monod opposed vehemently any explanation of life based on life-specific forces, laws, principles or the like. Instead, he claimed, life on Earth had its origin in a unique chance event, which has happened only once in the cosmos and will never be repeated. Provocatively, he even spoke of the "law" of chance, which had completely dominated the origin and the evolution of life. On the basis of the chance hypothesis, Monod developed a world-view in which man is depicted as a "gypsy" living at "the boundary of an alien world" and whose existence has neither meaning nor purpose [25, p. 172 f.].

Monod justified his hypothesis by appeal to the structures of biological macromolecules. As primary carriers of genetic information and physiological function, nucleic acids and proteins epitomize the essence of living matter. Monod, therefore, believed that the search of life-specific laws must focus on the sequence pattern of the macromolecules, because these patterns—together with the physical and chemical environment—determine the molecules' biological properties. Consequently, the origin of the sequence patterns of biological macromolecules was the point of departure of Monod's analysis. From this, he finally concluded that the known sequence patterns show no evidence for the presence of law-like or goal-directed behavior of living matter. He considered this finding to be the proof that genetic information and biological function emerged from a purely random synthesis of their building blocks.

With the hypothesis of the random origin of life, Monod wanted to combat all teleological world-views that purported to trace the existence of life back to some purposeful design. From an epistemological point of view, however, Monod's hypothesis is just as unsatisfying as the theories which it was supposed to refute. Algorithmic complexity theory immediately makes it clear that the randomness of the sequence pattern of biological macro-molecules—in contrast to Monod's claim—does *not* allow any conclusions about their origin [18]. Not even the randomness as such can be inferred conclusively from such patterns. Moreover, any statement that assigns to chance the rank of law is unscientific, not only because it misuses the concept of law, but also because it perverts the task of science—i.e., to find rigorous explanations on the basis of genuine laws. Finally, modern research into the evolution of biological macromolecules demonstrates that the origin of life is based on *both* chance *and* the law-like behavior of matter (Sect. 6.9).

Let us now consider the teleological hypotheses on which Monod's criticism was focused. A common feature of all these hypotheses is that they assert that the material organization of living matter is the result of a goal-directed principle. However, there are quite different ideas as to what this principle is. They range from the assumption of life-specific forces and laws ("vitalism") to the idea that all objects and phenomena of Nature, and even the universe itself, are animated ("animism"). However, the talk of such principles always remains vague as long as these do not have any tangible meaning. Nevertheless, one may try to solidify the core of the teleological argument, namely the existence of an overarching principle controlling the origin and evolution of life.

To do so, we will focus on the idea of life-specific law. As in the case of the chance hypothesis, we can clarify the epistemological status of this idea by applying the theory of algorithmic compressibility [18]. This has the advantage that it allows a straightforward interpretation of the idea of law-like behavior without necessarily referring to the traditional idea of laws of Nature.

Let us consider, for example, the genetic blueprint of a living organism. It can be represented as a binary series. If such a sequence is considered to be the result of a life-specific law, then it must be compressible. Otherwise, the binary series would be a random one. We already know, however, that the non-existence of a compact algorithm cannot be proven. Consequently, theories assuming the existence of life-specific laws cannot be refuted. This makes it incumbent upon those who peddle the existence of such laws to demonstrate the validity of even a single one—a demonstration that so far is lacking.

Thus, neither the chance hypothesis nor the teleological hypotheses yield any solution to the problem of the origin and evolution of life. The chance hypothesis is unsatisfactory because it traces biological evolution exclusively back to the action of blind chance and thus transforms the origin of life, as Monod himself pointed out, into a kind of lottery. The teleological hypotheses are inacceptable because they are immune against all attempts to refute them. In this respect, teleological explanations violate the demarcation criterion that separates science from pseudo-science (Sect. 2.7). For this reason, teleological hypotheses have survived to this day, despite the enormous advances and successes of the modern life sciences, which do not require or make use of them.

The above considerations rest on an information-theoretical understanding of the essence of a scientific law. It makes use of the fact that, in principle, all empirical data can be represented by binary sequences. From the viewpoint of algorithmic complexity theory, a set of data follows a rule or a law if the data can be compressed, and thus simplified, by a compact algorithm. Thus, from an information-theoretical point of view, scientific laws are algorithms that allow the compression of empirical data.

Even though the algorithmic approach leads only to an operational understanding of what we call a scientific law, it nevertheless highlights a significant aspect of scientific thinking that the physicist and philosopher Ernst Mach once called the "economy of thought" [26, p. xi]. Accordingly, the objective of scientific research is to be regarded "as a minimal problem, consisting of the completest possible presentment of facts with the *least possible expenditure of thought*" [26, p. 490].

The principle of economy of thought is also known in the philosophical literature as "Ockham's razor". This term alludes to the Franciscan monk William of Ockham, who in the fourteenth century, in connection with the so-called "problem of universals", demanded that one should not make unnecessary assumptions in explaining reality; rather, the number of explanatory assumptions should be pared down with, as it were, a sharp razor, and everything should be left aside that is not strictly needed for the explanation.

This economy of thought aims at simplicity of scientific representation and thus belongs to those guiding principles that are decisive for the unification of scientific theories. Indeed, science regards it as the hallmark of a successful concept if this can explain more facts with the help of simpler laws than a competing concept can. If one continues the argument, the question arises as to what a final, self-contained scientific description of the world might look like. Although this question hardly seems amenable to rigorous analysis, an astonishingly simple answer can be given on the basis of complexity theory [18, chapter 10].

For this purpose, one has to complete consistently the information-theoretical interpretation of the concept of natural law. This leads us to the following consideration: A comprehensive physical theory of the world, called "world formula" for short, must find its conclusion in a universal theory which—in principle—describes all conceivable physical phenomena. Maybe, for example, such a theory will one day emerge from the fusion of gravitation theory and quantum theory.

Let us assume in a thought experiment that we already have a final and complete theory describing all the phenomena of physics—those of macro-physics as well as those of microphysics, including quantum physics. From a formal point of view, the world formula would be the most compact algorithm possible that could allow calculation of everything that is of physical interest. Such a universal algorithm would come close to the Universal Mind envisaged by Laplace at the climax of the mechanical age (Sect. 3.7).

As the final, most compact and most straightforward algorithm of the physical world description, the world formula would be not compressible any further. However, apart from the fact that we can never attain any certainty as to whether we are actually in possession of such a universal theory, or whether it would bring research in physics to an end, the above analysis reveals a highly peculiar result: the sought-for world formula would, by definition, be the most compact algorithm possible for a physical explanation of the world.

It could be used to derive any physical regularity. At the same time, however, the world formula itself would be the epitome of irreducibility, complexity, and randomness—how paradoxical that result would be!

5.5 Circular Reasoning

Wherever we encounter paradoxes, the limits of our cognition become visible. Yet paradoxes are by no means the only boundaries of human thought. Vicious circles form another barrier. This is the case when a statement that is to be justified is already presupposed in the argument.

A typical example is the hermeneutic circle, according to which one can only understand something if one has already understood something else (Sect. 3.4). At first glance, one could argue that the hermeneutic circle is insofar unproblematic as it merely documents the relativity of all understanding. However, the hermeneutic circle is not intended to justify the relativism of all understanding, but rather the exact opposite: to document the fact that real understanding is absolute and precedes any relative interpretation of the world.

In the common interpretation of philosophical hermeneutics, the process of understanding is not seen as an open chain of acts of comprehension that continues endlessly in its relative frame of reference, but as a closed circle that forms the immovable horizon of all understanding. However, if the act of understanding refers back to the understanding subject itself, then the hermeneutic circle would indeed become a vicious one. Philosophical hermeneutics tries to avoid this by giving the mode of understanding some ontological background and by conceiving of it as an expression of our "being-in-the-world" per se. Hans-Georg Gadamer, one of the founders of philosophical hermeneutics, calls this aspect the "fusion of horizons" in the act of understanding [27, p. 317].

The fact that any form of understanding is itself embedded in a superordinate context of understanding seems to turn each understanding, inevitably, into a vicious circle. An illustrative example of this is the central thesis of Sect. 7.4, "Only knowledge can control knowledge", which refers to the responsibility of science. At first glance, this thesis seems to be a circular statement. However, that will only be the case if the knowledge that is to be controlled is already supposed to be complete and definitive. In reality, however, human knowledge is never complete; it is only provisional and can continuously be expanded and corrected, thereby renewing itself without falling into a vicious circle of self-control. A loss of control only occurs if knowledge is considered to be absolute.

One can avoid such problems by opening up a new and superordinate level of reasoning for the item that is to be justified. The rational justification of the concept of rationality is an example of this. Since rationality cannot be justified on the basis of itself, one needs an external reference frame providing the context of reasoning. Very often, the context of justification is the practical level at which basic terms of our intellectual world, such as knowledge (Sect. 1.1), truth (Sect. 2.9), meaning (Sect. 6.8) and others become connected with the meaningful and successful actions of man.

Nicholas Rescher [28], for example, states at the beginning of his book *Rationality* a definition according to which rationality consists of the intelligent pursuit of appropriate goals (Sect. 4.10). Following this definition, one could also concede that animals possess something like rationality, because animals can certainly pursue goals efficiently. If we want to exclude this interpretation of the concept of rationality, we will have to tighten up the reference system, and make it more precise.

To do this, we might consider an "intelligent" action not merely from the general point of view of its efficiency. Instead, we can include in the notion of "intelligent" the presence of convictions, the assessment and evaluation of options for action, the selection and definition of appropriate goals and the like—aspects that cannot readily be supposed to be present in animals. However, this opens up further levels of reasoning, which now refers to the concept of intelligence. The example underlines our statement above, that terms are always embedded in a complex hierarchy of terms. It shows once more that the content and scope of a term are always context-dependent (Sect. 1.9).

Circular reasoning, however, is not only a problem of the definition of terms. It also occurs in connection with the clarification of important questions concerning the foundations of the sciences. The most impressive example of this comes from physics and refers to the justification of the structure of time. Let us examine this in detail, and look at some ideas on how to solve it.

We experience temporality as an asymmetry of the course of natural events. The past we perceive as factuality and the future as potentiality, with the present as the point of transition from the possible to the factual. Without structuring temporality into time modes "before" and "after", generalizations such as "thunder follows lightning" could not be made at all. Nonetheless, perceived causal relationships make up the basis of all experience.

In contrast, a temporally structured reality does not seem to be mandatory for our understanding of the physical world. Physical laws in which the difference between the past and the future is nullified include, for example, the laws of mechanics. Here, time has only the character of a parameter that functions as a mere counting measure. Since the mechanical laws are invariant with respect to the reversal of time, they can be used to calculate from a current state not only future states, but also all past states. This is precisely what enabled the Laplacian demon to display his clairvoyant abilities in the first place. At the same time, however, the strength of the mechanical laws is also their weakness, since the mechanical laws as such cannot explain the direction of the natural course of events.

On the other hand, everything points to temporality as a property of natural processes that is present *a priori*. This essential experience is expressed in physics by the second law of thermodynamics, which states that in Nature only those processes take place spontaneously in which entropy increases (Sect. 4.6). Assuming that the second law can also be applied to changes in the universe itself, it would become a universal law of development and possibly the key to an understanding of the structure of time.

This is not the place to examine the difficult question of whether and to what extent thermodynamics can actually be applied to cosmic dimensions. However, if we ascribe the irreversibility of natural events to an objectively given time-structure, then this has an inevitable consequence for physics: it is the reversibility of physical theories that demands an explanation, while irreversible theory is precisely what one would expect.

In fact, it can be shown that the asymmetry of time implies reversibility of the fundamental theories of physics. This results from a basic principle of quantum field theory, known as the TCP theorem and discovered by Gerhard Lüders and Wolfgang Pauli in the 1950s. According to the TCP theorem, the causality of natural processes would be violated if it were not possible to convert a natural process by the combined application of the following three operations into another possible process: (1) reversal of all motion (time, T), (2) reversal of all charges (C), (3) spatial mirroring about a point (parity, P). This, in turn, means that the compatibility between the elementary natural laws on the one hand and time-structured reality (including the causality principle upon which it rests) on the other comes at the price of "symmetrically swapping" the past and the future with one another. In other words: The causality principle enforces, owing to the TCP invariance, the reversibility of the fundamental laws of physics.

With this recognition, the focus of the physical justification of time's structure is shifted back to the second law of thermodynamics. The second law, however, was derived from experience. Yet any experience presupposes the validity of the causal principle which, for its part, presupposes what the second law seems to verify, namely the structuring of time into past, present and future. Obviously, a circular argument is emerging here.

Despite this dilemma, there have been repeated efforts to find an adequate explanation for the structure of time. Best known is the sophisticated attempt by Ludwig Boltzmann to derive the entropy law from the laws of statistical mechanics, and thereby also to justify the structure of time (for details see [24, chapter 7]). Although the mechanical laws themselves are reversible, Boltzmann finally came up with an elegant way of circumventing the problem of mechanical reversibility by introducing the concept of "exceptional" initial conditions [29].

According to Boltzmann's model, the entropy of the universe increases because, in the early phase of its creation, the world was still far from a state of equilibrium, i.e. a state of maximum entropy. The increase in entropy, which determines the direction of time, could then be explained by the approach of the universe to its "normal" state, namely the state of equilibrium. When all deviations from the equilibrium state are balanced, so that equilibrium is reached, the directionality—and thus the temporality—of all events would be nullified.

Boltzmann pursued the idea that irreversibility is a consequence of the particular conditions under which the laws of mechanics work in Nature. This approach underlines once more the central thesis of Sect. 4.8, according to which the initial conditions give the natural laws their relation to reality in the first place. In the present case, the initial conditions cause the natural laws to imprint a structure upon reality, a structure that we experience as the direction of time. The initial conditions are "exceptional" in so far as they carry information about the early state of the universe.

Alongside the "initial-state hypothesis", Boltzmann suggested an alternative scenario for the universe, the so-called "fluctuation hypothesis". While the initial-state hypothesis assumes that the world emerged from a highly improbable initial state and has since been moving toward thermodynamic equilibrium, the fluctuation hypothesis starts from the equilibrium state. Accordingly, the universe is already in a state of balance, while our "local" world, in which we live, is at present in an entropy fluctuation—a huge one, but locally restricted. The dissipation of this fluctuation, according to Boltzmann, then also leads to an increase in entropy, which we perceive as an asymmetry of time.

The difference between the two hypotheses is illustrated in Fig. 5.5, which shows the (hypothetical) course of the entropy of the universe. The initial-state hypothesis provides a foundation for the structure of time by the

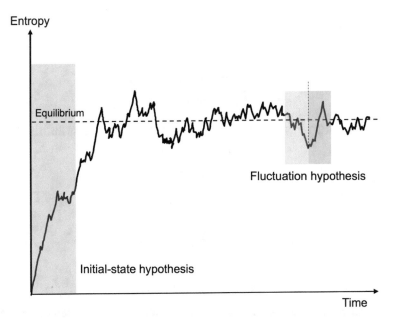

Fig. 5.5 Hypothetical time course of entropy in the universe. Applying thermodynamics to the universe, the "initial-state hypothesis" assumes that the development of the universe began under exceptional initial conditions and has since been—as it is now—approaching thermodynamic equilibrium, i.e. the state of maximum entropy. The increase in entropy determines the direction of time. Thus, in the equilibrium state, the passage of time as we perceive it would be annulled. However, the statistical interpretation of the entropy function allows an alternative explanation of time's structure: According to the "fluctuation hypothesis" we are living in the wake of a large fluctuation in the universe's entropy, and time's structure results from the tendency to restore thermodynamic equilibrium. Both hypotheses, proposed by Ludwig Boltzmann, correlate the structure of time with an increase in entropy

assumption of exceptional starting conditions. The fluctuation hypothesis, in contrast, attempts to relate the structure of time to deviations from a maximum of entropy that has arisen because of statistical fluctuations. In both cases, the universe is considered as approaching thermodynamic equilibrium, either from an initial distant state (initial-state hypothesis) or after a large fluctuation away from equilibrium (fluctuation hypothesis).

Let us look at the fluctuation hypothesis in more detail (Fig. 5.6). In this case, the universe is regarded as an isolated system, completely symmetrical in time and space. This means that there are no preferred directions in time or space. Moreover, it is presumed that the universe is in thermodynamic equilibrium, that is, in a state of greatest possible disorder. However, there are fluctuations around the equilibrium state—similar to the fluctuations that

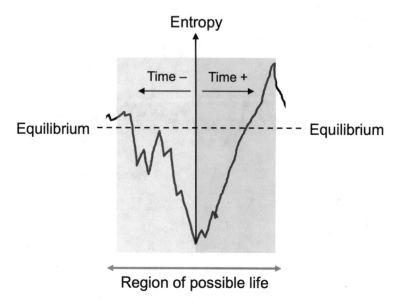

Fig. 5.6 Boltzmann's fluctuation hypothesis. The fluctuation hypothesis is an alternative to the initial-state hypothesis. It derives the structure of time from a gigantic, but local entropy fluctuation of the universe. The model assumes that the universe is in thermodynamic equilibrium and completely symmetrical in time and space. In this case, the probability that we find ourselves in a phase of decreasing entropy (left flank of the fluctuation) is just as great as that of our living in a stage of increasing entropy (right flank). The time structures correlated with the change in entropy behave like mirror images of each other. Boltzmann concluded from this that the definition of "past" and "future" is based on mere convention

occur in a chemical equilibrium (see also Fig. 3.10). Thus, there are regions of time and space in which the degree of disorder deviates from its greatest possible value. Given the cosmic vastness of time and space, it seems entirely possible, as Boltzmann argued, that entropy fluctuations in the universe may occur on the scale of entire stellar systems. Fluctuations of this huge magnitude Boltzmann denoted as "individual worlds".

Yet how are we to explain the fact that precisely we humans inhabit a time-structured "individual world" and that we can perceive the passage of time? An immediate answer would be that life is in any case only possible in a world that is far from thermodynamic equilibrium. Only under those conditions can the differences between states of matter be great enough to allow life to arise. Such a situation is given in the two flanks of a huge entropy fluctuation. Seen in this way, it would be no great surprise to discover that we live in just such an individual world.

The entropy curve, sketched in Fig. 5.6, is symmetrical. Consequently, the probability that we find ourselves in a phase of decreasing entropy is just as great as that of our living in a stage of increasing entropy. However, either of the two processes—although they run in opposite directions—can define the direction of time. Therefore, we may justifiably wonder, how can one then speak of a past and a future at all? Boltzmann countered this objection with the argument that the definition of the direction of the time would be a mere matter of consensus: "For the universe, the two directions of time are indistinguishable, just as in space there is no 'up' or 'down'. However, at any point on the earth's surface, we call the direction toward the center of the earth 'down'; in just the same way, an organism that happens to be in a particular time-phase of an individual world will define the direction of time as being the one that is opposed to less probable states rather than more probable ones (the former would be the past or the beginning, the latter would be the future or the end)." [3, p. 396; author's transl.]

Let us return once more to the initial-state hypothesis. As mentioned, the model posits that at its inception the universe was in a state of minimum entropy, i.e., in a physical state that was far from thermodynamic equilibrium. From this, Boltzmann concluded that the entropy of the universe must be steadily increasing and will continue to do so until thermodynamic equilibrium is reached, as predicted by the second law. Accordingly, the structuring of time into past, present and future would be a direct consequence of the irreversible development of the universe.

Thus, the initial-state hypothesis rests fundamentally upon the assumption that the universe "is still very young" [30, p. 140; author's transl.]. Only under this precondition could the universe have developed in the way described by Boltzmann. However, the assertion that the universe has developed at all can only be verified through its physical footprints. In fact, there are numerous examples of such footprints, which allow conclusions on the history of the universe. A well-known example is the cosmic background radiation, which can be interpreted as the electromagnetic "echo" of the birth of the universe (the "big bang"). However, such observations are only significant as records of the development of the world if we also presuppose that we can differentiate objectively between *actual* events and *merely possible* events. However, this precondition is precisely what the initial-state hypothesis was supposed to justify in the first place.

It seems impossible to provide an ultimate justification of temporality within the framework of physics. This leads to a dilemma. On the one hand, we experience temporality in the direction of natural events, expressed in physics by the second law of thermodynamics. On the other hand, there is no doubt that any experience already presupposes the structuring of time. For Immanuel Kant, who penetrated the presuppositions of human perception to a depth that no other philosopher had attained, time belongs to the *a priori* forms of human intuition through which we perceive the world and which make any experience, as such, possible at all (Sect. 2.5).

We come up here against the limits of physical explanation, because ultimate justifications are no longer possible without running into circular reasoning. The only way to escape this dilemma is to test the models for their consistency. The logical basis of a consistency proof can be described as follows. If two statements A and B are logically equivalent, then their consistency means that one can deduce both B from A and also A from B. This conclusion in itself is not yet circular. It becomes a vicious circle only in the case that one (wrongly) claims that the truth of A (or of B) has been proven by this procedure. It is rather the case that verification of consistency can only lead to the weaker conclusion that *either* both A and B are true *or* both are false.

Concerning the structure of time, one has to demonstrate that, given the validity of the second law, time can have the structure that it is supposed to have. The physicist Carl Friedrich von Weizsäcker, for example, attempted to establish such consistency by sharpening the concept of probability [31]. He argued that the theory of probability could only be applied meaningfully to the future, i.e. "possible" events. In contrast, it would be meaningless to ask about the likelihood of "factual", i.e. past, events.

On the basis of this thought, von Weizsäcker believed that one could reconcile the irreversibility of natural processes with the time-symmetric laws of mechanics. Therefore, he proposed that in the statistical interpretation of entropy, the concept of probability should only be employed for the calculation of *actual* transitions—that is, transitions toward the future. From this, it follows at first that there will be an increase in entropy in the future. However, since every past moment was once the present, the increase in entropy follows for everything that at that time belonged to the future—that is, also for a time-span that today belongs to the past. Accordingly, the asymmetric application of the probability concept guarantees the consistency of the second law of thermodynamics with the laws of reversible mechanics.

5.6 The Impossibility of a Perpetual Motion Machine

There are many limitations of science that result from its inability to cross the boundaries of technical feasibility. However, we are not referring here to limitations that result from technical deficiencies in state-of-the-art science and technology, but rather to limits that are set by the laws of Nature themselves and thus cannot be broken by human action. Even though we can use the laws of Nature to realize technological goals, we cannot change the scope of the laws of Nature through our actions. This has various reasons. Technical feasibility, for example, is limited from the start by natural constants such as like the velocity of light or the absolute zero—i.e., the lowest attainable value—of temperature.

In this section, we will focus on the laws themselves. At the outset, it is important to emphasize that the laws of Nature are not uniform in character. Rather, they can be divided into two classes, universal and special (or systemic) laws. The universal laws, like the law of gravity, are neither spatially nor temporally limited, nor are they bound to particular forms of matter.

By contrast, special laws, such as Kepler's laws of planetary motion, only apply under certain conditions. Although special laws are a consequence of universal laws, they are restricted to particular configurations of matter. For example, Kepler's laws apply to a system of planets revolving around a central star. On condition that the masses of the planets are small compared with that of the central star, and that the planets do not come too close to each other in their orbits, Kepler's laws can be derived from the law of gravity. The derivation follows the basic pattern of a physical explanation, whereby in this case the explanandum is not an event, but a special (i.e. Kepler's) law that can be deduced from the universal laws if certain boundary conditions are specified (for details see [32, p. 306 ff.]).

Another example of a special law is Ohm's law, which states the relationship between voltage, current and resistance in an electric circuit (Sect. 3.9). It applies only to systems consisting of a voltage source and an electrical conductor. Using our earlier turn of phrase, we can also say that in such an electrical circuit matter is subject to certain boundary conditions—forms of its organization which confer upon the general laws of electrodynamics the particular form of Ohm's law.

Let us now turn our attention to the fundamental laws of physics, in particular to those laws that set limits upon technical feasibility. To these belong the first and second laws of thermodynamics. The first law postulates the conservation of energy, the second law the increase of entropy. Both laws differ from other fundamental laws of physics in that they are "exclusion laws", i.e., laws that rule out the construction of a certain kind of machine, called a perpetual motion machine.

The term "perpetual motion machine of the first kind" is used to describe a machine that keeps itself in constant motion without consuming energy (Fig. 5.7). The experience that such a device cannot be built is the subject of the first law of thermodynamics, which posits the conservation of energy: energy cannot be generated from nothing and it cannot be annihilated.

The "perpetual motion machine of the second kind" is also an exotic machine, of which we again can say that our experience shows it to be impossible. For example, it is not possible to build an ocean liner that sails across the world's oceans and draws the necessary energy from the water by extracting heat from it. Or, in general terms: It is impossible to build a machine that produces power by cooling its environment down. Such a device would violate the second law.

In the end, the idea of a perpetual motion machine leads to the metaphysical question of whether any "something" can be created out of "nothing" (*creatio ex nihilo*) at all. This issue already occupied the philosophers of antiquity. Epicurus, an antagonist of Plato and his school, had argued that nothing could arise out of nothing and nothing could become nothing.

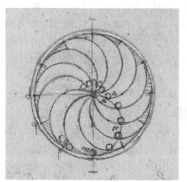 

Fig. 5.7 Drawing of a perpetual motion machine by Leonardo da Vinci. Through the centuries, man has tried to realize the dream of a perpetual motion machine. In the Renaissance, various attempts were made to build such a device. The figure shows a draft sketch of a perpetual motion machine by Leonardo da Vinci (left) and a 3D illustration (right) based on his sketches. Leonardo da Vinci was presumably the first to recognize that a perpetual motion machine cannot be realized. [Left: Victoria and Albert Museum, London. Right: Science Photo Library/Alamy Stock Photo]

Accordingly, the law of conservation of energy almost seems like a modern confirmation of Epicurus' idea.

With the impossibility of building a perpetual motion machine of the first or the second kind, two principles of experience were formulated in physics, each showing an insurmountable limit placed upon technical feasibility. It took some time for scientists to realize that two fundamental laws of Nature are hidden behind these principles, and that the numerous attempts to build a real, physical perpetual motion machine had been doomed to failure from the outset.

Even if the idea of machines that run indefinitely could not be realized, it nevertheless inspired, above all, the creativity of engineers of the mechanical age. Thus, at the end of the eighteenth century, much effort was dedicated to putting the idea of self-moving and self-controlling machines into practice. The unbeatable pioneer in this field was Jacques de Vaucanson. He developed, among other things, the first fully automatic loom, which—anticipating modern computer technology—could be programmed with punch cards. Above all, Vaucanson became famous for his quirky constructions such as the mechanical flute player or the artificial duck. The latter could not only cackle and flap its wings, but it could also take in food, and it even possessed an artificial gastrointestinal tract. When in 1770 the Swiss precision mechanic Abraham Louis Perrelet succeeded in constructing a pocket watch with an automatic winding mechanism (*montre perpétuelle*), one seemed to be very close to the goal of permanent self-movement.

The fact that all attempts to build a perpetual motion machine have failed bears out the validity of the second law. Nonetheless, it must be emphasized once again that the second law is based exclusively on experience. It only tells us what kinds of processes are *not* possible. Since it cannot be derived from classical physical theories, its true nature has always remained somewhat unclear. Nevertheless, there have been various attempts to clarify the scope and the possible limits of the second law. This also includes a thought experiment that was developed by the physicist James Clark Maxwell in 1871 and which later acquired the name "Maxwell's demon".

Maxwell believed that the second law of thermodynamics is only in accordance with our experience as long we consider macroscopic systems consisting of vast numbers of molecules. Here, deviations from the second law would not be visible to us, which invokes the impression that the second law has unlimited validity. In this way he drew attention to the possibility that the second law might appear in a new light if one were able to advance to a detailed, physical molecular description of the highest conceivable precision.

Since Maxwell's thinking was deeply rooted in the mechanical paradigm of the nineteenth century, he illustrated this thought by the idea of a hypothetical being—very similar the Laplacian demon—that was able to observe and measure the position and velocity of every single molecule.

At the end of his book on the *Theory of Heat*, Maxwell wrote: "One of the best established facts in thermodynamics is that it is impossible in a system enclosed in an envelope which permits neither change of volume nor passage of heat, and in which both the temperature and the pressure are everywhere the same, to produce any inequality of temperature or of pressure without the expenditure of work. This is the second law of thermodynamics, and it is undoubtedly true as long as we can deal with bodies only in mass, and have no power of perceiving or handling the separate molecules of which they are made up. But if we conceive a being whose faculties are so sharpened that he can follow every molecule in its course, such a being, whose attributes are still as essentially finite as our own, would be able to do what is at present impossible to us. For we have seen that the molecules in a vessel full of air at uniform temperature are moving with velocities by no means uniform, though the mean velocity of any great number of them, arbitrarily selected, is almost exactly uniform. Now let us suppose that such a vessel is divided into two portions, A and B, by a division in which there is a small hole, and that a being, who can see the individual molecules, opens and closes this hole, so as to allow only the swifter molecules to pass from A to B, and only the swifter molecules to pass from A to B, and only the slower ones to pass from B to A. He will thus, without expenditure of work, raise the temperature of B and lower that of A, in contradiction to the second law of thermodynamics." [33, p. 308 f.]

Let us briefly summarize the basic features of Maxwell's thought experiment (Fig. 5.8). The system under consideration is thermodynamically isolated, i.e., it exchanges neither matter nor energy with its surroundings. To begin with, the system is in thermal equilibrium. By unmixing "hot" and "cold" molecules, the demon generates, step by step, an ever-increasing temperature gradient in the system; this does not involve his doing any work, as the sliding door operates without friction. Since the whole process is accompanied by a decrease of entropy, the second law is violated. The actions of the demon induce a systematic deviation from equilibrium—in contrast to random statistical fluctuations, referred to in the preceding section.

As a fictional being, Maxwell's demon has no material properties. Any other information processing device could replace him. In other words, Maxwell's thought experiment is in fact a blueprint for a perpetual motion machine of the second kind, a machine that can operate as a thermal engine

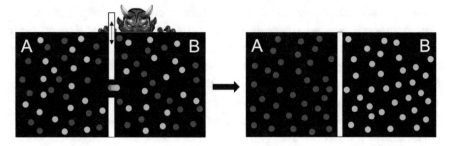

Fig. 5.8 Maxwell's demon is a thought experiment questioning the universal validity of the second law of thermodynamics. Two chambers A and B, which are separated by a partition, contain equal numbers of gas molecules of the same kind, which are in thermodynamic equilibrium at some given temperature. The higher the average energy of the molecules, the higher the temperature of the gas. For simplicity, let us denote all molecules with a velocity higher than the average value as "hot" (red spheres) and all with a lower velocity as "cold" (blue spheres). Moreover, there is a small hole in the partition that can be opened and closed by means of a frictionless sliding door. A hypothetical intelligence (the "demon") observes and measures the exact position and velocity of every single molecule (left). Moreover, he can sort the particles according to their speed (or "temperature") by opening the slider for "cold" molecules only if they are moving from A to B and for "hot" molecules only if they are moving from B to A. All other movements between A and B he blocks by closing the door. Over time, the "hot" molecules become concentrated in compartment A and the "cold" molecules in compartment B (right). [Demon: © Dreamstime.com]

without dissipating energy. The heat reservoir needed to make the engine work results from the action of the demon, who separates "hot" from "cold" molecules.

It is clear that Maxwell's demon possesses his supernatural capability only because he has information about the molecules, and this information allows him to get round the second law. The physicist Leó Szilárd was the first person who scrutinized this aspect of Maxwell's thought experiment [34]. He concluded that the amount of information that Maxwell's demon had to invest in order to circumvent the second law is exactly equal the decrease of entropy caused by the demon's manipulations. Thus, the entropy balance is maintained by the compensation of *entropy loss* through *information gain*, the latter denoted as "negentropy" (cf. [35]). The second law remains fully valid.

Before Szilárd presented his solution of "Maxwell's demon", Ludwig Boltzmann had already pointed out that entropy and information are closely related to each other (Sect. 3.6). Boltzmann considered entropy to be a measure of the loss of information that occurs if one replaces the detailed description of all molecules in an equilibrated system by their average kinetic energy [36]. Statistical mechanics, in turn, teaches us that the temperature of an ideal gas is directly proportional to the kinetic energy of the gas molecules. Thus, temperature is an *intensive* quantity. Its value does not depend upon the actual size of the system (the number of gas molecules), but only upon their *average* energy. Thus, there must be another quantity, complementary to temperature, that is *extensive*—one that does depend on the extent or amount of gas present; otherwise our knowledge of the total thermal energy of the system would be incomplete. This extensive quantity is the entropy.

The entropy function thus provides us, as Boltzmann recognized, with a measure of the loss of information that occurs in statistical physics when one forgoes exact knowledge of the key thermodynamic data of each particle within the statistical ensemble. Boltzmann was thus the first to set out in detail the dependence of scientific knowledge upon the means used to describe, and to express abstractly, the system being considered.

In Maxwell's time it was not foreseeable that his thought experiment would one day introduce information as a new concept into physics. Even more, beside matter and energy, information has nowadays become the third pillar of physics. This has brought about a radical change of thinking in science, and it can be expected that this will open up new ways of looking at and solving persistent problems. It is possible that the strange phenomena of quantum physics, which have bothered physicists for a long time, may one day find a firm grounding in the concept of information—in the same way as information has now become the guiding principle for understanding the riddles of living matter.

In biology too, the effects of the second law of thermodynamics have long been, and continue to be, topics of lively discussion. Among other things, the central question here is that of whether life, with its sophisticated and highly ordered structures and functions, really could have originated by natural evolution from non-living matter. Is natural evolution capable of reversing the direction of the natural happening prescribed by the entropy law? In other words, does evolution violate the entropy law, according to which the course of natural events always leads to states of higher entropy, i.e. greater disorder? How "creative" is evolution?

When at the end of the nineteenth century a scientific controversy arose about these questions, the evolutionary biologist Ernst Haeckel, one of the leading scientists of his time, declared self-confidently: "The universe itself is a perpetual motion machine." [37, p. 202] It had, he supposed, neither a beginning nor an end. Instead, the universe behaved like a machine that is in eternal motion and continually produces new structures. Consequently, Haeckel believed that the entropy laws were disproved by the bare fact of cosmic evolution. He went even further, claiming that the law of conservation of energy applied only to individual processes, but not to the world as a whole.

What is often overlooked, even today, is the fact that the second law applies only to "isolated" systems, i.e., systems that exchange neither energy nor matter with their environment. Living beings, in contrast, are open systems. They can compensate for their internal entropy production by tapping into a supply of "free energy" (the physicists' term for energy that can be converted into useful work). This process serves metabolism (Sect. 5.7). Thus, there is no contradiction at all between the existence of highly ordered systems, like living systems or crystals, and the second law of thermodynamics.

However, as already emphasized, living beings are characterized primarily by their functional, and not by their structural order (Sect. 1.4). Their functions, in turn, are encoded in the genetic information which, moreover, regulates and controls all essential life processes. The central role played by genetic information in the living organism is another reason why the entropy law is today receding into the background as regards the physical and chemical understanding of life. Its place is being taken by the problem of the emergence of information, which is challenging physics and chemistry: Can meaningful information, as the basis of organic life, originate from information-less matter? Did early evolution work like a "perpetual motion machine", generating information from nothing?

Within the framework of the Darwinian theory of evolution, it is taken for granted that evolution automatically leads to a higher development of life and thus to constant expansion and enrichment of genetic information. This hypothesis has in the past been backed up by plausible arguments. So far, however, these arguments have indeed been based on plausibility alone. Only more recently has a theoretically sound justification emerged, opening up an exciting chapter of current basic research in the border area of physics, chemistry and biology ([38], see also Sect. 6.10).

Last but not least, modern information theory also throws light on the question of whether and to what extent the biological world admits creation from nothing. Darwin himself avoided this question, by shifting the actual source of evolutionary progress to the environment, which selectively evaluates the genetic diversity caused by mutation. However, from an information-theoretical point of view, the question arises as to how much information must have been present in the environment at the beginning of the process, so that evolution could lead to the formation of purposeful, functional structures. The fact that we can only speak of information in relation to some other information suggests that information cannot originate from "nothing". In other words, the environment as the recipient of genetic information can only fulfil its evaluating function in evolution if it itself is already in possession of background information.

This consideration leads to the general question of how much information is needed to understand some other information. With this question, which includes the difficult concept of "understanding", we seem to reach the limits of the scientific approach. If all understanding depends, in the manner hinted at above, on previous knowledge on the part of the recipient, then it must be something subjective, for which no objective measure can be found. Therefore, it is all the more surprising that, contrary to expectation, the general question posed above can be answered not only objectively but also quantitatively ([39], see also Sects. 6.8 and 6.9).

The idea behind this approach can be exemplified by human language. Let us consider the following sentence: "The world is computable." Obviously, this sentence carries meaningful information. The sequence of its letters, however, is random, in the sense that it cannot be generated by employing a simple algorithm (Sect. 5.3). We do not know of any rule that would allow the subsequence "The world is…" to be extended meaningfully in any way. There are numerous conceivable continuations, but none of them could be derived from the subsequence itself. Meaningful information cannot be compressed without at the same time also changing its content, to a greater or lesser degree—or even destroying it. From an algorithmic point of view, strings that carry some message are sequences of maximum complexity. This statement is obviously correct, although it cannot be proven in a strict sense. However, we have already seen in Sect. 5.4 that there are plausible arguments why meaningful information necessarily has a syntactic structure that is aperiodic and thus random.

Let us now look more closely at the process of understanding. To do so we divide the overall process into two steps. The first step is the perception of a message, i.e., its mere registration by the receiver. The subsequent step will then be the interpretation by the receiver. We ask: how much information *at least* is involved in the overall process?

To find the answer, it suffices to analyze the first step. Assume that a sender transmits to a receiver some meaningful information, encoded, for example, in a sequence of binary digits. This sequence is then, according to the previous assumption, one of maximum complexity. Because it cannot be compressed any more, the recipient must register the sequence in its full length before he can start on the actual process of evaluating it. In other words, the mere registration of meaningful information, preceding the act of interpretation, already contributes to the receiver's advance information. The syntactic structure of the information received is, so to speak, the template on which the actual process of understanding takes place. It already defines the amount of information which is necessary for understanding, regardless of how much additional information the recipient needs for handling the registration and the subsequent step of interpretation. Thus, the advance information required to understand a piece of information must be at least of the same complexity as the information to be understood.

These are the minimum requirements for understanding information. They determine the scope of the context in which understanding is possible. Remarkably, one can come to this conclusion without even having to define and explain the complicated concept of understanding. The sole decisive fact is that any understanding of information is inevitably preceded by the recipient's registration of the information, and that this alone already determines the minimum amount of information that is necessary for its understanding. This result, which we have reached here with the help of a plausibility consideration, can be justified by using the concept of the Turing machine. In this case, two Turing machines are considered which "communicate" with each other, whereby the output of one machine is the input for the other machine and vice versa. Such a self-contained system of transmitter and receiver cannot automatically increase its program complexity. In other words: there can be no machine that generates information out of nothing. We can also express this result as the impossibility of a perpetual motion machine of a third kind [40]. The important question of whether evolution can be considered to be a perpetual motion machine, generating genetic information from nothing, we will investigate at the end of Chap. 6.

5.7 Fluid Boundaries

The limits of science discussed so far were of a logical or epistemological character. They delineate, among other things, the edge of the cognitive reach of science. However, there are also limits of science that result from methodological prerequisites. We remember that physics, for example, is based mainly on the methods of abstraction, separation, simplification and idealization. These are necessary instruments of structuring the multitude of appearances, events, objects and processes that constitute reality.

Even more fundamentally, the methodological preparation of reality is always preceded by another important step, namely the conceptual containment of the subject of research. This step, however, is not unproblematic, because reality is a complex fabric of appearances with fluid transitions and boundaries. It is obvious that this challenges our conceptional thinking in a number of ways (Sect. 5.10).

The efforts to find an adequate definition of life are paradigmatic of this. Though we know a multitude of different forms of life, it seems impossible to develop a general concept of life that includes not only necessary but also sufficient criteria for drawing a sharp borderline between living and non-living matter. This is all the more surprising because all known life forms have emerged from a common origin. In any case, one would expect to be able to grasp the collective essence of living beings, even though they have taken different paths in evolution.

The problem of defining life is not only of academic interest. It also has enormous practical and ethical significance for our modern life-world, which is shaped not least by the recent advances in genetic engineering and reproductive medicine. New biotechniques raise a multitude of philosophical questions that call for an adequate definition of the living. In genetic engineering, for example, one needs to know in which respects bacterial life differs from plant life, plant life from animal life and animal life from human life. In reproductive medicine, questions arise concerning the beginning of human life, such as "at which developmental stage of the fertilized egg human does life begin?" or "when does personal life arise?". Last, and above all, comes the fundamental question of how living matter differs, if at all, from non-living matter. Is it possible to find a meaningful definition of life which goes beyond earthly life and can be applied, for example, to identify unknown forms of matter as "living" matter?

The fact that there is no simple answer to the question "What is life?", can be explained, though only superficially, by the circumstance that living matter is extraordinarily diverse in its manifestations. At a deeper level, however, the difficulties stem from the fact that the boundaries between living and non-living are fluid. This excludes from the outset the possibility of an exhaustive definition of life. Instead, one must accept that there are numerous definitions, covering in each case only partial aspects of the living.

A definition of life frequently encountered is based on the following three features: metabolism, self-reproduction and mutability. However, one immediately sees that these criteria are not sufficient to distinguish clearly living from non-living matter. There are natural objects to which these properties apply, but which do not fit into our overall picture of living beings. Crystals are the best example. They grow because their molecular building blocks attach themselves to the surface of a crystal nucleus. In this way, the structure of the miniature crystal is repeatedly reproduced, until it finally becomes visible in the crystal's shape. Indeed, the growth of a crystal is the prime example of the self-reproduction of a non-living structure.

There are further parallels between crystals and living beings. Properties such as metabolism and variability can also be observed in crystals. From a physical point of view, the metabolism of a living being is the turnover of "free" energy, i.e. energy that can perform work. The energy supply is necessary to keep the complex life processes going and to prevent the decay of the organism into the "dead" state of thermodynamic equilibrium. The organism gains free energy by converting high-energy substances into low-energy ones. Thus, metabolism is nothing other than a permanent energy flow by which the organism is kept in working order.

Something comparable can be observed in the growing crystal. Here, however, the flow of energy takes place in the opposite direction, i.e. energy is not supplied to the structure; on the contrary, the growing structure yields up energy to its surroundings. Thus, the molecular building blocks can only fit into the ordered structure of the crystal lattice if the energy of motion of the building blocks is dissipated. This can be done, for example, by cooling the crystallization solution, thereby reducing the kinetic energy of the molecules.

Even the phenomenon of variability (mutability) can be observed in crystals. However, its causes are entirely different from those underlying the mutability of living beings, viz, random changes in their genetic material. With crystals, variability results from the fact the growth of a crystal is never perfect. Occasionally, errors occur because foreign atoms occupy some positions in the crystal lattice, or because some positions remain empty. In the visible crystal, which consists of a vast number of atoms, there is an almost unlimited number of possibilities for distribution of such "blanks" over the lattice structure. However, since unoccupied positions affect the individual shape of the crystal, the richness and variety of structures are inexhaustible. Similar forces are at work in snow crystals. The uniqueness of a snowflake's structure, for example, is due to the particular atmospheric conditions under which it crystallized (Fig. 5.9).

Because of the astonishing parallels between crystals and living beings, it is understandable that crystal formation was regarded until well into the twentieth century as an analog for the formation of living organisms. In his studies of inorganic life, entitled *Kristallseelen* ("Crystal Souls"), Ernst Haeckel even put crystals on a par with living beings [41]. He saw the close relationship between crystals and living beings as a confirmation of his "monistic" world-view, according to which there is no essential difference between inanimate and animate Nature. Although the analogy between crystals and

Fig. 5.9 The uniqueness of snow crystals. Snow crystals are known for their richness of forms. Although they can be classified into certain basic forms, every single crystal is unique, because of the particular conditions under which it originates. A snow crystal appears when water vapor turns directly into ice without liquefying first. The crystallization starts with the formation of a small hexagonal plate. If more water condenses onto the surface of the nascent ice crystal, then new branches sprout from its six corners, and it finally grows to more complex shapes such the one shown in the picture. What ultimately appears depends on the detailed prevailing atmospheric conditions: temperature, humidity etc. Since the smallest differences in the growth conditions of an ice crystal lead to different forms, there are myriads of possible structures. Crystals were long considered to be an analogy for the living organism, as their growth seemed to resemble the biological process of self-reproduction. Today this analogy is only of historical interest. It can at best be applied to viruses, which are at the border between non-living and living matter. [Image: Rubelson, Wikimedia Commons]

organisms had initially been based merely on external features, the physicist Erwin Schrödinger later imbued it with deeper meaning when he interpreted chromosomes as "aperiodic crystals" (Sect. 5.4).

Modern research, however, has greatly weakened the idea of an analogy between crystals and organisms. Nevertheless, it has also demonstrated that, as Haeckel correctly concluded, there is no sharp borderline between the non-living and the living. Viruses, for example, are at the threshold of life. They have features of both living and non-living objects. Within their host cells, they show the typical properties of a living system. Although they do not have their own metabolism, they can intervene in the metabolism of their host cells. reprograming it for the production their offspring. Viruses can reproduce very rapidly and can also mutate just as quickly. The latter feature serves the viruses as a highly effective means of masking themselves and thus protecting themselves from external attacks. Outside their host cells, however, viruses behave like inorganic molecules. They can not only be crystallized, but

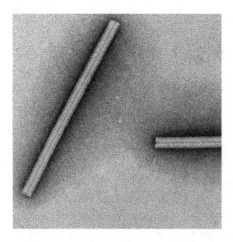 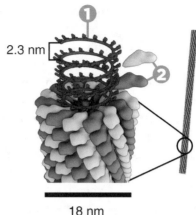

2.3 nm

18 nm

Fig. 5.10 The tobacco mosaic virus (TMV) as an example of the fluid boundary between animate and inanimate matter [42]. The left-hand figure shows a color-enhanced transmission electron micrograph of particles of tobacco mosaic virus; the right-hand figure is a schematic representation of the structure of such a virus, at higher magnification. A TMV particle consists of two classes of biological macromolecules, a nucleic acid (1) carrying the genetic information, and 2130 identical protein units (2) arranged spirally around the nucleic acid to form a stable, protective envelope for the genetic material of the virus. After penetrating its host cell, the virus sloughs off its protein coat. This releases its genetic material, and the metabolism of the host cell becomes reprogrammed to produce new virus particles. As complex as this process is, it can be replicated and studied in the test tube. Dissociation and reconstitution experiments of this type were first performed successfully in the mid-1950s [43]. They demonstrated that the transition from non-living to living matter is reversible. [Left: Science History Images/Alamy Stock Photo. Right: Thomas Splettstoesser, Creative Commons]

also broken down into their molecular components and reassembled into infectious particles, with loss of their original individuality (Fig. 5.10).

Viruses make it clear that a definition of life based on self-reproduction, metabolism and mutability is not sufficient to demarcate living matter unambiguously from non-living matter. Since we are dealing here with a fluid border, the possibility of an objective delimitation is already excluded for logical reasons [24, chapter 3]. A complete definition of the term "life", on the other hand, would only be possible if there were a sharp border between non-living and living. In this case, however, the definition would have to fall back on at least one feature that is specific to a living being. Only through this feature can the fundamental otherness of living matter, in contrast to non-living matter, be expressed at all. This, in turn, means that life can only be defined comprehensively by including an irreducible feature of life. In short, a sufficient and complete definition of life is only ever available at the price of tautology.

Conversely, assuming as a working hypothesis that the transition from the inanimate to the living is a fluid one, a logically sound definition of life is possible, but now at the price of incompleteness. One is faced here with a paradoxical situation. The reductionist research strategy in biology aims at the complete description of living matter by physicochemical terms and laws. This goal, in turn, implies that the transition from non-living to living matter is continuous, or at least quasi-continuous. This, however, means that the concept of life must inevitably remain incomplete.

The difficulties encountered in defining life demonstrate once more that the methods of the exact sciences—such as separation, abstraction, simplification, idealization and the like—are necessarily accompanied by a conceptual narrowing-down of the object of research. Finally, the definition of life also depends on the question being asked of the living system under analysis.

To give an example: From a physical point of view, all living systems must avoid falling into the (dead) state of thermodynamic equilibrium. This means that a living system can only maintain its material and functional order if the entropy production within the system is compensated for by a supply of free energy, i.e. energy working against the tendency toward equilibration. The living system gains free energy by metabolism, the turnover of energy-rich to energy-deficient material. Only through metabolism can the living organism counteract the degradation of its order structures, as a consequence of the entropy law. Thus, from a thermodynamic viewpoint, the definition of life will emphasize—beside self-reproduction and mutability—above all *metabolism* as a decisive feature of living matter. At the same time, these criteria seem to provide the most straightforward definition that physics can give the phenomenon of life at all.

However, if we change the perspective and consider the living being as a system that is instructed and driven by information, then we can also reduce the definition of life to the formula [38]:

$$\text{Life} = \text{Matter} + \text{Information}$$

This definition highlights an aspect shared by all living beings, including viruses. They have a piece of specific information, laid down in their genome, which governs and controls all their basic life processes. In physical terms, the genetic information is anchored in the specific physical boundary conditions under which the natural laws operate in living systems (Sect. 4.8). These boundaries define the system's organization in so far as they restrict the physical and chemical processes that are *in principle* possible to those that *actually* do take place in living beings.

The gradual nature of the transition from non-living to living matter has the consequence that every definition of life necessarily contains a normative element. This immediately becomes clear if we consider the content of genetic information. The content defines, as we said above, the functional organization of a living being. Functionality, however, is also a flexible term. We can easily see this if we use the efficiency of a machine as a measure of functionality. Thus, a machine can in principle possess an efficiency between zero and a hundred percent. Here, it is a relatively arbitrary matter to state what degree of efficiency will qualify any given machine to be regarded as functioning: it is just a question of the norm that we apply to the notion of functionality. An ideal machine that achieves an efficiency of a hundred percent, and could set a standard in this respect, cannot exist, since under the laws of physics energy can never be completely converted into work.

Let us summarize: It appears a little strange that the scientific definition of life has a certain degree of arbitrariness. However, this is a direct consequence of the fact that the boundaries at the interface between non-living and living matter are fluid ones. Consequently, any definition of life depends on the scientific perspective from which the phenomenon of life is to be examined. Many acrimonious debates on the concept of life could be avoided if sufficient attention were paid to this problem.

5.8 The World of "As-Well-As"

Let us consider the question "What is life?" from another point of view. We remember (Sect. 1.9) that the physicist Richard Feynman once claimed that it is not crucial for a scientist whether and how to *define* the objects of investigations, but rather how to *measure* them [44]. This statement expresses the typical view of positivistic science, according to which only those things that are objectifiable and quantifiable can become the subject of research. Correspondingly, the exact sciences do not claim to penetrate to the essence of the objects of their research. Instead, their interest focusses on such phenomena as can be prepared following the methodological guidelines of the reductionist research program. This restriction also applies, not least, to the life phenomena.

The first scientist who critically questioned the reductionist research program in biology was Niels Bohr, in the early phase of quantum physics [45]. Referring to the basic difficulties associated with the process of observation and measurement in quantum physics, Bohr arrived at the conclusion that a similar problem could occur regarding the physical analysis of the living matter. In quantum physics, to the surprise of the scientific community, it had turned out that the object observed and the experimental arrangement to observe it cannot be separated from each other. They form an indissoluble unit (Sect. 1.10). For this reason, it is no longer possible to determine what actually happens with a quantum object during a physical measurement. Instead, the object appears as the sum of all possible measurements that can be carried out on it [32, chapter 6]. In other words, it depends on the experimental measuring arrangement whether, for example, light shows the characteristics of a wave or those of a particle. Each aspect expresses an essential feature of light, although they are mutually incompatible from the viewpoint of classical physics. However, they can never contradict one another, because it is only through the choice of measuring arrangement between two mutually exclusive options that the one or the other feature of light emerges.

The wave-particle dualism of quantum physics is one of the most astonishing discoveries of modern science. It means that at the submicroscopic level one encounters a world of "as well as", in which the feature of a physical phenomenon that is observed depends upon the choice of measuring device itself. This leads to the unusual and elusive idea that physical reality has an ambiguous character, and only through observation and measurement does it become what we call "objective" reality. For this very reason Bohr, Heisenberg and other quantum physicists believed that physical theory should include only observable quantities. However, this demand did not meet with unanimous approval among physicists. On the contrary, it triggered a stormy debate on the appropriate interpretation of quantum phenomena, which continues to this day (Sect. 3.6). This debate is mostly about the metaphysical question of the essence of physical reality, i.e. precisely the kind of questions that positivist science tries to avoid.

The Copenhagen interpretation of quantum physics, as initiated by Bohr and others, is based on an instrumental understanding of physics that centers around the physical measurement process and its possible interactions with the object under observation. This explains why Bohr made his physical understanding of life phenomena entirely dependent on the conditions of their observation and measurement. Moreover, in this question Bohr adopted the position of the atomic physicist—that is, he assumed that a complete physical description of the living organism must lead to full knowledge of its atomic composition and states. This goal, Bohr pointed out, can only be achieved by breaking the organism down into its components and following the life phenomena experimentally in all their atomic details. This breakdown, however, would destroy the organism and its functions. Thus, the object that was to be investigated becomes lost.

These considerations led Bohr to conclude that "non-living" and "living" are two complementary states of matter, in the same sense as waves and particles are to be regarded as complementary manifestations of light [45]. With this picture in mind, he adopted an epistemological position regarding life phenomena which one may describe as holistic, if not even vitalistic. Although Bohr rejected the traditional idea of vitalism according to which there exists a life force acting beyond physics, he did not want to rule out the possibility that the remarkable stability of the living organism against external disturbances is due to the effect of a life-maintaining principle.

For a long time, Bohr's reflections about *Light and Life* from 1933 had a considerable impact on physicists' thinking regarding living matter. At the beginning of the 1960s, however, Bohr felt compelled to revise his views on this issue [46]. In the meantime, the progress of biophysical research had proceeded so rapidly that it had become difficult to believe seriously in the irreducibility of life. The reproductive self-preservation of living organisms, which only a few years before had seemed to evade any physical explanation, now proved to be a physicochemical property of biological macromolecules, and thus became accessible to apparently unrestricted experimental analysis.

It has furthermore become clear that for a physical understanding of living matter it is not necessary to trace life phenomena down to the behavior of single atoms. The basic life processes such as heredity, metabolism and the like take place at the level of molecules and macromolecules and not, as Bohr initially had assumed, in the atomic domain. Moreover, a complete nuclear description of the phenomena of life would in any case be impossible, because the living organism is an open system that continuously exchanges energy and matter with its environment. Thus, as regards the fate of single atoms, there is anyway no fixed border between living and non-living matter.

Bohr's idea of transferring the principle of complementarity to living matter is nowadays only of historical interest. It elucidates Bohr's views on the issue of complementarity, but it failed to make any decisive contribution to solving, or even to clarifying, the riddle of life—even if it occasionally appears in a new guise as an argument against biological reductionism.

The actual meaning of the complementarity principle is that it enforces a radical rethinking regarding our understanding of reality. It demonstrates that, in the atomic domain, the classical notion of an object is no longer applicable. Instead, one has to realize that our conventional terms and ideas only fit our daily world of experience, and they fail to describe the microcosm adequately.

Therefore, the difficulties in satisfactorily interpreting the appearances of the sub-microphysical world are correspondingly great. We continually run the risk of trying to explain something incomprehensible by using the inadequate expressions of human language to describe what actually can be described only in mathematical terms. Bohr chose a fitting comparison for this dilemma, as Heisenberg reported in his memoirs [47]. The quantum physicist, Bohr said, finds himself in the situation of a dishwasher who is faced with the task of cleaning dirty glasses with dirty water and dirty dishcloths.

Quantum physics seems incomprehensible to us, above all, because it no longer follows the logic of "either/or" with which we are familiar. Instead, it leads us into a world of "as-well-as", that is so difficult to comprehend. Therefore, it is all too understandable that scientists have been looking for a new logic, called quantum logic, that could describe quantum-physical phenomena in a logically flawless form. The new type of logic is conceived of as a propositional logic, in which the truth value of statements has only a certain probability. However, quantum logic is in so far controversial as it is tailored to a domain of experience that can only be reached through technical devices, and this contradicts the traditional understanding according to which logic is valid independently of any form of experience.

5.9 The Fuzziness of Truth

In addition to the problem of *ambiguous* truth values, there is also the problem of *vague* truth values. The latter always occur if it is not possible to draw a clear boundary between true and false statements. Already in ancient times, the Megarian philosopher Eubulides of Miletus had pointed out the logical aspects of the problem of vague truth when he asked his contemporaries the catch question: "How many grains form a heap?". One grain is, without any doubt, not a heap. Even two grains are not a heap. The conclusion that one grain more or less does not make a difference between heap and non-heap is generally accepted, regardless of the number of grains considered. However, repeating the above argument again and again, even one billion grains would finally not yield a heap. This conclusion, however, would be nonsense. Who could deny that a billion grains do not make up a heap?

The irritating feature of the "sorites" paradox (from the Greek word *soros* for heap) is the fact that a chain of correct conclusions leads to a wrong result. Now, one could try to defuse the paradox by merely denying the truth content of some underlying assumptions. Thus, for example, it has been argued that the initial statement ("one grain is not a heap") is wrong. Another objection has claimed that our judgment on the conclusion is wrong because there are no heaps at all. Of course, these are extreme points of view, and ultimately they make little sense.

Let us now look at the standard explanation, which is considered to be the most convincing solution of the heap paradox. This explanation is based on a logical inference rule, called *modus ponens*. It states: If "*A* is true" and "From *A* follows *B*" both hold, then *B* is also true. However, the *modus ponens* is no longer admissible as soon as the statement *A* is not right a hundred percent of the time. In this case, the vagueness of premise *A* is transferred to the conclusion *B*. Since, in the heap paradox, *B* is the premise for a subsequent conclusion *C*, the truth value of the premises inevitably decreases with each further conclusion as on a sloping road. Thus, in the end, one comes from a "fairly" true proposition, namely "one grain is not a heap", to the utterly false assertion that even "one billion grains are not a heap".

This is not the place to go deeper into the logical problem of the heap paradox. A survey of these can be found elsewhere [48, chapter 3]. For us, it is only crucial to realize that the truth value of statements is often vague and that this blur in a logical conclusion can be amplified with every step of an argument until it finally leads to a completely wrong conclusion. Many acrimonious debates in society and as well as in daily life would be avoidable if the parties involved were aware of the nature of vague statements and the self-reinforcing effect that they can have.

A simple example can show that transitions between true and false statements are often continuous ones. Let us look at a long, bar-shaped panel whose left end is black and whose right end is white (Fig. 5.11). In between, its color is a continuous series of shades of grey. At the left end of the grey scale, the statement "The bar is black" is entirely true. If we now go from left to right, the truth content of the statement "The bar is black" decreases continuously. Therefore, it is quite justified to describe a conclusion as entirely true, or as entirely false, or as "fairly" true, or as "fairly" false, whereby the terms "fairly true" and "fairly false" lie close together. Where, in this case, we should draw the line between true and false is a matter of convention, and thus a normative question. With the usual two-valued logic, according to which statements are either true or false, such fluid transitions cannot be adequately grasped; to do this, one needs a logic that takes account of the fact that statements can be more or less true, or more or less false.

A logical system that operates with more than two truth values is called multi-valued or "multivalent" logic. A simple example of this is trivalent logic, which beside "true" and "false" also includes the truth value "undetermined". Beside, one differentiates between logical systems that operate with a finite number of truth values and those that use an infinite number of truth values. An example of the latter, in which "true" and "false" appear as limit values of statements that allow an infinite number of truth-values in between, is the "fuzzy logic" developed by the mathematician and computer scientist Lotfi Zadeh. This type of logic has found a full field of application, especially in the engineering sciences.

It seems that multivalent logic, to which also quantum logic belongs (Sects. 3.6 and 5.8), is more suitable for describing reality than is bivalent (true/false) logic. One reason for this may be the fact that reality appears to us, to a large extent, as a continuum in which transitions between phenomena are more or less fluid. Therefore, statements about reality can usually no longer be described merely as true or false. Instead, life wisdom enters here, according to which the truth often lies in between—a typical case for the application of multivalent logic.

Nevertheless, there are reservations about multivalent logic based on the same argument as in the case of quantum logic. Both are criticized for being too closely adapted to the conditions of reality, which, it is asserted, ultimately gives them a descriptive character. In the view of the critics, however, logic is an aprioristic law of thought, i.e. its function is prescriptive. Moreover, it is pointed out in this context that the metalanguage in which the different logical systems are examined is in turn based on classical, i.e. two-valued, logic.

The problem of fuzzy truth values is significant, not least for philosophical theories of truth. Let us take as an example the discourse theory of truth proposed by Jürgen Habermas (Sect. 2.9). According to this theory, truth emerges in social discourse, provided that this follows specific rules. However, if the truth of statements is unclear or indefinite for fundamental reasons, the blur is inevitably transferred to the discourse. Thus, for example, ambiguities may be amplified according to the pattern of the heap paradox, leading the discourse finally into a dead end and calling into question the aspired-to development of valid arguments.

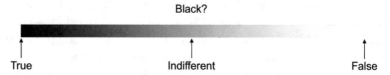

Fig. 5.11 Floating truth values. At the left end of the panel, the statement "the panel is black" is true. At the right end, this statement is false. In between, there is any number of intermediate truth values, which continuously decrease from absolutely true to absolutely false, as on a sloping path

Beside logical fuzziness, there is a further problem that seems to question the discourse theory of truth. This problem is related to the fuzziness of information exchange in human communication. It is evident that mutual understanding between two partners in a dialog is only possible if they have standard background information. In the ideal case, they must even have background information of the same complexity, to communicate with each other at the same semantic level (Sect. 5.6). In reality, however, no dialog partners ever have precisely the same background information. Instead, there is an asymmetry in this respect: One of the partners will always have background information that is more complex and exceeds the background information of the other partner. This difference is fundamental and cannot be evened out in the course of communication. It is merely seemingly eliminated by the dialog partners accepting a certain degree of blurriness in their mutual understanding, which in turn is manifested in the vagueness of their statements.

Moreover, the limitations of the forms of linguistic expressions lead to a degree of fuzziness in interpersonal dialog. Since linguistic expressions have a discrete structure, they can only depict reality approximately, because the actual world appears as a continuum. To avoid vagueness in communication, the dialog partners have to adapt their linguistic terms to factual reality with the highest possible precision. This means that they must sharpen the scope and content of their linguistic expressions by continual, collective effort. If pursued to perfection, such a procedure would make the attainment of intersubjective understanding a never-ending process. Consequently, in the real world, partners in dialog content themselves with a greater or lesser degree of mutual comprehension.

5.10 Limits of Conceptual Thinking

Language is not only a means of communication; it is also the basis of our world-understanding (cf. Chap. 1). Thus, one may suspect that the limits of human language also limit man's knowledge of the world. The fact that terms are always limited in scope and content seems to support this suspicion. As mentioned above, the world of experience appears to us as a world of fluid transitions, i.e. more or less as a continuum. The attempt to grasp the continuum of reality by utilizing the discrete terms of language is about as futile as the attempt to capture the water of the Atlantic Ocean with a fishing net.

With similar arguments, the philosopher Heinrich Rickert tried to point out the limits of concept formation in science [49]. According to Rickert, the natural sciences are forced to use a network of terms with an unusually wide mesh, because they aim to represent reality with the highest degree of generality. For this purpose, the natural sciences would have to increase the scope of their terms constantly, in order to subsume as many phenomena of reality as possible under general concepts. However, Rickert emphasized that the more comprehensive and broader such concepts become, the more they lose their empirical content and thus their closeness to reality.

In contrast to the natural sciences, according to Rickert, the aim of the historical sciences is the individualization, i.e. the most precise description possible, of the object of cognition. Unlike the natural sciences, he stated, the historical sciences create an ever-finer net of terms because they seek to grasp as many details of reality as possible. However, to reach a common understanding, historical sciences have to emphasize certain phenomena from the immense abundance of unique and individual appearances. Yet this does not happen with the aim of terminological generalization, but rather because of the meaning that the phenomena in question have for the life-world of man. The objective yardstick for selection is therefore given in the general cultural values of humankind, i.e. those values that all people share. This line of reasoning led Rickert to try to establish the historical sciences as cultural sciences – a point of view that still has many followers today. The roots of this dualistic understanding of science, which go back to Wilhelm Windelband, have been described in Sect. 4.2.

According to Rickert, it is this kind of conceptual structuring of reality that constitutes the essential difference between the natural sciences and the cultural sciences. He argued that the natural sciences throw a net of terms over reality that is as wide-meshed as possible, to grasp the supra-individual and general structures of real life. The cultural sciences, in contrast, use a dense web of terms that allows depiction of the fine structure of reality. The former move further and further away from reality, by striving for an ever more general picture of reality. The latter approach reality by drawing an ever more specific view of it.

The conceptual boundaries that Rickert drew within the scientific landscape served a dual purpose. On the one hand, they were intended to delimit the scope of the exact sciences; on the other, they were to emphasize the importance of the cultural sciences for our world-understanding. However, Rickert linked his science dualism with the untenable idea that scientific cognitions based on general concepts ultimately have less value for understanding reality than do the insights of cultural studies that resort to individualizing terms. Rickert thus attempted to define the value of scientific knowledge primarily on the basis of concept formation. The substantial difference existing between a scientific explanation, supported by natural laws, and the mere description of facts faded out in his considerations.

From today's perspective, the sharp dividing line between nomothetic and idiographic sciences is in any case dissolving. As outlined in Chap. 4, the unique and historical structures of reality are also increasingly coming into the focus of nomothetic explanations. The limitations imposed upon representation of fact by the discrete concepts of human language do not impose any substantial limits upon scientific knowledge. They merely lead—and Rickert's analysis indeed demonstrates this—to a loss of detailed information, comparable to the loss of information that occurs with the transition from "analog" to "digital" forms of representation.

A clock illustrates the difference between these two forms of representation. With a conventional analog clock, the hands move continuously over the dial, so that in principle an infinite number of values of time can be read. With a digital clock, on the other hand, only discrete time–values are displayed. Like the color spectrum of a rainbow, elements of analog information merges seamlessly into one another, while digital information consists of a limited number of finite gradations.

The description of reality through language corresponds to the digital form of representation. Therefore, it can only depict reality like a raster or graticule image. However, this raster varies from language to language, so that there will be overlaps as well as mismatches between the meaning zones of certain words (see [50]). Moreover, language is not a rigid network of terms with a mesh size that is given once and for all; on the contrary, it is evolving in respect of its cognitive functions.

The mathematician and philosopher Richard von Mises has compared language to a tool that, like other human tools, is continually being adapted to the meet new realities and challenges: "Every existing tool is tried out, at some time or another, for new, not originally intended, purposes and thereby undergoes certain changes directed toward these purposes, which then lead to the creation of a new type of tool. In the same way, one tries continually to express by means of the existing linguistic elements (word roots, word forms, word and sentence combinations, modes of expression) every new experience. For adaptation to the new situations, however, one often has to introduce little changes, which sometimes last, replacing the old form or, perhaps, adding to it. In this fashion the language is continually modified and enriched by adaptation in the 'change of usage'." [51, p. 21]

Von Mises believed that the principle of the progressive change of language use could also be applied to the development of scientific thoughts. He followed here an idea of Ernst Mach, who was one of the first to point out that the evolution of thoughts consists in a continual adaptation to facts. Von Mises, therefore, did not want to draw a sharp line between the creation of language and the emergence of science. On the contrary, he in fact regarded the creation of language as the embryonic stage of science.

The idea that the progress of our knowledge rests on the continual change of the use of language and thinking is an involuntary reminder of the image with which François Jacob described progress in biological evolution [52]. According to Jacob, biological evolution "tinkers" with existing biological structures until random changes and natural selection lead to structures that are either better adapted to their functions or can take on new functions. In higher organisms, the tinkering process is accelerated by genetic recombination, because the latter gives birth, time and again, to new and innovative combinations of the genetic material. In this way, evolutionary progress emerges by change and adaptation of the biological tools to the demands of the environment.

There are many indications that human language has adapted—by continuous changes in language use—to reality. This means that the use of language has adapted to the givenness of reality and not vice versa, as some philosophers believe. Without the progressive evolution of the use of language, we would still find ourselves in the situation of Stone Age man, trying to handle reality with only one and the same tool. It is the critical use of language, and the constant development and enrichment of its means of expression, that finally brought about the advancement of human knowledge and led mankind out of its Stone Age existence. Thus, it is not helpful for a critical understanding of the world if some philosophers declare language to be an absolute starting point of our world cognition. Such a philosophy is uncritical, and indeed counterproductive, because it forces human language into a stiff corset of world cognition that obstructs the path to understanding language itself, its origin and its development.

No doubt, it is a vital task of philosophy to point out the limitations of human cognition. However, it is an equally important task of critical thinking to distinguish apparent from real boundaries and thus to separate the wheat from the chaff. Otherwise, the door is wide open to arbitrariness of thinking. The philosophy of Jean-François Lyotard [53] is an outstanding example of this. He turned the cognition goal of modern science upside down. He declared that the undecidable, the unpredictable, the uncontrollable, and thus non-knowledge, must be at the center of postmodern science. Here again, we encounter the slogan "Anything goes" in its most radical form (Sect. 3.10).

However, the insights into the limitations of human knowledge must not be perverted to a pseudo-science, one which justifies its findings exclusively from "paralogy"—i.e., from contradiction, anti-method and absurdity—and for which, in the end, real cognition only consists in dissent with the known. As important as the insights into the limits of science are: real progress of human knowledge is determined by the perspectives that serve an affirmative and positive understanding of the world.

References

1. Tarski A (1944) The semantic conception of truth. Philos Phenomenol Res 4(3):341–376
2. Schelling FWJ (1980) Ideen zu einer Philosophie der Natur als Einleitung zum Studium dieser Wissenschaft (1797). In: Schelling. Schriften von 1794–1798. Wissenschaftliche Buchgesellschaft, Darmstadt. English edition: Schelling FWJ (1988) Ideas for a Philosophy of Nature: as Introduction to the Study of this Science (transl: Harris EE, Heath P, introduction Stern R.) Cambridge University Press, Cambridge
3. Hölldobler B (1971) Communication between ants and their guests. Sci Am 224 (3):86–93
4. Küppers B-O (2010) Information and communication in living matter. In: Davies P, Gregersen NH (eds) Information and the Nature of Reality: From Physics to Metaphysics. Cambridge University Press, Cambridge, pp 170–184
5. Musil R (1995) The Mathematical Man (transl: Pike B, Luft SD). In: Pike B, Luft SD (eds) Musil R: Precision and Soul: Essays and Addresses. University of Chicago Press, Chicago, pp 39–43 [Original: Der mathematische Mensch, 1913]
6. Frege G (1967) Begriffsschrift, a formula language, modelled upon that of arithmetic, for pure thought (transl: Bauer-Mengelberg S) In: Heijenoort J van (ed) From Frege to Gödel, a Source Book in Mathematical Logic, 1879–1931. Harvard University Press, Cambridge/Mass [Original: Begriffsschrift, 1879]
7. Frege G (1960) The Foundations of Arithmetic: A Logico-Mathematical Enquiry into the Concept of Number (transl: Austin JL). Harper & Brothers, New York [Original: Die Grundlagen der Arithmetik, 1884]
8. Frege G (2016) Basic Laws of Arithmetic. Oxford University Press, Oxford [Original: Grundgesetze der Arithmetik, 1893, 1903]
9. Russell B, Whitehead AN (1910–1913) Principia Mathematica, vol 3. Cambridge University Press, Cambridge
10. Fraenkel AA, Bar-Hillel Y, Levy (1973) A Foundations of Set Theory. Elsevier, Amsterdam
11. Weyl H (1946) Mathematics and logic. A brief survey serving as a preface to a review of "The Philosophy of Bertrand Russell". Am Math Mon 53:2–13
12. Hilbert D (1984) On the infinite. In: Benacerraf P, Putnam H (eds) Philosophy of Mathematics: Selected Readings. Cambridge University Press, Cambridge, pp 183–201 [Original: Über das Unendliche, 1926]
13. Reid C (1996) Hilbert. Springer, New York
14. Hilbert D (1902) Mathematical problems. Bull Am Math Soc 8(10):437–479

15. Gödel K (1931) Über formal unentscheidbare Sätze der Principia Mathematica und verwandter Systeme I. Monatsh Math Phys 38:173–198
16. Lucas JR (1961) Minds, Machines and Gödel. Philosophy 36:112–127
17. Penrose R (1989) The Emperors's New Mind: Concerning Computers, Minds, and the Laws of Physics. Oxford University Press, Oxford
18. Küppers B-O (1990) Information and the Origin of Life (transl: Woolley P) MIT Press. Cambridge/Mass [Original: Der Ursprung biologischer Information, 1986]
19. Bennett CH (1982) The thermodynamics of computation—a review. Int J Theor Phys 21:905–940
20. Solomonoff RJ (1964) A formal theory of inductive inference. Inf Control 7:1–22 (Part I); 7:224–254 (Part II)
21. Kolmogorov AN (1965) Three approaches to the quantitative definition of information. Probl Inf Transm 1:1–7
22. Chaitin GJ (1966) On the length of programs for computing finite binary sequences. J AMC 13:547–569
23. Schrödinger E (1964) What is Life? (Combined reprint of "What is Life, 1944" and "Mind and Matter, 1958"). Cambridge University Press, Cambridge
24. Küppers B-O (2018) The Computability of the World. How Far Can Science Take Us? Springer International, Cham
25. Monod J (1971) Chance and Necessity: An Essay on the Natural Philosophy of Modern Biology (transl: Wainhouse A). Random House, New York [Original: Le hazard et la nécessité, 1970]
26. Mach E (1883) Die Mechanik in ihrer Entwicklung historisch-kritisch dargestellt. F.A. Brockhaus, Leipzig. English edition: Mach E (1919) The Science of Mechanics. A Critical and Historical Account of its Development (transl: McCormack T). The Open Court Publishing. Chicago
27. Gadamer H-G (2004) Truth and Method. Continuum, London/New York [Original: Wahrheit und Methode, 1960]
28. Rescher N (1988) Rationality: A Philosophical Inquiry into the Nature and the Rationale of Reason. Oxford University Press, Oxford
29. Boltzmann L (1897) Zu Hrn. Zermelo's Abhandlung "Ueber die mechanische Erklärung irreversibler Vorgänge". Ann Phys 60:392–398
30. Hund F (1979) Geschichte der physikalischen Begriffe. Bibliographisches Institut. Mannheim
31. Weizsäcker CF von (1980) The Unity of Nature. Farrar, Straus & Giroux, New York [Original: Einheit der Natur, 1971]
32. Scheibe E (2001) Between Rationalism and Empiricism: Selected Papers in the Philosophy of Physics (ed: Falkenburg B). Springer, New York 2001
33. Maxwell JC (1871) Theory of Heat. D. Appleton, New York
34. Szilárd L (1929) Über die Entropieverminderung in einem thermodynamischen System bei Eingriffen intelligenter Wesen. Z Phys 53(11–12):840–856

35. Brillouin L (1962) Science and Information Theory, 2nd edn. Academic Press, New York
36. Boltzmann L (1964) Lectures on Gas Theory. University of California Press, Berkeley [Original: Vorlesungen über Gastheorie, 1896, 1898]
37. Haeckel E (1934) The Riddle of the Universe. Watts & Co, London [Original: Die Welträtsel, 1899]
38. Küppers B-O (2016) The Nucleation of Semantic Information in Prebiotic Matter. In: Domingo E, Schuster P (eds) Quasispecies: From Theory to Experimental Systems. Springer International, Cham, pp 67–85
39. Küppers B-O (2013) Elements of a Semantic Code. In: Küppers B-O, Hahn U, Artmann S (eds) Evolution of Semantic Systems. Springer, Berlin/Heidelberg, pp 67–85
40. Küppers B-O (1996) Der semantische Aspekt von Information und seine evolutionsbiologische Bedeutung. Nova Acta Leopold 72(294):195–219
41. Haeckel E (1917) Kristallseelen: Studien über das anorganische Leben. Alfred Kröner, Leipzig
42. Butler PJG, Klug A (1978) The assembly of a virus. Sci Am 239(5):62–69
43. Fraenkel-Conrad H, Williams RC (1955) Reconstitution of active tobacco mosaic virus from its inactive protein and nucleic acid components. Proc Natl Acad Sci USA 41(10):690–698
44. Feynman RP, Leighton RB, Sands M (1963) The Feynman Lectures on Physics, vol 1. Addison-Wesley, Reading/Mass
45. Bohr N (1933) Light and Life. Nature 131:457–459
46. Bohr N (1963) Light and Life. Revisited. ISCU Rev 5:194–199
47. Heisenberg W (1971) Physics and Beyond: Encounters and Conversations (transl: Pomerans AJ). Allen and Unwin, London [Original: Der Teil und das Ganze, 1969]
48. Sainsbury RM (2008) Paradoxes. Cambridge University Press, New York
49. Rickert H (1986) The Limits of Concept Formation in Natural Science: A Logical Introduction to the Historical Sciences (abridged edition, transl: Oakes G). Cambridge University Press, Cambridge [Original: Die Grenzen der naturwissenschaftlichen Begriffsbildung, 1896]
50. Hjelmslev, L. 1970. Pour une sémantique structurale (1957) In: Essais Linguistiques. Éditions de Minuit, Paris, pp 105–121
51. Mises R von (1968) Positivism: A Study in Human Understanding. Dover Publications, New York [Original: Kleines Lehrbuch des Positivismus, 1939]
52. Jacob F (1977) Evolution and Tinkering. Sci New Ser 196(4295):1161–1166
53. Lyotard JF (1984) The Postmodern Condition: A Report on Knowledge (transl: Bennington G., Massumi B). University of Minnesota Press, Minneapolis [Original: La condition postmoderne: Rapport sur le savoir, 1979]

6

Perspectives: Designing Living Matter

6.1 We Must Know, We Shall Know

The overwhelming advance of science and technology in modern times has engendered a world-view often referred to as scientism. In the most comprehensive sense, scientism is regarded as a cultural current that tries to extend science, its ideas, concepts, methods, practices etc. to all essential aspects of human existence and society. Accordingly, science and technology are regarded as decisive driving forces for cultural progress. In a narrower sense, scientism is only a synonym for rational and analytical thinking in a knowledge-based society. In that sense, scientism is merely a generic term linking the main philosophical currents within science: positivism, reductionism, materialism, naturalism and others.

Today, knowledge and the competences that it engenders occupy a prominent place in our "knowledge society", in which everything seems to be centered on the guiding principle of all-embracing scientism. This is mainly because science and technology are bridging, increasingly closely, the traditional boundaries between the living and the non-living, the natural and the artificial. The driving forces of this development are the rapidly advancing life sciences combined with progress in information technologies. These disciplines determine not only the perspectives of science in the long run but also the practical impact of science on human life.

Scientism reached its first peak in the eighteenth century. We remember that at the height of the age of mechanism Pierre-Simon de Laplace had proclaimed the provocative idea of a Universal Mind which could calculate all past and future states of the world with the help of the mechanical laws

© Springer Nature Switzerland AG 2022
B.-O. Küppers, *The Language of Living Matter*, The Frontiers Collection,
https://doi.org/10.1007/978-3-030-80319-3_6

provided that he has complete knowledge of the world's present state
(Sect. 3.7). Nothing, Laplace asserted, remains hidden from the Universal
Mind in the transparent world of mechanism. Even the thoughts, feelings and
actions of man were in principle foreseeable and predictable, as the argument
ran, for according to the mechanistic understanding of the world even spiritual phenomena are nothing more than material processes and therefore
subject to the laws of mechanics.

The mechanical paradigm finally gave rise to a thoroughly materialistic
world-view. However, materialism was not generally accepted, even among
scientists. Already in the second half of the nineteenth century, there arose a
violent controversy about this question within science. The debate was triggered by a lecture on the limits of scientific cognition, given in 1872 by the
physiologist Emil du Bois-Reymond at the *Assembly of German Naturalists
and Physicians* [1].

Du Bois-Reymond, who was one of the leading scientists of his time,
addressed in his lecture seven world-riddles, or shortcomings of science,
which he claimed were unsolvable within the framework of the materialistic
world-view: (1) the essence of matter and force, (2) the origin of movement,
(3) the emergence of life, (4) the apparently deliberate, purposeful quality of
Nature, (5) the origin of simple sensations, (6) rational thinking and (7) the
problem of free will. Some of these problems, such as the essence of matter
and force and the origin of movement, du Bois-Reymond considered to be
unsolvable in principle, because he believed them to transcend our possibilities for world-understanding. Other problems, such as the emergence of life,
he regarded as solvable, but only on the premise that matter has started
moving at all.

Not least, du Bois-Reymond saw in mental phenomena a fundamental
border that the natural sciences could not cross. Mind, he claimed, could
never be understood within a mechanistic conception of Nature, since sensations, feelings and thoughts could not be traced back to the physical and
chemical properties of matter. Du Bois-Reymond was deeply convinced that
mental phenomena were immaterial in nature and therefore eluded the
mechanical paradigm of science. In reality, he stated, the human spirit could
not be anything more than a weak image of the Laplacian Universal Mind.
This Mind, he argued, was subject to the same limits of cognition as we are,
and for it the seven riddles were as unsolvable as for humans. Since du
Bois-Reymond assumed an ascending interdependence between these
world-riddles, which together appeared to him to be a cohesive, fundamentally insoluble complex of problems. So he ended his lecture with his now

famous dictum *Ignoramus et ignorabimus*—we do not know and we shall never know—any solution to these world-riddles.

Du Bois-Reymond's critical attitude to the mechanistic science of his time seems all the more remarkable as he himself was a prominent supporter of the mechanistic understanding of Nature. He regarded it as the ultimate goal of the sciences to trace every event of the material world back to the movement of atoms. More than this, he saw in the rising natural sciences of the nineteenth century actual progress in human culture. He even went so far as to claim that "the history of science is the real history of mankind" [1, p 134; author's transl.]. Thus, it seems all the more surprising that, in his *Ignorabimus* lecture, du Bois-Reymond contradicted his own scientific ideal, the mechanistic conception of Nature.

Du Bois-Reymond was well aware that his assertion "we shall not know" was a provocation to the scientific optimism of his time and would inevitably lead to emphatic contradiction. He himself referred to his assertion as "Pyrrhonism in a new guise" [2, p. 6; author's transl.]. Pyrrhonism, also called scepticism, goes back to Pyrrhon of Elis and denotes a philosophical movement in antiquity, which fundamentally doubted the possibility of true knowledge of reality.

The objections expected by du Bois-Reymond were not long in coming. Above all, it was the evolutionary biologist Ernst Haeckel who became du Bois-Reymond's most powerful opponent. Like no one else, Haeckel had campaigned for the spread of Darwin's theory of evolution toward the end of the nineteenth century. In fact, Darwin's central idea of the evolution of life by natural selection seemed to confirm fully the mechanistic paradigm of modern science. Ludwig Boltzmann, one of the leading scientists of the time, was convinced that evolutionary theory would be the crowning glory of the mechanistic age. In a lecture given in 1886, he highlighted Darwin's achievement with enthusiastic words: "If you ask me for my innermost conviction, whether it will one day be called the iron century or the century of steam or electricity, I will answer without hesitation that it will be called the century of the mechanical conception of Nature, the century of Darwin." [3, p. 28; author's transl.]

That is why Haeckel felt it presumptuous that du Bois-Reymond had declared the origin and development of life to be an unsolvable mystery. Haeckel furthermore stood uncompromisingly for a monistic doctrine of Nature, according to which even mental phenomena are nothing other than material processes. In anticipation of the modern idea of evolutionary epistemology (Sect. 2.8), Haeckel believed that one could also explain the cognitive abilities of the human brain by evolution, i.e., as adaptations of human beings to the requirements of their living conditions. Moreover, Haeckel strictly rejected du Bois-Reymond's assertion that there are unsolvable world problems. Except for the problem of free will, which Haeckel claimed to be a dogma, because it was based on deception and did not exist at all, he considered the remaining world mysteries either to have been solved in the context of the materialistic conception of Nature or at least to be solvable in principle [4].

More generally, the assertion "ignoramus et ignorabimus" provoked a massive counter-reaction from most scientists. Haeckel, to cite him once more, commented that the word "ignorabimus" conveyed only false modesty in view of the scope of the natural sciences. In reality, he asserted, it expressed downright presumption. It raised the assertion "we shall not know" to a principle and thus to a verdict, postulating for all time the unsolvability of some of the fundamental issues of science.

Haeckel accused du Bois-Reymond of adopting an agnostic attitude because he considered certain things as being unknowable in principle. This criticism is quite understandable, as du Bois-Reymond's view was fundamentally opposed to science's awareness of progress in the nineteenth century. In a lecture broadcast in 1930, the mathematician David Hilbert even described the assertion "we shall not know" as "foolish", as in his view there was no such thing as an unsolvable problem at all. He concluded his lecture by alluding to, and deliberately misquoting, du Bois-Reymond with the words: "We must know, we shall know." [5, p. 963; author's transl.]

Du Bois-Reymond's world mysteries are a mixture of physical and metaphysical issues which necessarily lead to a confusion of categories. Thus, the problem of the origin of all things, which involves the philosophical question "why is there anything and not nothing at all?" cannot be answered within the framework of the empirical sciences. It cannot even be answered within the framework of philosophy. Since du Bois-Reymond coupled those questions with issues of the empirical sciences, he also shifted the concrete problems of science to the realm of metaphysics and vice versa. Modern science, however, does not admit issues that have the character of "mysteries". Instead, all great questions of current science can be broken down into clear-cut subproblems of empirically answerable questions which either have already been solved or else can reasonably be expected to be solvable. Thus, the empirical sciences today are striving to reduce the field of metaphysics further and further, even though it is not possible to banish metaphysical questions entirely from science.

Today we know that there are indeed unsolvable problems, not least—yes, even—in pure mathematics. The most famous of these is the 23rd problem on the list of mathematical problems that Hilbert proposed as requiring solution. However, this problem proved to be unsolvable in principle (Sect. 5.2). The proof of this, first given by the logician Kurt Gödel, demonstrated that truth and provability are by no means congruent concepts, as Hilbert had implicitly assumed. Nevertheless, even in this case, it makes no sense to speak of a world mystery. On the contrary, the unsolvability of Hilbert's 23rd problem has led to profound insights into the essence of self-referentiality and has thereby sharpened and enriched our scientific thinking.

Looking back from today's perspective, one can say that Hilbert's maxim "we must know, we shall know" has become the undisputed guiding principle of modern scientific research that, despite the knowledge of its limits, persistently endeavors to make the incomprehensible comprehensible, the unpredictable predictable and the immeasurable measurable.

6.2 How the "Impossible" Becomes Possible

The physical chemist Manfred Eigen, who became well known for his investigations of extremely fast chemical reactions, described impressively in his Nobel lecture how the boundaries of scientific research, which often at first appear to be insurmountable, can yet be overcome—if one has the right idea [6]. Before Eigen developed his revolutionary techniques, rates of chemical reactions were measured according to a standard procedure: the two reactants are mixed, and the course of the reaction is followed by means of a suitable indicator. However, this procedure runs up against a difficult problem. There are chemical reactions for which the mixing of the reactants takes many times longer than the reaction itself. An example of such an extremely fast reaction is the neutralization of an acid by a base. When Eigen started his investigations, such reactions were considered to be immeasurably fast.

How was this problem to be overcome? Eigen's ground-breaking idea was not to initiate the chemical reaction by mixing the reaction partners and then following their equilibration, but rather to start the reaction in the equilibrium state itself. Eigen then disrupted the chemical equilibrium and followed, employing suitable instruments, its restoration. Of course, here too, the disturbance must take place more quickly than the restoration. An abrupt disruption of a chemical equilibrium can be caused by administering a very short pulse of energy. One can do this, for example, by a sudden change in temperature or pressure, or by an abrupt change of an electrical field, as occurs during the discharge of a capacitor. Such disturbances, to which the system reacts by restoring its chemical balance, can be applied extremely rapidly, so that the subsequent adjustment of the equilibrium can still be measured.

This example shows us that one does not necessarily overcome an obstacle in science by continually trying to jump over it. A better way is often to circumvent the obstacle skillfully. Naturally, this requires a high degree of creativity. Yet it was precisely this approach that led to the successful measurement of what previously had been regarded as "immeasurably fast" reactions.

The methods developed by Eigen are a prime example of how an at first seemingly unsolvable scientific problem can in the end be solved. This opened the gateway to a new field of research, which subsequently yielded deep insights into the mechanisms of chemical reactions. Thus, for example, the elementary steps of the transfer and processing of genetic information, and the regulation of biocatalysts, could be investigated experimentally with unprecedented accuracy.

Above all, the life sciences have demonstrated the importance of the development of new research techniques. Only the constant refinement of observation methods has made possible the transformation of biology from a merely descriptive to an exact and experimental science (Fig. 6.1). Although magnifying glasses had already made structures visible that were difficult to see with the naked eye, it was only the development of light microscopy that allowed biology to acquire insight into the hitherto hidden world of cells.

The invention of electron microscopy in the twentieth century finally opened up a further, previously invisible area of living matter, namely the field of cell organelles and their substructures. In recent times, it has even been possible to circumvent a physical limit of light microscopy, employing fluorescence, and in that way to extend classical microscopy into the range of molecules without having to perform invasive interventions into the cells. This technique, developed by the biophysicist Stefan Hell, is called nanoscopy [7]. It is a further impressive example of how bright ideas can overcome the boundaries of physics.

The new techniques of brain research are also exciting. While at first cerebral activities could only be tracked by recording brain waves, employing electroencephalography, the action and excitation patterns of the brain can now be mapped directly on a computer screen by using magnetic-resonance imaging. Such investigations, which require no organic interventions into the brain, have provided first conclusions about the fundamental processes of consciousness: hearing, speaking, reading, feeling and the like.

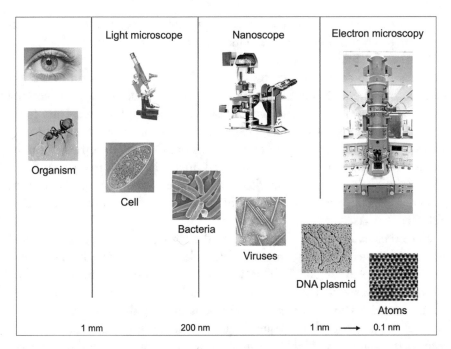

Fig. 6.1 Observation techniques used in biology. The possibilities for making biological structures visible, at all levels of biological organization, is paradigmatic for the overcoming of the natural boundaries of human perception by science. The figure shows length scales and spatial resolution limits of visual inspection (the human eye), of light (optical) microscopy, of far-field optical nanoscopy and of electron microscopy. Nanoscopy, developed in recent years from fluorescence microscopy, extends the resolution of light microscopy well beyond the theoretical limit, calculated by Ernst Abbe in the nineteenth century, of half the wavelength of the light used (~ 200 nm). In contrast to electron microscopy, this technique has the advantage that it can be applied without invasive interventions into the biological material. [Images: Wikimedia Commons]

Beside the continuous improvement in observation techniques, numerous chemical and physical methods have been designed to investigate brain functions in detail. Thus, the "patch-clamp technique", developed by the neurobiologists Erwin Neher and Bert Sakmann, does not only allow one to measure the flow of current in a single nerve cell; it also allows analysis of the exchange of substances through the tiny molecular channels in the cell wall [8, 9]. The paramount importance of this technique for brain research is quite comparable to the development of light microscopy, giving access to the microstructures of living beings.

Of outstanding importance for modern biology are, last but not least, X-ray methods for structure determination of biological macromolecules and techniques for finding out their sequences. These have made it possible to elucidate the molecular structures of biomolecules and to determine the genetic information of numerous organisms. A few decades ago, it seemed almost inconceivable that the genome of even the simplest organisms could ever be deciphered. In the mid-1970s, the techniques of molecular biology allowed only the sequencing of short sections of DNA or RNA molecules; the sequence determination of only a hundred nucleotides took many months of laboratory work. At that rate the sequence determination of the human genome, which consists of several billion nucleotides, would have taken millions of years.

Nevertheless, even in the early 1980s, the virologist Renato Dulbecco had already envisaged the utopian goal of completely deciphering man's genetic construction plan by 2005. This idea led to the creation in 1989 of an international research network, the so-called "Human Genome Project", in which numerous major research institutions around the world participated [10]. Indeed, the goal of sequencing the human genome, which at first seemed to be mere wishful thinking, was almost completed within less than ten years. The Human Genome Project benefited not least from a steadily intensifying race between a state-funded research network and the private company *Celera Genomics* founded by the scientist and entrepreneur Craig Venter, which expected to get a leading market position in biotechnology by sequencing the human genome.

In the end, the private company was a step ahead in this race; however, at the expense of completeness and accuracy. On 26 June 2000, the deciphering of the human genome was officially announced in the presence of the President of the USA. Even though only 80% of the genome had been sequenced by then, so that at best a first working draft was available, the event was celebrated worldwide with superlatives. Sometimes it was compared to the moon landing, sometimes to the invention of printing. Some spoke of a Copernicanian revolution; others thought the event to be in every respect unique and incomparable.

The initial enthusiastic expectation that the comprehensive analysis of the human genome would make it possible to track down the causes of genetic diseases, and one day to cure them by appropriate gene therapies, was followed by disillusionment. It soon became clear that there was still a long way to go between the mere availability of sequence data and an understanding of the functions encoded by the genome (Fig. 6.2). In the course of the sequence analysis, the human genome had been broken down into innumerable fragments, which, however, could only become informative if they were considered as elements of a structure representing a meaningful whole. For this purpose, it is necessary to clarify the complete organization of the genome, i.e. the position, function and coordination of all the genes within it.

As already described (Sect. 1.5), the fabric of functions encoded in genetic information unfolds step by step, in endless feedback loops between the genome and its physical and chemical environment. However, nothing seems to biologists to be more inscrutable than this entangled process. To understand the expression of the genome in its essential features, it is first necessary to locate the genes on the genome and to weight them according to their functional significance.

Bioinformatics deals with precisely this task. It aims to develop the mathematical tools that will enable us to order the enormous amount of data provided by the experimental genome analysis, and to correlate the items of information with one another. Here, everything revolves around the question of which function an information fragment has in the context of all the other information fragments, and how this functional context may possibly change over time. This is like the proverbial search for a needle in a haystack. Nowadays, such questions are approached experimentally by using "gene chip" technology. Thousands of DNA fragments are prepared in the laboratory and fixed on a microscopically small carrier (the "chip"). These serve as probes to filter out the respective complementary, active, gene sequences in the biological cell. With this sophisticated technology, the activity of tens of thousands of genes can be traced in a single experiment.

Naturally, the complexity of the genome is reflected at the level of the molecular function carriers, the proteins. The totality of proteins encoded by the genome is called the "proteome", by analogy to the genome. Proteins perform a wide range of functions in the living cell, mainly affecting and regulating metabolism. To gain detailed insight into their individual functions, one has not only to determine which proteins are activated by the genome at which time and in what quantity, but also to find out how they interact within the context of other proteins.

The amount of experimental data that molecular biology has accumulated in the meantime is so immense that a new problem of understanding emerges. It raised the question of how the countless individual pieces of information can be put together to form a coherent picture that corresponds

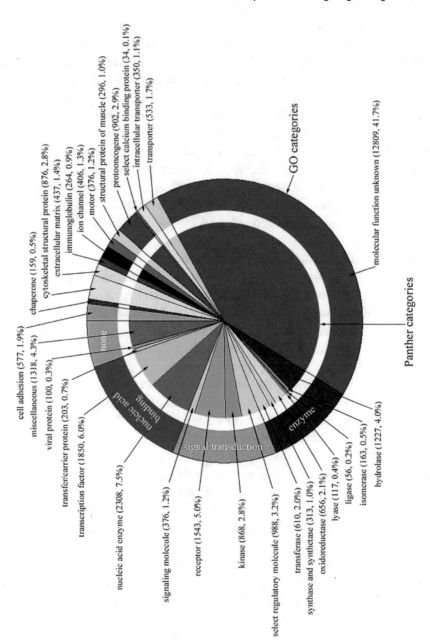

◄**Fig. 6.2** Human genome, categorized by function of each gene product (as known in 2001). The diagram documents a milestone toward an understanding of the human genome. It shows the results of the analysis of 26,383 human coding genes, classified according to the functions of their transcribed proteins. For each category of gene, the number of genes is shown along with the percentage of all genes. The outer circle shows the assignmet to molecular function categories in the Gene Ontology (GO), the inner circle shows the assignment to molecular function categories within *Celera's* "Panther" scheme. The map was published in 2001 by Craig Venter and his team [11]. At that time, only 58% of the gene products could be classified. In the years following, the map was completed and refined step by step. The sequence determination of the human genome is an outstanding example of how science can make possible the seemingly impossible. This is demonstrated by the explosive development of sequencing methods in a relatively short period. In the Human Genome Project, it took more than 10 years and $3 billion to sequence a single human genome. In 2006, it still took several months and cost €10 million to sequence a large genome. Today, it is possible to sequence a complete human genome for less than €1000 in under a day

to a living system as a whole. Problems of this kind are dealt with in a new field of biological research, referred to as systems biology.

At the current state of research, it is not yet possible to unravel the complex life processes involved and to trace them back to a sequence of simple mechanisms and causal chains, in the sense of the reductionist research program. Because of these difficulties, speculation arose as to whether entirely new explanatory approaches, going beyond the previous concepts, may be necessary for understanding the phenomena of life.

It comes as no surprise that, in the course of this discussion, an increased interest in holistic or organismic views has again arisen. However, the experimental progress of biology calls into question neither the reductionist research program nor the mechanistic explanatory approach. On the contrary; if we do not want to settle for a mere description of living matter, we must—even though this seems to be hopeless at the moment—inevitably strive for a detailed mechanistic analysis of the molecular events that take place within it. Only this kind of analysis can ultimately depict and explain the network of cellular processes in all its causal dependencies. The most promising theoretical concept, one that may in time come to guide and support experimental research, will be "molecular linguistics" (see Introduction).

The new techniques of systems biology support this kind of analysis by filtering out the biologically relevant processes from the multitude of conceivable interaction patterns. This procedure already reduces the complexity of possible functional sequences by many orders of magnitude. Systems biology is supported in this challenging task by the concepts of network theory, information theory and others. These are structural sciences, which have the appropriate abstraction level to analyze and to model the complex networked processes that occur in living matter (Sect. 4.9).

6.3   Genetic Information and Communication

Given the complexity of life processes, some commentators have considered the tremendous effort involved in the human genome project to be pointless from the outset. However, this attitude toward genetic research is unscientific, for two reasons. On the one hand, the first step toward a basic understanding of the living organism necessarily requires knowledge of its genetic information. On the other, it has become apparent that, despite all the complexity of living matter, there are also a number of uniform and straightforward principles that control life processes.

One of the surprising insights, for example, relates to the number of human genes, which is considerably lower than had been expected. Earlier, it had been thought that at least one hundred thousand genes were involved in determining the structure of the human organism. The detailed sequence analysis, however, showed that the human genome consists of only twenty to twenty-five thousand genes. This number is comparatively small, considering that even the fruit fly or the threadworm already have between ten and twenty thousand genes.

It is also amazing that the mouse's genome has almost as many genes as the human one. Even more surprising is the discovery that a comparatively simple organism like the rice plant possesses considerably more genes than humans do. Moreover, there seems to be a close genetic relationship between humans and lower organisms such as baker's yeast, the worm or the fruit fly. Many of man's hereditary endowments are also found in lower organisms, albeit in slightly altered form. Apparently, the essential functions of living organisms such as growth, cell division and metabolism are controlled by similar genetic programs.

Yet the most surprising result of the sequence analysis of the human genome is the fact that, so far, a large part of it could not be associated with any function at all. Molecular biologists once used to refer to the genome's "functionless" DNA segments as junk DNA. However, it is difficult to imagine that such a large proportion of the DNA in the human genome should only be waste material. Therefore, it is supposed that some unknown function may be hiding behind the "junk" (or more precisely: non-coding) DNA after all. Maybe, too, the functional gaps in the human genome are an experimental field for evolution. Moreover, no convincing explanation has so far been found for the finding that non-coding DNA contains broad similarities to the genome of certain retroviruses. Retroviruses are RNA viruses whose genetic material is transcribed into DNA after infection of the host cell; DNA containing the viral sequence is then permanently integrated into the host cell's genome. In the Introduction to this book it is argued that all these puzzling findings may find their solution in the idea according to which the so-called "junk DNA" actually represents the grammar of genetic language that regulates gene expression (cf. also Sect. 6.9).

The fact that we have gained such deep insight into the organization of the genome is due mainly to the coordinated worldwide activities that took place within the human genome project. Beside, the analysis of the human genome has also propelled forward the genomic analysis of numerous other organisms. To mention only two examples: In 1995, the complete genome of the dangerous pathogen *Haemophilus influenzae* was deciphered. One year later, the genome of brewer's yeast was published, an organism that is closely related to the human cell. Thus, as a by-product of the human genome project, milestones were set in the development of genetics that no-one would have dared to dream of two decades earlier.

These successes were significant for basic biological research for two reasons. On the one hand, simple organisms such as bacteria or yeast have been among the most important model organisms in biology for decades, and the molecular basis of their life processes can now be studied. On the other, they provided a test not only of the efficiency of the new sequencing techniques, but also of the practical application of a global research network. Just like research into the atomic nucleus, research into the human genome requires enormous experimental effort, which can only be brought about by large research groups in an international cooperation. A new feature of genome research, however, was the fierce competition between state institutions and private companies, which at times threatened to undermine the quality of the research results.

Beside the genomes of man, chimpanzees, chicken, mice, threadworms, fruit flies, rice plants and other higher organisms, at the time of writing the genetic information of well over ten thousand micro-organisms is now known. The sequence data are managed in a central "gene bank", which is continually being expanded and up-dated. The gene bank is of great importance in this respect because it can be used to replace or supplement conventional gene maps with molecular maps of much higher resolution.

Of course, the difficult task of interpreting the data still lies ahead of us. The mere sequence data provide as little information about the life-dynamics of an organism as a list of telephone numbers tells us anything about life in a city. The sequence data only become meaningful and useful when we understand the functional relationships controlled by the genes and thereby learn to understand the essence of living matter.

Here is the point, however, at which the question of the scope of genetic determinism arises. To what extent is the organism determined by its genetic program at all? Giving genetic determinism a radical interpretation, according to which everything comes from the genome, seems to leave no free space for either the organic or the mental development of an individual. In the light of modern biology, however, it is becoming increasingly clear that an interpretation invoking such a strong contradiction is not only misleading but also plain wrong.

Such misinterpretations of genetic determinism come about because they are based on a concept of genetic information that implicitly presumes information to have an absolute character. However, there is no information in an absolute sense. As already outlined, every item of information is relative. It acquires sense and meaning only in relation to some other information. This idea is an indispensable prerequisite for any communication between a sender and a receiver of information. As a matter of course, the inherent contextuality of information also applies to the information in the genome.

The information in the genome, too, unfolds its meaning only in an appropriate environment, namely that provided by a cell. Only this environment gives genetic information, which is in itself ambiguous, a clear meaning at all (cf. [12]). However, since the physiological context changes with each step of gene expression, the actual content of the genetic information only becomes fixed in the course of continuous interaction with the environment (Sect. 6.9). This procedure can best be compared to a dialog, in which the meaning of words and sentences is to be determined by communication. Thus, genetic determinism must be understood as a "generative" determinism based on genetic language (cf. Introduction).

The regulation of gene expression, in short "gene regulation", thus becomes the key for understanding the sophisticated, information-driven interactions of and among biomolecules. The molecular interactions are mechanical in nature, i.e., the sequences of processes follow physical and chemical laws according to a genetically anchored causal scheme, but they do not interlock like cogwheels in clockwork. The causal scheme that they follow we may describe as the "grammar" of genetic language (Sect. 6.9).

At first glance, this mechanism seems to contradict the idea of a clockwork-like mechanism. On closer inspection, however, the apparent contradiction appears in a different light. It is true that clockwork-like regularities have been the epitome of a purely mechanical process, for centuries in fact. However, a mechanical process need not necessarily be inflexible, even if it is based entirely on a mechanical mode of action. Self-regulating mechanisms, for example, always need a certain degree of freedom, i.e. flexibility, to perform their self-regulatory function at all.

Self-regulation in living matter demonstrates this. Here, the mechanism of self-regulation is instructed by genetic information. The fact that genetic information has the structure of a language also allows "communicative" exchange of information between the gene products to take place, and this regulates all process sequences. Despite this, self-regulation has a purely mechanistic explanation. The metabolism of lactose in bacteria exemplifies this (Fig. 6.3). In this case, gene regulation is controlled by continual feedback, based on reaction and response. In detail, specific metabolic products serve as control units which, through tthe action of a regulatory protein ("repressor") either block or unblock the gene section ("operon") of the bacterial genome involved in lactose metabolism. The operon has for the regulation mechanism of lactose metabolism a function similar to that of a sentence in human language (cf. Table 1.1). Thus, the metabolism of lactose illustrates the importance of language-based communication of biomolecules in living matter.

The analogy between genetic information and human language is apparent. It is also entirely plausible. Living matter is based on numerous causal chains that are hierarchically organized. This is also reflected in the hierarchical organization of the genome. Therefore, the expression of the genome cannot occur by translation of the entire genetic information in one fell swoop. Because of its hierarchical organization, genetic information must be expressed as required. However, this requires perpetual communication controlling each step of gene expression. In other words, the dynamization of

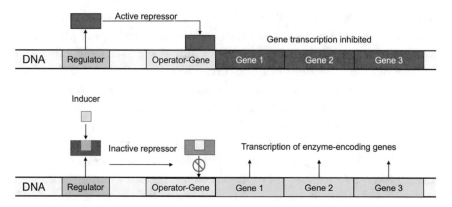

Fig. 6.3 Regulation of lactose metabolism in bacteria. The regulation of lactose metabolism in bacteria, shown here in its basic form, is an example of the sophisticated communication among biomolecules of the living cell. Upper figure: The metabolism is regulated by a protein, termed a "repressor", which functions as a switch. When no lactose is present, the enzymes needed for lactose metabolism are not required, and the repressor blocks the operator gene, preventing the reading of the DNA (from left to right in this diagram) and thus shutting metabolism down: genes 1, 2 and 3, which encode these enzymes, are not expressed. Lower figure: If a lactose molecule ("inducer") arrives, it binds to the repressor and inactivates it. The repressor then no longer blocks the operator gene, with the result that the DNA is read, so that genes 1, 2 and 3 are turned on, the respective enzymes are produced, and lactose metabolism can take place. In this way the cell economises with its resources, producing enzymes only when they are required. The gene unit shown, responsible for the entire, regulated process, is termed an "operon"; it consists of the operator gene and several genes that are funtionally involved in the lactose metabolism

genetic information by gene expression must follow rules that correspond to a linguistic mechanism. In this sense, gene expression is driven by language.

It must be emphasized, however, that human language and the language of genes are not congruent. Human language has peculiarities that are to be traced back to the evolutionary development of man. It is adapted to functions that are different from those of genetic expression. The one is tailored to interpersonal communication and serves the organization of the human life-world. The other serves the intermolecular communication and controls the organization of living matter. Accordingly, the two forms of language are based on different communication machineries. In other words, despite the common features, one must not place the "language of structures", operating in living matter, on a level with the structure of human language. Rather, the latter must be interpreted as the highly complex and specialized product of evolution–although its basis is in line with its genotypic origin.

Therefore, it has become a goal of modern biology to learn the language of living matter and to modify the genetic text, or even to write entirely new versions of it. Such attempts have led to new fields of research aiming at the modification, reprogramming and generation of genetic information. They were accompanied by the development of the new technique of genetic engineering, somatic cell nuclear transfer and artificial evolution, allowing us to override the natural irreversibility of organic development and evolution.

6.4 Reprogramming Gene Expression

Until the middle of the twentieth century, the term "biotechnology" was still mainly associated with fermentation, i.e. the transformation of substances by microorganisms such as yeast, fungi and bacteria. Nowadays, new techniques have emerged that go far beyond the biotechnical procedures of the past. Beside artificial evolution, discussed in the next section, modern genetic engineering and reprogramming of gene expression stand in the foreground of development.

Genetic engineering aims at the modification and recombination of genetic material. The first steps toward this technique were taken around 1970 by the invention of recombinant DNA technology (described in Sect. 7.1). Following upon this invention, the goal-directed modification of the genomes of microorganisms ("white" genetic engineering), of plants ("green" genetic engineering) and of animals ("red" genetic engineering) has led to numerous applications in agriculture, pharmacology, environment protection and other fields. A famous example is Dolly, the cloned sheep, whose lineage is outlined in Fig. 6.4.

Today, the most efficient procedure allowing the precise modification of DNA is gene editing by CRISPR-Cas9. This is the designation for a new tool of genome editing invented by Jennifer Doudna and Emanuelle Charpentier [13]. This procedure has not only enhanced the accuracy of genome editing but also significantly reduced the time required for it. It is of groundbreaking importance for basic genetic research as well as for the further development of the technique of genetic engineering itself. Its possible applications in medicine are expected to provide entirely new insights into the genetic predispositions accompanying genetic diseases and their possible therapy by genome-editing.

Here, however, we will focus on the breathtaking progress that has been made in the life sciences by the various procedures of reprogramming gene

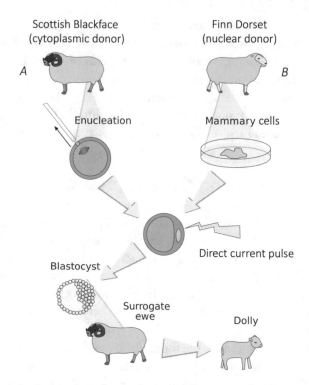

Fig. 6.4 Reproductive cloning of organisms allows one to make a copy of an organism true to its original. The scheme shown refers to the pioneering experiment of cloning a lamb from somatic cells in the mid-1990s [16]. Since then, numerous animals have been cloned–most recently, a pair of primates (long-tailed macaques) [17]. Reproductive cloning is based on the transfer of the nucleus of an adult somatic cell of organism *B* to the enucleated egg cell of organism *A*. The genetic information contained in the cell nucleus of *B* is then expressed in the enucleated egg cell *A*. If *A* and *B* are from the same species, the result will be an identical clone of the nuclear donor *B*, even if *A* and *B* are different breeds. By reproductive cloning, the genetic individuality of an organism gets lost. [Image: Squidonius, Wikimedia Commons]

expression that are collectively referred to as "somatic cell nuclear transfer". This name stands for a technique by which a nucleus from a fully differentiated cell undergoes complete genetic reprogramming when it is introduced into an enucleated oocyte (cf. [14]).

Somatic cell nuclear transfer was first successfully applied for the reproductive cloning of genetically identical animals (Fig. 6.4). In this case, the nucleus of an adult somatic cell of an organism *B* is introduced into the enucleated oocyte of an organism *A*, whereupon the genetic information contained in the cell nucleus of B is expressed in the cytoplasm of cell A. If A and B belong to the same species, the result of the ontogenetic development is an identical clone of the organism B.

If the oocyte and the donor nucleus are from different species, then hybrids can form across species boundaries (Table 6.1). Cross-species cloning is expected to provide essential insights into the interaction of the cell nucleus with its cytoplasmic environment. As argued in the previous section, only the context gives the–in itself ambiguous – genetic information an unambiguous meaning. Since the genome also modifies its own context, with each step of expression, the process closely resembles the communication between sender and receiver of a piece of information [15].

If the recipient cytoplasm and the donor nucleus are from the same species, they are based on a common mechanism of gene expression. This is, so to speak, the common basis of mutual understanding between sender and recipient of genetic information. In other words, genetic information is—in the sense of genetic determinism—a necessary and sufficient condition for the reproductive maintenance of an organism, as genetic information (only) becomes operational in "dialog" with its cellular environment.

The situation is different, however, if genetic information is expressed in a foreign cellular environment, for example in the cytoplasm of another species. In this case, the sender (genome) and the receiver (cytoplasm) of genetic information are not adapted to each other, since they have different

Table 6.1 Interspecies cloning. Some examples of interspecies cloning, as carried out up to 2007 (for the complete and detailed list see [18]). Since the term "species" is not sharply defined, terms for subgroups and transitional groups such as subspecies and breed are used. The taxonomic classification of the above list is based on a definition that regards species as a group of organisms that could interbreed naturally and produce fertile offspring

Taxonomic relationship	Recipient oocyte	Donor cell	Year
Interclass	Cow	Chicken	2004
	Rabbit	Panda	2002
Interorder	Cow	Dog	2005
	Cow	Whale	2004
Interfamily	Cow	Sheep	1999
	Goat	Tibetan antelope	2007
Intergenus	Cow	Goral	2006
	Wildcat	Leopard cat	2006
Interspecies	Cow	Zebu	2001
	Goat	Ibex	2005

evolutionary histories. Presumably, this will be reflected in the mechanisms of gene expression as well. In consequence, the evaluation standards of gene expression may be expected to shift the more, the larger the taxonomic distance between the genome and the cytoplasm.

The influence of cytoplasmic factors on gene expression has been investigated in various ways, including for example by cross-species nuclear transfer experiments in fish. For this purpose, a hybrid fish was cloned by transferring the nucleus of a common carp into the enucleated egg cell of a goldfish (Fig. 6.5). The experimental results suggest that the interaction between the nucleus and the maternal factors is not only crucial for early developmental steps, but also that it exerts an influence on later development.

This and similar experiments demonstrate that the cytoplasmic context plays an active part in genome expression. At the same time, they confirm the importance of communication in living matter, as outlined in Sect. 6.3. Therefore, the technology of reprogramming gene expression, which

Fig. 6.5 Cross-species cloning of a hybrid fish (right) from a common carp (middle, which provided the nucleus) and a goldfish (left, which provided the egg). The experiment, carried out by the biologists Yong-Huan Sun and Zuo-Yan Zhu, served to clarify the influence of cytoplasmic factors on the development of organisms [19]. In the case shown, all hybrids were identical to the nucleus-providing common carp in respect of exterior phenotypic characteristics (form, color, etc.), while no characteristics of the egg-providing goldfish were visible. However, an X-ray investigation showed that the vertebral numbers of the cloned fish were typical of the goldfish and distinctly different from those of the common carp (for details see [19]). [Image: courtesy of Yong-Huan Sun and Zuo-Yan Zhu]

intervenes directly in the dialog between the genome and its cellular context, may also help to clarify basic questions of theoretical biology. These questions include that of the nature of genetic language. Moreover, it is quite possible that the elucidation of gene expression will also allow far-reaching conclusions to be drawn about the mechanism of evolution. There is much to suggest that the context-dependent development of the individual organism in particular, and the context-dependent evolution of living matter in general, follow the same principles. It will be the task of theoretical biology to work out in detail the common features and the differences between ontogenetic and evolutionary development.

The practical implications of somatic cell nuclear transfer are apparent. For example, this technique may provide an auspicious approach for the reconstruction of endangered species. Above all, however, it may help to establish animal models that are useful for biomedical research. Moreover, with the simultaneous breakthrough in stem-cell research, the technique of reprogramming genetic systems has acquired entirely new perspectives.

Stem cells are characterized by their nearly unlimited capacity to divide and to differentiate into cells of virtually all types. However, this only applies to the so-called "totipotent" stem cells, which include fertilized egg cells up to the third stage of division, i.e. the eight-cell stage. This stage is followed by differentiation, which restricts the development of these cells. From the eight-cell stage onwards, they are, therefore, referred to as only "pluripotent" cells (Fig. 6.6).

An embryonic stem cell is, as it were, the original form of a differentiating cell; from it, in principle, each of the approximately two hundred cell types of the human organism can emerge. Such stem cells can be cultivated endlessly, so that research has a virtually inexhaustible reservoir of differentiatable cells at its disposal. Particularly advanced are investigations with stem cells of the mouse. Here, it has already been possible to grow active heart-muscle cells, nerve cells and the like.

One aim of stem-cell research is to cultivate human tissue and organs in vitro, with a view to using them for regenerative therapeutic purposes. This perspective may sound like a distant dream of the future. However, a vital step toward this goal has been the discovery by John B. Gurdon [20] and Shinya Yamanaka [21] that somatic cells can regain their capacity to differentiate into any bodily cell type. The reverted cells are termed "induced pluripotent stem" (iPS) cells. The possibility of reprogramming mature cells into pluripotent stem cells was still considered unthinkable a few decades ago. However, experiments with iPS cells have now shown us how to reverse the

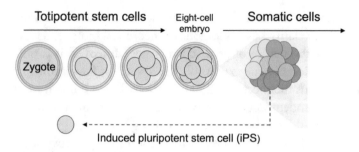

Fig. 6.6 Reversal of cell differentiation by reprogramming embryonic stem cells. After the eight-cell stage of embryogenesis, the process of cell differentiation becomes irreversible. The discovery that somatic cells can regain their capacity to differentiate into any cell type allows reversal of the natural process of differentiation by biomedical methods

process of biological differentiation, a process that in normally growing organisms really is irreversible.

The groundbreaking results of embryonic stem-cell research, along with the new methods of somatic cell nuclear transfer and gene-editing, are of outstanding significance, as they allow us to modify and design living matter to a hitherto unimagined extent. Even though the recombination of genetic material has been the basis of traditional breeding methods from time immemorial, the instruments of modern biology and reproductive medicine have endowed scientific progress with an entirely new quality. Consequently, the biosciences are at the front-line of the cultural and social upheavals that have been triggered by modern science.

We need hardly emphasize that this development poses a serious ethical challenge to society—one which demands of science a high willingness to take responsibility. However, the guidelines for this task cannot be laid down by an invariant, metaphysically based value system. Quite on the contrary: it has become clear that a number of the ethical problems once thought to be fundamental have been sorted out by science itself, since the rapid progress of biotechnology has revealed ways to circumvent them. Stem-cell research is an example of this, and it impressively verifies the central thesis, which will be outlined in Chap. 7, according to which knowledge itself can only be controlled by progress in knowledge. Therefore, the value system of today's knowledge society must rely upon a knowledge-based canon of values which, like science itself, is open for change and is continuously adapted to the objective conditions of progressing knowledge.

6.5 Artificial Evolution

The versatile applications that biotechnology offers are sometimes dismissed as pure wishful thinking. However, the deciphering of the human genome should have taught us that what is supposedly impossible can move into the realm of the possible more quickly than we might have imagined in our wildest dreams.

Unattainable wishful thinking also seems to lie behind the idea that one day man will be able to plan biological evolution and steer it according to his goals. If, however, this idea could indeed be realized, then it would undoubtedly have drastic consequences for our understanding of Nature. At the moment when science gains control over evolution, the barriers between the natural and the artificial would disappear without trace.

However, the idea of artificial evolution is not as distant from reality as it seems at first glance to be. Optimization strategies based on the model of natural evolution are already being applied in solving technical problems. They are an essential element of bionic technology (Sect. 7.2). Admittedly, this is only an "analogy" technique, in which the problem-solving capacity of natural evolution is merely copied. Only if such experiments could be carried out with biological material could one speak of artificial evolution in the proper sense.

Given the enormous complexity of living systems, one might doubt that such experiments are possible at all. Moreover, the time factor is likely to be decisive here, as the evolution of natural systems usually takes place on time scales that exceed all experimentally accessible time scales by many orders of magnitude. However, the speed of evolution can be influenced from the outside. As the population geneticist Ronald Fisher discovered, the speed of evolutionary development is proportional to genetic diversity. The latter, in turn, depends on the mutation rate with which the genetic material of a population changes during its continuous reproduction. Thus, to achieve a higher speed of evolution, one can increase the mutation rate artificially with the help of mutagens. However, noticeable acceleration of evolution can only be achieved for organisms that form large populations and reproduce very quickly, such as bacteria and viruses.

In experiments to test this, bacteria and viruses indeed show all the characteristics that we would expect of Darwinian evolution. Nevertheless, such experiments are not really suitable for pursuing any goal of artificial evolution. Because the organisms on Earth have already undergone a long phase of optimization, they are very limited in their capacity for further development. If one wishes to initiate artificial evolution with developmental possibilities that are still largely open, then it is expedient to start from the molecular precursors of life, because only at this (very low) level of complexity can the degrees of freedom of evolution be exploited to the full.

For example, evolutionary experiments with biomolecules might be carried out to develop drugs with predetermined properties. In such a case, one would start with a random mixture of biomolecules and filter out specific functionally active molecules by employing selective methods to optimize them according to the principles of natural selection. It would also be conceivable to set up experiments in which biological macromolecules are synthesized directly from their molecular components, to expose them to artificial evolution in the test-tube. All these considerations presuppose, however, that the principles of natural selection and evolution already operate in the pre-biological domain, and also that the corresponding processes can be simulated in the test-tube.

Precisely this has been achieved, thanks to a pioneering experiment carried out by the molecular biologist Sol Spiegelman in the mid-1960s [22]. Spiegelman and his colleagues were investigating the properties of a class of viruses which molecular biologists term "bacteriophages", or "phages" for short, as they attack bacteria. In their experiment with RNA phage $Q\beta$, they discovered a virus-specific enzyme that exclusively propagates the genetic material of the virus. Figuratively speaking, the enzyme runs like a sewing machine along the viral RNA molecule, making complementary ("negative") copies of the viral RNA by stealing the RNA building-blocks already present in the infected bacterial cell. In a second replication step, the "negative" is copied and thereby turned back into a "positive" copy. This mechanism, based as it is on the principle of complementary base-pairing, is called "cross-catalytic replication" (Fig. 6.7).

However, since errors occur at random during the copying process, some copies always differ from the original viral RNA, the so-called master sequence. Such mutations are the reason why the genetic material of a replicating virus is no longer homogeneous when it has passed through several reproduction cycles. Instead, it consists of the master sequence and the resulting mutant distribution (together these make up the so-called "quasispecies"). Therefore, a reproducing virus RNA is always surrounded by a mutant spectrum, like a comet that is followed by its comet tail.

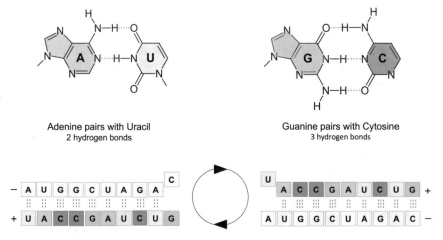

Fig. 6.7 Cross-catalytic replication of nucleic acids (see also Sect. 1.5). Above: By way of hydrogen bonds, the nucleobases A and U and the nucleobases G and C join up to form base pairs that are geometrically almost identical. The base pairs differ, however, in the strength of their binding. GC-pairs are hold together by 3 hydrogen bonds, AU-pairs only by 2 hydrogen bonds. Below: Complementary base-pairing is the basis for the replication of a nucleic acid molecule. By joining up with complementary nucleotides, the nucleic acid to be copied first generates a "negative" copy (the "minus" strand) and in a second replication step it is converted back into a "positive" copy (the "plus" strand). In the chromosomes, for example, the two complementary DNA strands are joined up into a double helix resembling a screw (see Figs. 4.7 and 4.16). In reproduction the DNA strands become separated, like the halves of a zipper, and each strand is then joined up to a newly formed complemetary strand

Spiegelman's experiments caused a big stir when he succeeded in isolating biochemically all the components that are involved in the reproduction of the virus. At first, he was only interested in proving that the genome of the virus can be reproduced in the test-tube at will. However, the idea soon arose to put the system under selection pressure, so as to filter out the variants in the viral RNA's mutant spectrum that reproduce the fastest.

Such an experiment is started by inoculating a nutrient solution with the RNA of the Qβ virus (Fig. 6.8). The nutrient solution contains all the ingredients necessary for cell-free replication of the viral RNA. After a short incubation period, a small sample of the reaction solution is transferred into a fresh nutrient solution to initiate a new phase of replication. This procedure is repeated several dozen times. The repeated transfer of a part of the reaction solution into a fresh culture medium leads automatically to the selection of those RNA mutants whose replication rate is higher than the average rate of all the variants present. However, since the repeated selection of rapidly replicating molecules continually raises the average replication rate, it is to be expected that the mutant which has the highest replication rate of all the variants will ultimately be selected, to the exclusion of all the others. Obviously, in cell-free replication, the structural features of RNA (Fig. 6.9) are the crucial factor regarding the molecules' replication behavior.

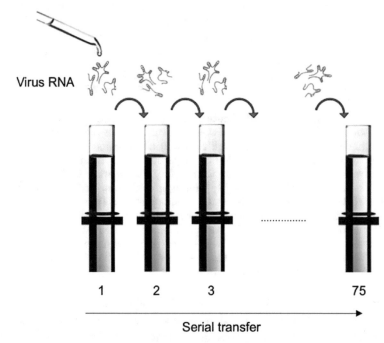

Virus RNA

1   2   3   75

Serial transfer

Fig. 6.8 Darwinian selection in the test-tube. First experiments of this kind were carried out with the genetic material of a bacterial virus [22–24]. They are based on a "serial transfer" technique. A reaction solution (test-tube 1) is inoculated with a virus RNA. After incubation for a certain time, a sample of the reaction solution is transferred to a fresh nutrient solution (test-tube 2). This procedure is repeated several times. The dilution automatically exerts selection pressure upon the RNA population, favoring the variants that replicate fastest for details see text). In the pioneering experiment, first carried out by Spiegelman and his colleagues, the RNA of the bacterial virus Qβ was subjected to 74 transfers [22]. As an end product, an RNA variant was isolated that had lost much of its genetic information. At the same time, however, this RNA variant was able to reproduce itself many times faster than the original RNA could. Today, such experiments can be performed by laboratory robots (cf. Fig. 6.11)

In Spiegelman's laboratory the expected outcome of the experiment was achieved after only a relatively small number of transfers (74 transfers, in fact). The real surprise, however, was the result of the subsequent product analysis. It showed that, under the luxurious conditions of its test-tube world, the genome of the virus had "thrown away" all the information that would hinder its rapid reproduction. Importantly, this included the genes that are needed for the natural infection cycle of the virus, but which are not required for cell-free reproduction in the test-tube. These genes were completely removed from the viral genome by mutation and selection. The shorter an RNA sequence is, the less time it needs for reproduction, and the more copies are produced in a given time. Therefore, any random shortening of the chain length led to a selective advantage over other, competing RNA sequences. At the end of the experiment, the remaining RNA variant had lost more than 80% of its genetic material, so that it only possessed the bare ability to reproduce itself and had lost all its other abilities.

The result of the serial transfer experiment actually contradicts our intuitive understanding of evolution, according to which biological complexity increases in this process. Here, instead, complexity is decreasing. All the information that the virus needs for replication in vivo is eliminated in the test-tube experiment. Up to now, no one has succeeded in initiating an evolution experiment leading to enrichment of genetic information. What was observed in Spiegelman's experiment was only a mere adaptation, and not a development toward a higher level of complexity, as the Darwinian idea of "evolution by adaptation" suggests.

To study the evolutionary process in more detail, Spiegelman and his team isolated one of the smaller RNA variants created in his transfer experiments and determined its nucleotide sequence. They then again subjected this variant to the selection pressure of a transfer experiment by adding a dye to the reaction solution. Certain dyes, such as ethidium bromide or acridine orange, are known to interfere with or prevent the replication of nucleic acids because they interfere with the nucleic acid's folded, tertiary structure. In this experiment, however, the RNA species developed resistance against the dye, and the resistance could be traced back to three point mutations in the RNA's nucleotide sequence (Fig. 6.10).

The "naked" RNA molecules emerging by evolution in vitro did not carry any information that would have enabled them to reproduce autonomously.

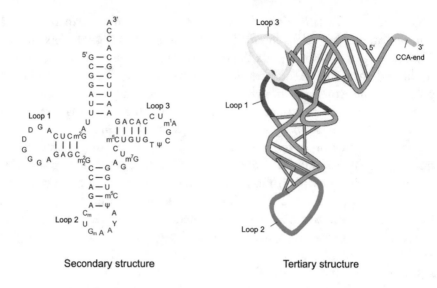

Secondary structure   Tertiary structure

GCGGAUUUAGCUCAGDDGGGAGAGCGCCAGACUGAAYAΨCUGGAGGUCCUGUGTΨCGAUCCACAGAAUUCGCAACCA

Primary structure

Fig. 6.9 Structural classification of RNA. The figure shows the structure of phenylalanine transfer RNA (tRNA^{phe}) which is involved in protein biosynthesis. The molecule consists of 76 nucleotides, of which some are chemically modified. Left: The folding of the nucleotide sequence ("primary structure") of RNA by intramolecular base-pairing is termed its "secondary structure". For large RNA structures the secondary structure can be modeled by an algorithm that simulates the optimum folding of the primary sequence, by using thermodynamics and auxiliary information. The secondary structure considers only the backbone of the RNA and the folding of the chain due to base-pairing, while it neglects all other conformational properties such as the interactions between the loops and other factors that contribute to the overall structure, the "tertiary structure". Right: By the secondary structure's interaction with its physical and chemical environment, the molecule's three-dimensional "tertiary structure" arises [25]

Instead, they needed the help of an enzyme to catalyze their reproduction. Therefore, in these experiments selection could only aim at influencing the dynamical features of the RNA molecules, features determined exclusively by the molecules' structure. Accordingly, RNA molecules with random changes in their nucleotide sequence that reduced or prevented the dye's binding had a selective advantage in the adaptation competition.

The structural diversity of RNA molecules is presumed to have been an essential factor in the early evolution of life. At the beginning there was no biotic environment that could have served as a frame for adaptation in the evolution of genetic information. In that phase, only the physical and

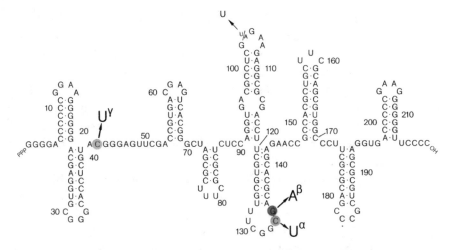

Fig. 6.10 Adaptation of RNA molecules by natural selection. The replication of RNA molecules is inhibited by certain dyes, which burrow into the RNA's folded structure and impede or even prevent reproduction. The RNA molecule shown here has been subjected to a serial transfer experiment in the presence of the dye ethidium bromide [24]. After several transfers, a variant was isolated that was (to a certain degree) dye-resistant. This variant differed from its parent sequence by three point mutations, which appeared not simultaneously, but successively (in the order indicated, from α to γ). Furthermore, the nucleobase 104, which in the parent RNA was variable, became the fixed nucleobase U. RNA molecules that originate by cross-catalytic reproduction under selection pressure exhibit high symmetry in their nucleotide sequence, reflecting the fact that natural selection of RNA aims at optimizing the molecule's structural features under test-tube conditions

chemical environment determined the dynamic, and thus the selection behavior of molecules, as simulated in Spiegelman's test-tube experiment. In the latter, the targets of selection were exclusively the structural properties of autocatalytic molecules, and accordingly evolution was restricted to the "syntactic" level of the RNA's nucleotide sequence.

However, because RNA structures are enormously complex and distinct from one another, they provide a nearly inexhaustible reservoir for natural selection in the real world. For the RNA variant shown in Fig. 6.10, there are already 10^{130} combinatorially possible alternatives of its nucleotide sequence. Thus, from a purely chemical point of view, a correspondingly large variety of possible RNA structures is conceivable, all of which could enter into selection competition with each other. These are more (possible) RNA structures than there are molecules in the entire universe.

In the 1980s, Sidney Altman [26], Thomas R. Cech [27] and others made the surprising discovery that RNA molecules much smaller than Spiegelman's

RNA variants may have catalytic properties like those of enzymes (cf. Fig. 6.26). Such RNA molecules are called "ribozymes". Today we know that ribozymes catalyze a number of chemical reactions, including their own reproduction. The discovery of ribozymes led to a paradigm shift in molecular biology, abolishing the sharp distinction between information carriers (nucleic acids) on the one hand and function-bearers (proteins) on the other. Ribozymes are of particular importance for the understanding of the origin of life, as they represent the link between chemical evolution and molecular self-organization (see Sect. 6.9).

The test-tube experiments, simulating the "evolution" of biological macromolecules, refuted the long prevailing idea according to which the principle of natural selection is a specific characteristic of living systems—a dogmatic claim that many scientists had persistently defended for a long time. Even in the 1960s, the biophysicist Ludwig von Bertalanffy asserted: "And even if complex molecules like nucleoproteins and enzymes are considered as being 'given', there is no known principle of physics and chemistry which, in reactions at random, would favor their 'survival' against their decay." From this, he concluded: "Selection, i.e., favored survival of 'better' precursors of life, already presupposes self-maintaining, complex, open systems which may compete; therefore selection cannot account for the origin of such systems." [28, p. 82] Remarkably, this claim comes from the same period in which Spiegelman carried out his pioneering experiments. Shortly afterwards, Eigen was also able to prove from a theoretical point of view that selection is indeed a general principle of Nature, which, under certain conditions, already appears among self-reproducing molecules [29].

Within the framework of these investigations, Eigen also developed the concept of an evolution machine. This machine was first conceptualized as an idealized model for the physical and chemical studies of molecular evolution [23, 29, 30]. For the theoretical foundation of biological evolution it plays a role similar to that of the Carnot engine in the physical foundation of the second law of thermodynamics, or that of the Turing machine in the mathematical theory of computability (Sects. 4.4 and 5.3). In the 1990s, the concept of an evolution machine was also put into practice in new technology (Fig. 6.11).

The idea of using an evolution machine to address questions about the origin and evolution of life focuses our attention on two possible approaches to such questions. On the one hand, we may ask what the historical conditions were under which life originated. This would require a detailed reconstruction of the physical and chemical conditions that prevailed on Earth and which ultimately led to the formation of living systems—a

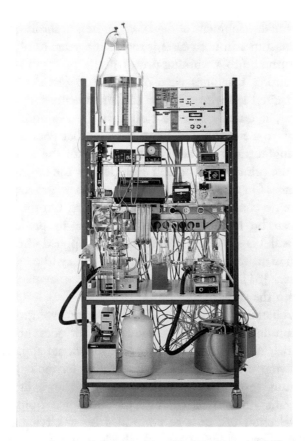

Fig. 6.11 Prototype of an evolution machine. The machine shown here, developed by Eigen and his co-workers in the 1990s, is today on display in the German Mueum, Bonn. Nowadays, advanced evolution machines have become an indispensable part of modern biotechnology, aimed primarily at the development of pharmacological substances. [Photo: courtesy of the German Museum Bonn]

formidable challenge. On the other hand, more realistically, we can ask what the general principles are that once brought about the transition from non-living to living matter. To investigate this we can use experimental systems that are free of historical specifics. The evolution machine meets precisely this description.

As pointed out above, the concept of an evolution machine was at first a theoretical construct for the study of the evolution of biological macro-molecules. However, the early experiments of Spiegelman already prompted the idea that the serial-transfer technique could be automated, turning the abstract model of an evolution machine into a practical reality. This step is comparable to the development of digital computers according to the pattern of the Turing machine. In both cases it took a few years until the theoretical idea could be turned into a working prototype.

Initially, evolution machines were only used to test experimentally the theoretically acquired knowledge about the fundamentals of molecular evolution. However, it quickly became clear that the evolution machine also pointed the way to a new kind of biotechnology that goes beyond conventional genetic engineering. Artificial evolution has for the first time made it possible to use the principles of natural evolution for the targeted synthesis of biological reagents. On the one hand, substances can be generated in this way that would have no chance of survival in free Nature. On the other, existing biomolecules can be further optimized according to predefined criteria. Gerald Joyce and his co-workers have recently applied the technique of directed evolution to redesign a ribozyme that can synthesize its own evolutionary ancestor [31]. From such experiments, we may expect completely new insights into the early history of life.

Evolutionary biotechnology is applied molecular evolution under controlled and reproducible laboratory conditions. It does not require any "natural" material, nor does it depend on the functional apparatus of living organisms or living cells. The advantages of this technology are clear: in contrast to natural evolution, artificial evolution allows the processes of reproduction, variation and selection to be uncoupled from one another. Furthermore, selection can now be controlled by specifying a direction of development. Artificial evolution has today become an innovative technology, able to induce evolutionary development in an abiotic medium.

6.6 Pathways to Artificial Life

Evolution experiments with biomolecules have refuted a traditional doctrine of biology according to which only living matter is capable of evolution. This belief managed to entrench itself in the minds of biologists, although none other than Charles Darwin himself had already speculated about the possibility of prebiotic evolution. In February 1871 Darwin wrote to the botanist Joseph Hooker: "It is often said that all the conditions for the first production of a living organism are now present, which could ever have been present. But if (and oh! what a big if!) we could conceive in some warm little pond, with all sorts of ammonia and phosphoric salts, light, heat, electricity, &c., present, that a protein compound was chemically formed ready to undergo still more complex changes, at the present day such matter w[ould] be instantly absorbed, which would not have been the case before living creatures were found." [32, vol 3, p. 18]

The idea that in an energy-rich chemical "primordial soup" some kind of evolution could begin to operate was taken up by the biologists Aleksandr Oparin and John Scott Haldane at the beginning of the twentieth century and expanded into a possible scenario for the origin of life. A few decades later, Stanley Miller, a student of the chemist Harold Urey, was able to simulate such a chemical scenario in a glass flask [33]. The experimental result demonstrates that amino acids, the building stones of proteins, may originate spontaneously under prebiotic reaction conditions (Fig. 6.12). The test, however, must not be misunderstood in the sense that it actually reflected conditions on the primordial Earth. The experiment rather demonstrated that a spontaneous synthesis of amino acids from abiotic substances is possible *in principle*.

There seems to be no way in which we can reconstruct the specific historical conditions under which chemical evolution took place in the early phase of the development of life. Everything allowed by chemical and physical laws will happen, provided that the corresponding reaction conditions also prevail. However, since the precise conditions on the early Earth are largely unknown, many possible scenarios are conceivable. Consequently, this has opened the door for numerous speculations. However, it is the primary task of science to investigate natural processes independently of the contingent conditions of their historical givenness. Therefore, the scientific understanding of the origin of life must focus primarily on the law-like relationships that would govern the formation of primitive life forms. In this respect, the Miller–Urey experiment has undoubtedly played a pioneering part.

Numerous subsequent investigations have verified the idea that complex organic molecules, right up to biological macromolecules, can emerge even in an abiotic environment. Even though many questions regarding chemical evolution are still unanswered, no serious doubt remains that the chemical building blocks of the living organism could arise spontaneously under suitable reaction conditions and, as Darwin suspected, evolve further.

The fact that the processes of chemical evolution can be simulated in the test-tube seems to bring us closer to an old human dream: the generation of life in a retort. In late antiquity and the Middle Ages, the idea of the *homunculus*, an artificially created miniature human, was widespread. Especially the alchemists felt magically attracted by this idea. In a book published in 1527 and entitled *De generatione rerum naturalium*, attributed to Paracelsus, there is even a recipe for producing a *homunculus*. It is based on the idea, prevailing at that time, that only male seed is life-giving material, and

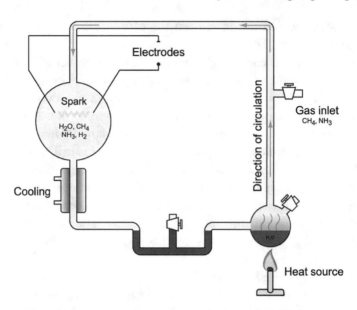

Fig. 6.12 Miller–Urey experiment. The figure shows the experimental arrangement that was used for the first time in the 1950s to simulate the prebiotic synthesis of amino acids. Water is brought to the boil in a glass flask (lower right). The water vapor drives a reducing gas mixture of methane (CH_4) and ammonia (NH_3) into a chamber (upper left). Here, spark discharges between two tungsten electrodes induce various chemical reactions. The reaction mixture then runs through a cooler (lower left), where it condenses. The non-volatile reaction products are collected in a U-tube next to the water reservoir (centre, bottom). The volatile products are re-distilled out of the water reservoir and, together with the water vapor, again exposed to electrical discharges in the reaction chamber. The non-volatile products, on the other hand, are concentrated in the U-tube. [Image: Carny, Wikimedia Commons]

therefore life can also develop outside the mother's womb. For this purpose, according to the recipe for the *homunculus*, it is only necessary to mix a little semen, urine and blood and to keep the reaction mixture in a sealed flask at a constant temperature for a few weeks while supplying proper nutrients.

Nowadays, we would not even remotely come up with the idea that living organisms can originate merely by mixing together the corresponding organic substances. Nevertheless, the ideas of the alchemists do point in the same direction as the modern idea of a "prebiotic soup" as the starting point for the development of life. That there is no impassable boundary between living and non-living matter became evident in the nineteenth century, when the chemist Friedrich Wöhler succeeded in synthesizing "organic" urea from "inorganic" ammonium cyanate. In our time, with the far-reaching discoveries of molecular biology, the recognition that the boundaries between the animate and the inanimate are fluid has became a certainty (Sect. 5.7).

Since the groundbreaking experiment of Miller and Urey, research has been able to draw a more accurate picture of the emergence of life, dividing the entire process into three major phases. Thus, beside the initial stage of chemical evolution, there must have been a phase of molecular self-organization during which the biological macromolecules—proteins and nucleic acids—organized themselves into a self-sustaining and self-reproducing unit. Such a self-reproductive unit may be considered to be the forerunner of the living cell. The third phase in the transition from non-living to living matter is the Darwinian phase of biological evolution. Following this picture, one necessarily arrives at the conclusion that there have been numerous steps on the way from molecules toward the first living beings. All these steps together make the transition from non-living to living matter appear as a quasi-continuous process.

The concept of the natural self-organization of matter and evolution of life provides a consistent scientific working hypothesis for explaining the origin of life. It also allows the problem of artificial life to be addressed. Given the fluid transitions at the threshold of life, the first question to be asked is: Which concept of life are modern approaches to artificial life to be based on? We have already seen that any definition of life necessarily has a normative character if the transition between non-living and living matter is a fluid one. Consequently, the meaning of life always depends on the specific aspect addressed by the particular questions the researchers are asking. In short: There are many degrees of freedom in defining life, and this ambiguity is reflected in the ambiguous use of the term "artificial" life.

In order not to restrict the concept of life too much at the outset, we will use the definition given in Sect. 5.7: Life = Matter + Information. This is the most abstract definition of life that one can imagine. It goes beyond our intuitive understanding of terrestrial life, since it characterizes, as outlined in Sect. 4.8, the deep structure of organized matter. For this reason, the above formula provides the best framework for systemizing the different concepts of artificial life. It allows, for example, "artificial life" to be based upon highly diverse forms of material organization. It also leaves a margin for the application of various boundary conditions ("information"), ones that might imprint on matter a functional structure bearing characteristic features of living beings.

Let us consider two examples illustrating this. The first refers to artificial life forms based on "electronic matter". In this case, the computer serves as a virtual laboratory for the generation of structures that we may term electronic or digital life (cf. [35]). Examples of this are the early attempts of Aristid Lindenmayer to simulate the growth of multicellular organisms by employing so-called "generative" algorithms [36]. For this purpose, Lindenmayer introduced a mathematical formalism ("L-system") that is based on the recursive application of a defined set of rules with certain specifications. The digital life forms generated in this way look deceptively like plants, for example (Fig. 6.13). Such computer simulations demonstrate that simple algorithms may already generate intricate patterns that reflect perfectly the elegance and beauty of natural objects (see also Sect. 3.1). Scientifically, these experiments were aimed at the mathematical modeling of possible mechanisms underlying the development of organic structures.

Another most fascinating version of digital life is based on a computer game invented by the mathematician John Horton Conway. The game simulates life processes like the rise, fall and alternations in a population of organisms ([37]). The game starts from a homogeneous space partitioned into cells, like a chessboard (ideally, an infinitely large one), divided into rows and columns (Fig. 6.14). Each grid square represents a "cell" that may adopt one of two states: occupied or empty. Each cell has eight neighboring cells. Their states may be referred to as "alive" or "dead". At the beginning, the first generation of "living" cells is arbitrarily placed on the grid.

The dynamic behavior of a cell follows the rules described in the legend of Fig. 6.14. By applying these rules, specific aspects of living populations are simulated, such as isolation, cooperativity and overpopulation. Although the rules of the game are simple, they nevertheless lead to very complex behavior. Thus, the cells may move over the grid; they may disappear or multiply; they may remain unchanged; or they may oscillate. Figure 6.15 shows a typical result of the game, in the first and the 50th generation.

This is not the place to go further into the details of the game. Instead, let us consider the game's background and its relationship with the problem of artificial life. The "cells", which develop according to the rules of the game, are reminiscent of the cells on the tape of a Turing machine (Fig. 5.2). Their states also change in time according to specific transition rules, and they also develop in discrete steps. The cells of the game, however, have the property that their further development depends not only on their own state but also the states of their respective neighboring cells. Dynamical systems with such features are called cellular automata. In modern informatics, they serve in the modeling of information processing machines.

Fig. 6.13 Virtual life. Photorealistic simulation of the development of the plant *Mycelis muralis* (from [34, p. 91]). The structures, generated by a context-sensitive L-system (see text below), constitute an early example of the computer-aided creation of "digital life forms". Such experiments were designed primarily to investigate the possible mathematical structures underlying the processes of development and pattern formation in biology

The foundations for the modern theory of automata were laid by the mathematicians Stanislaw Ulam, John von Neumann and others in the 1950s. At that time, the question also arose as to whether one can conceive of an automaton that can reproduce itself. Soon after, von Neumann succeeded in demonstrating with mathematical rigor that this is in principle possible. Since self-reproduction is a characteristic feature of living systems, cellular automata inevitably came into view in theoretical biology. Since then, numerous cellular automata have been designed, simulating artificial life. Conway's life game is a vivid representation of this approach.

It is amazing that the application of a set of relatively simple rules already leads to such complex dynamical behavior of the cellular elements as is the case for Conway's life game. Conversely, it may seem strange that simple patterns that reproduce, change and develop on a computer's screen are already denoted as artificial life. However, as mentioned, any scientific definition of life is a matter of the perspective from which the phenomenon of life is investigated. There is no uniform definition that comprises necessary and sufficient criteria which might allow one to draw a borderline between artificial and natural life. Even for natural life, one has to adjust and specify the definition anew from case to case.

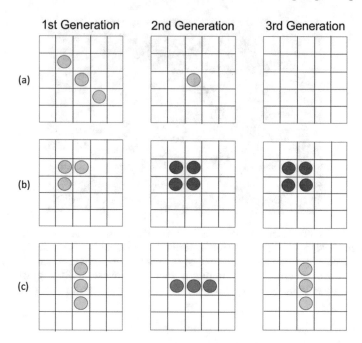

Fig. 6.14 Conway's life game. The rules define survival, death and birth. (1) Survival: a cell survives to the next generation if two or three of its adjacent squares are occupied as well. (2) Death: a cell dies if either more than three or fewer than two adjacent squares are occupied. (3) Birth: an empty square becomes occupied if exactly three of its adjacent neighborhood squares are occupied. Three examples of the development of a three-membered configuration, in each case applying the rules three times, are shown. Configuration (a) is unstable. It dies out after the second generation. Configuration (b) goes over from the 2nd generation into a stable block of four cells. Configuration (c) oscillates; it is termed a "blinker". Other structures, like the "glider", move across the grid, passing through a repeating sequence of shapes on the way. The various configurations are classified according to their dynamical behavior. They have also been given names like "beehive", "traffic light" etc

The greater the degree of abstraction from natural conditions, the more the concept of life disintegrates into specific variants deviating from the natural forms of life. This problem is inevitably reflected in the quite different usages of the term "artificial life". This designation, however, does not imply that the structures generated by specific algorithms are really alive. Instead, these investigations aim at modeling aspects of life, in order to attain a better theoretical understanding of the underlying mechanisms. An example of this is provided by the algorithmic models that are designed to simulate biological development.

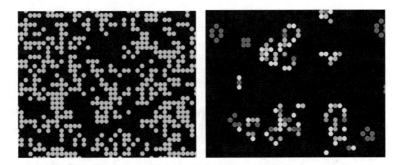

Fig. 6.15 Computer simulation of Conway's life game. Snapshot of digital life forms as they typically develop in Conway's life game. The computer simulation starts from a random distribution of occupied squares (left). By application of simple rules, a variety of cell clusters arise that show complex, dynamic behavior such as we know from real living populations. The screenshot of the computer display after 50 generations (right) shows some typical structures, which develop or disappear or which pass into a stable configuration in the course of the game

There are also research strategies that tackle the challenge of generating artificial life from quite different scientific and technological perspectives. They can be classified, for example, according to their material and technical properties (Table 6.2). One very promising route is the attempt to start from a living being and to minimize it to such an extent that it just maintains its essential vital functions, metabolism and reproductivity. An experiment of this kind was carried out a few years ago by a group of researchers around Craig Venter [40]. The microorganism *Mycoplasma genitalium*, one of the smallest autonomously reproducing organisms, served as model (Fig. 6.16). Through targeted mutations, with which individual genes could be switched off, it was possible to make a rough estimate that only about a half of the genetic material is required to maintain the life functions of this organism under laboratory conditions. Such experiments undoubtedly constitute an essential step on the way to artificial cells, since they provide information about the minimum requirements of single-celled microorganisms, albeit only indirectly. Moreover, minimal organisms are very close, and similar, to natural life.

A different path toward synthetic life was taken some years ago by Vincent Noireaux and Albert Libchaber [39]. Inspired by von Neumann's automata theory, they constructed a vesicle bioreactor operating like a cell. To realize such an artificial cell, they isolated from the intestinal bacterium *Escherichia coli* a cell extract and provided it with an artificial cell wall. As it turned out,

Table 6.2 Forms of artificial life. Depending on the scientific and technological perspective, there are quite different approaches toward artificial life. (1) Digital life, bound to algorithms, only simulates life. (2) Synthetic life is based on *de novo* construction of life from organic compounds (Sects. 6.4 and 7.2). (3) Minimal life originates by reducing organic life to its minimum requirements. It is the closest to natural life. (4) Hybridized life forms can be generated either by genetic engineering or by merging biological with technical material. (5) Bionic life forms imitate the material properties, technical solutions and evolutionary strategies of real life (Sect. 7.2). There is no real borderline between these different concepts of artificial life, because the transition between non-living and living matter is fluid. Instead of "minimalizing" a living being by deconstructing its material complexity, one can go the other way around and reconstruct living beings from molecular components. Such a restitution experiment was carried out by Heinz Fraenkel-Conrad and Robley Williams with viruses in the mid-1950s [38]. They succeeded in breaking down simple virus particles into their basic molecular building blocks (nucleic acids and proteins) and then reassembling these into infectious virus particles, which however had lost their original individuality (for details see Sect. 5.7). Historically, this was the first step toward "synthetic" life

Classification	Property	Example
(1) Digital life	Simulated	Cellular automaton [36]
(2) Synthetic life	Biochemical	Artificial cell [40]
(3) Minimal life	Biochemical	Synthetic genome [40, 41]
(4) Hybridized life	Organic	Cross-species [19]
(5) Bionic life	Technical imitation	Robots

the synthetic cell could sustain genetic transcription and translation for several days. Even though the synthetic cell could not reproduce itself, it nevertheless represents an innovative starting point for building minimal cells that, in turn, allow detailed molecular and supramolecular studies of the mechanisms of self-reproduction at the threshold of life.

The diverse experimental approaches to the exploration of synthetic life provide essential support for the thesis of the quasi-continuous transition from inanimate to living matter. Since the biomolecules that are necessary for the construction of a living system can in principle also be produced chemically and subsequently optimized in evolution machines, it will one day be possible to create a great variety of synthetic life forms in the laboratory.

However, the synthesis of biological building blocks is only a preliminary stage on the way to complex artificial life forms. Only when these building blocks organize themselves to a functional whole will such artificial systems reflect organic life in every sense of the word. In biological systems, self-organization is ensured by genetic information. Thus, it will constitute the decisive step toward higher forms of artificial life when semantic information can also be implanted into synthetic biomolecules (Sect. 6.8).

Fig. 6.16 Minimal synthetic cells. The image shows a colored electron micrograph of *Mycoplasma mycoides* JCVI-syn3.0 cells, the first autonomously replicating bacterium controlled by a synthetic genome with just 473 genes. It was derived from *Mycoplasma bacteria* by a group of scientists at the Craig Venter Institute [41]. For this purpose, a genome based on the *Mycoplasma mycoides* genome was synthesized and then inserted into the shell of a *Mycoplasma capricolum* from which the genetic material had been removed. Such experiments are the starting point for investigating the function of living cells at their most fundamental level. [Image: Thomas Deerinck, NCMIR/Science Photo Library]

6.7   Toward Artificial Intelligence

The progress in researching artificial life also raises the question of the possibility of artificial intelligence. As with artificial life, the answer depends on clarification of the concept of "intelligence". It stands to reason to search at first for criteria referring to human intelligence. Ernst Cassirer, for example, considered man's ability to handle and manipulate symbols as a decisive criterion of intelligent behavior. For just this reason, he suggested that one should "define man not as an animal rationale, but as an animal symbolic" [42, p. 51].

However, Cassirer's interpretation seems questionable, because technical systems, such as computers, are also able to manipulate symbols. This objection raises a further question. Does a machine that can manage symbols have to be aware of this to be classified as intelligent? The Latin root *intellegere*, which means to "recognize", "understand" and the like, might suggest that conscious reflection accompanying action is an indispensable feature of intelligent behavior.

Such an interpretation is not mandatory. It is perfectly conceivable that a system could successfully exchange symbols with other systems and manipulate them without having insight into the symbols' meaning. Conscious reflection on the usage of symbols could become superfluous anyway, since the meaning of symbols only becomes evident through the effects that they exert on their recipient. In other words, from an external perspective, any system that can successfully manipulate symbols will appear to be intelligent.

Moreover, a definition of intelligence orientated exclusively toward human intelligence would neglect the fact that intelligent actions can also be seen, in different forms and gradations, in animals. The use of tools, the ability to solve problems arising from changing habitat, changes in strategic behavior, individual memory performances and the like, are clear indications of intelligent behavior of animals. Therefore, it is not possible to draw a sharp line between intelligent and non-intelligent behavior. Any such attempt inevitably runs into problems that are similar to those that arise in defining life.

Alan Turing tried to circumvent these problems by renouncing from the outset any attempt to narrow the concept of intelligence to specific features. Instead, he provided an operational description of intelligence that can be applied to machines—for example, computers. This procedure, known as the "Turing test", is performed in the following way [43]. A test person and two interlocutors (one of which is a computer) are communicating with each other (Fig. 6.17). A wall separates the interlocutors, so that they can neither see nor hear each other. Their communication is exclusively digital, using keyboard and computer screen. However, the test person is allowed to ask his interlocutors cunningly thought-out questions to find out which of the two is the computer. If after intensive questioning he is unable to identify his non-human interlocutor, the computer has passed the Turing test and is to be regarded as intelligent.

The Turing test is designed to check whether intelligent human behavior can be ascribed to a machine. It does not reveal anything about intelligence as such. In an experiment carried out some years ago, 11% of the test persons were unable to recognize that their interlocutor was a computer. This means, conversely, that by Turing's definition 89% of the subjects were more intelligent than the computer that answered them. One may expect, however, that with progressing computer technology, the relationship will shift more and more in favor of the machines, indicating their increasing intelligence. Some day, this development may even turn the Turing test upside down. Instead of measuring the intelligence of computers, the test could document the superiority of artificial intelligence over "natural stupidity".

In the Turing test, the answers of the computer are merely the electronic mirror image of the questions addressed to it. The test is based solely on an operational definition of intelligence, which does not include any forms of mutual understanding. Thus, a computer that passes the Turing test gives the false impression that it possesses creative intelligence. In reality, it is only reflecting intelligent behavior. Therefore, it would be interesting to find out whether a sufficiently complex computer could learn to understand what it is doing and whether it could also make its own decisions.

In this respect, any substantial progress in artificial intelligence will depend on how flexible and autonomous computers can become in the future. Various technological approaches that are based on biological models reflect this development toward autonomous computing. At present, three developmental lines stand out: (1) evolutionary computing, (2) organic computing and (3) biomolecular computing. However, this is only a rough classification, and fails to take into account the fact that these overlap in many respects.

Let us first consider evolutionary computing. Already in the early stage of computer technology, algorithms were develped that organized and optimized themselves, according to the model of biological evolution. They are summarized today under the general heading of "evolutionary algorithms". They include, beside evolutionary algorithms in the narrow sense, procedures such as genetic algorithms, genetic programming and evolutionary programming.

The generation of such algorithms follows in several steps by applying evolutionary mechanisms: recombination, mutation, selection and others. In this way, evolutionary algorithms can adapt themselves to given tasks. In science and technology, such algorithms are primarily used for optimization and search routines. With the development of adaptive algorithms, computers have become more and more autonomous, as they are increasingly able to determine the course of their computing operations themselves, organizing and optimizing their information processing internally.

Another information-theoretical approach to artificial intelligence, closely oriented toward living systems, is "organic computing". The aim of this is to transfer the information-driven mechanisms of biological self-organization, pattern formation, autonomous behavior and the like to technical operations. One of the most promising projects of organically structured information technology is the simulation of neural information processing in the brain (Fig. 6.18).

Artificial neural networks process, like their natural model, information on several tracks at the same time—a procedure that is denoted as parallel information processing. Compared with sequential information processing,

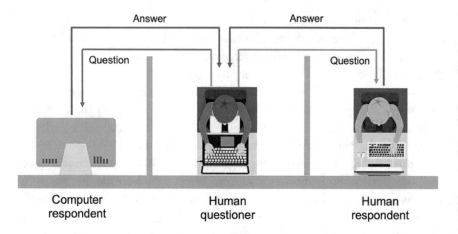

Fig. 6.17 The turing test. A test person is communicating with two hidden interlocutors: another test person and a computer. The first test person is allowed to ask any conceivable questions, which his interlocutors answer. If he is not able to work out which of the two interlocutors is the computer, then the computer has passed the intelligence test

which is done in strictly successive steps, parallel information processing has two significant advantages: On the one hand, it is much faster, because in this case tens of thousands of calculation steps can be carried out at the same time. On the other hand, parallel information processing reduces the adverse effects of errors, because any errors occurring are distributed among the parallel computing processes and are not propagated in the manner of a chain reaction.

Neural networks are designed to simulate realistically in the computer the performance of the human brain in functions such as the ability to learn or to recognise patterns. They are an essential part of a rapidly developing information technology, according to which machines generate knowledge by gaining information from the process of information processing itself. Machine learning is an exciting and fast-developing branch of artificial intelligence, and it has numerous applications. These range from biometrics, bioinformatics and image processing up to speech and text analysis, to mention but a few. Last but not least, machine learning has also become the basis for a new research area called "data mining", which pursues the discovery of knowledge in large databases.

A most impressive aspect of machine learning is "deep" learning. This technique is based on artificial neural networks which have a particular deep structure regarding the processing of information. The fundamental

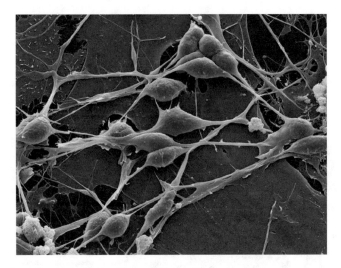

Fig. 6.18 Neural network. The human brain is the most complex organ created by Nature. It consists of about 10^{11} neurons (nerve cells), each of which is connected, on the average, with 10^3 other neurons. Shown here is a scanning electron micrograph of cortical neurons on glial cells (flat, dark grey) and their interconnecting dendrites. Glial cells provide structural support and protection for neurons. The extremely complex network of neurons gives the brain its tremendous power of data-processing. The simulation of the brain's information processing in the computer has inspired a new technology, termed organic computing. At present, the most advanced tool that this offers is machine learning, which enables machines to imitate brain functions such as learning or pattern recognition. [Image: David Scharf/Science Photo Library]

characteristic of deep learning is that the input information is processed over a cascade of intermediate steps, whereby each intermediate level provides the input information for the next step (Fig. 6.19).

The layer structure of information processing is an intrinsic property of the learning process, since any learning requires the continual re-evaluation of information. The intermediate steps between the input and the final output, the hidden layers, contribute to the refinement of the information processing. Deep learning has found a wide range of applications, ranging from natural language-processing, medical diagnostics and image identification up to autonomous driving of road vehicles.

Evolutionary and organic computing are first steps on the path toward an organic technology, which also includes other technological branches such as bionics and robotics (Sect. 7.2). Progress can be expected in this respect, not least from nanotechnology. This technology aims to miniaturize technical processes down to the molecular level.

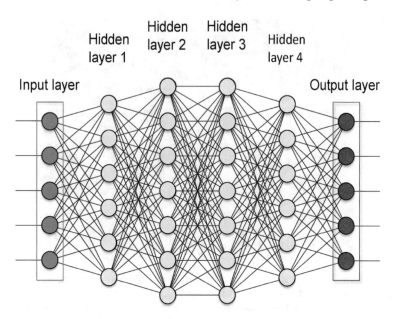

Fig. 6.19 Artificial neural network. Deep (machine) learning is a computer technique based on a multi-layered neural network. Between the input layer and the output layer, there are further intermediate (hidden) layers, in which the actual learning process takes place. For example, if the task is to recognize a human face, the raw data set of the image, consisting of a large number of pixels, is processed according to increasing levels of complexity. Layer 1: the machine identifies brighter and darker pixels. Layer 2: the machine learns to identify simple forms. Layer 3: the machine learns to recognize complex shapes and objects. Layer 4: the machine learns to associate shapes and objects with human faces. In this way, large data sets (input layer) can be categorized and processed according to their content as revealed in the output layer

The idea of nanotechnology was developed in the early 1960s by Richard Feynman in a lecture entitled *There's Plenty of Room at the Bottom* [44]. Feynman understood his talk as an "invitation to enter a new field of science". Beside many other ideas, Feynman, even at that time, discussed the possibility of building submicroscopic computers. His visionary thoughts, however, could only take on concrete form when—following the development of scanning tunneling electron microscopy at the early 1980s—atomic-force microscopy was developed, with which individual atoms and molecules can be selectively moved and assembled into molecular machines. Today, nanostructures, which are only a billionth of a meter ("nanometer") in size, play an important role in modern technology and, not least importantly, in medical and pharmaceutical research. Nanotechnology has also become a further pillar of biotechnology.

In the 1990s, the mathematician and computer scientist Leonard Adelman took up Feynman's idea of a molecular computer when he demonstrated experimentally that computing with molecules is indeed possible [45]. For his experiment, Adleman made use of the fact that the living cell already operates like a molecular computer, co-ordinating and controlling an enormous number of chemical reactions that occur simultaneously.

Using the tools of molecular biology, Adleman succeeded in building, in the test-tube, a "liquid" computer from DNA molecules and enzymes. This computer worked with literally hundreds of trillions of DNA molecules at once. Thus, the DNA computer is comparable to a massively parallel computing machine processing huge amounts of complex data. It demonstrated impressively its capacity by solving a difficult combinatorial problem, related to the so-called "Hamiltonian Path Problem" (for further details see [46]).

Computing with DNA is comparatively slow, as the rates of chemical reactions are far below the speed at which electrons move in the circuits of an electronic computer. In contrast, however, DNA computing can solve problems that require far more calculations than all calculators of the world could ever perform together. Thus, for example, in a DNA computer, billions of computing steps can be carried out at the same time. The best conventional computers are currently lagging far behind the benchmarks set by DNA computing. Since in a DNA computer information is stored in molecular form, the information density achievable exceeds that of traditional data storage by many orders of magnitude. One research team has even started to build a molecular Turing machine, which uses the filamentous structure of DNA as an "input tape" and particular enzyme complexes of the cell machinery as a monitor, carrying out defined "computational operations" on the molecular input tape (cf. [47]). The possibilities of computing with proteins have also been discussed [48]. Recently, even RNA computing in a living cell has been reported, marking a step toward programming cell behavior [49].

There seem to be no borders at the interface of biology and computer science (Table 6.3). The fact that living cells can be induced to carry out computations in the manner of computers brings us back to the question discussed at the beginning, "What is intelligence?". At present, artificial intelligence is still a "virtual" intelligence that serves to support human intellect in its central functions of recognizing, learning, communicating, calculating and the like. Yet we have to ask: How far can the development of artificial intelligence lead?

The philosopher John Searle has doubted that a computer, only following given algorithms, could be considered to be intelligent [50]. Inspired by the Turing test, he developed an alternative thought experiment that he believed would refute the idea of intelligent computers. He outlined a situation in which a person, alone in a room, can successfully edit a given text in a foreign language, let's say Chinese, without understanding the meaning of Chinese characters. The only precondition is that this person has a general background

Table 6.3 Forms of autonomic computation. Inspired by the idea of self-reliant (autonomic) computation, three approaches to artificial intelligence have evolved in recent times. (1) Evolutionary computing is the collective name for techniques based on algorithms that can adapt themselves to specified tasks by self-optimization [51]. The different methods partly overlap and allow a broad spectrum of applications. Autonomic computation employing adaptive algorithms is a decisive step toward higher forms of artificial intelligence. (2) Organic computing aims at the development of organically structured information technology. At present, the focus of organic computing is on artificial neural networks, which simulate the interactions between neurons in a nervous system. (3) Biomolecular computing is based on the capacity of biological macromolecules to store and process information. Biomolecular calculators, such as DNA computers, can not only solve mathematical problems, but they also provide, in combination with nanotechnology, new approaches to data-recording and data storage, medical diagnosis and other practical issues

Method	Specification	Applications
Evolutionary computing	Evolutionary algorithms	Communication engineering, Multidimensional optimization
	Genetic algorithms	
	Evolutionary programming	Agent-based modeling
	Genetic programming	Benchmark test
	Evolutionary optimization	Bionics (Sect. 7.2)
Organic computing	Artificial neural networks	Image recognition
		Speech recognition
		Natural language processing
		Autonomous driving
Biomolecular computing	DNA/RNA computing [52]	Data recording
		Medical diagnostics
		Cryptoanalysis
		Nanobiotechnology
	Computing with proteins [48]	

knowledge regarding the text and knows the syntactic rules for handling the signs. Searle tries to prove that such a test person would be well able to manage a document in Chinese—with such perfection that even a Chinese-speaker would not recognize that the text had been edited by a person who was not a native speaker of Chinese. Searle argues that this person would have passed the Turing test, despite not understanding a single word of the text edited. For the same reason, Searle concluded, a computer passing the Turing test only faked intelligence.

It is no wonder that the Chinese Room Argument has set off a heated debate on the question of whether artificial intelligence is possible at all. Searle's thought experiment has been strongly criticised, for example by Daniel Dennett [53]. However, one does not even have to go into the details of the Chinese Room Argument to see that it is misleading, not to say wrong, in equating intelligence with understanding.

Let us consider an example from physics. How can one still support Searle's argument when it is known that even scientists sometimes talk about scientific results without understanding them? Quantum theory is such a case. Its mathematical formulation describes correctly the phenomena of the submicroscopic world, even though we cannot understand quantum phenomena because of their lack of intuitive content (Sects. 1.10 and 3.6). Quantum theory is thus an example of successfully processing mathematical symbols, while applying only the rules of logical thinking, without understanding the meaning of the resulting theory. This discrepancy prompted Richard Feynman's legendary remark: "I can think I can safely say that nobody understands quantum mechanics." [54, p. 129]

As outlined above, any discussion on the possibilities of artificial intelligence depends on an appropriate definition of "intelligence". Turing, unlike Searle, was foresighted enough not to get involved in a debate on definitions. Instead, he based his test on an operational description of intelligence, looking only at the intellectual difference between a hidden computer and a human by asking whether the computer can give a human the impression that it too is a human.

In the Turing test, it does not matter whether the computer "understands" the information which it is sending out. For Searle, in contrast, understanding—whatever he himself "understands" by that—is an indispensable feature of intelligence. However, do we know whether a person with whom we are talking actually understands what he or she is telling us? Do, for example, quantum physicists understand the phenomena that they have discovered? This is a question that we do not usually think about. Since any understanding is relative, we are already satisfied if what a person is communicating to us makes sense to *us*.

As little as we can look into the mind of our interlocutor, just as little is our insight into the possible inner states of a computer. However, one thing is certain: cognitive capabilities depend on the number of neurons and their interconnections. Thus, intelligence is among other things a matter of the complexity of the neural organization of the brain. We therefore may ask at what level of complexity intelligent behavior begins that also includes understanding. In other words, how many neurons constitute a smart brain that also tries to understand what it is doing? This brings us back to Eubulides' famous question: "How many grains of sand make up a pile?" (Sect. 5.9). We have seen that we do not have a definite answer to this question. Instead, it suggests that the cross-over from non-intelligent to intelligent behavior is gradual, as is the transition between non-living to living matter.

Searle published his thought experiment at the early 1980s. Since that time, the computer-based techniques of artificial intelligence have developed so fast that Searle's objections to artificial intelligence have shrunk to the more or less trivial claim that artificial intelligence is not the same as the human mind. Even at that time, his argument seemed like a desperate attempt to resist the advance of scientific knowledge, as is typical of dogmatic principles of belief.

The compilation in Table 6.3 shows that modern information technology has long since outperformed the simple algorithmic approach to artificial intelligence. Machine learning in image recognition, speech recognition, language translation and the like are only a few examples of the technological progress in artificial intelligence. Given the rapid advancement of research and technology, one may say—to quote Ludwig Wittgenstein freely—that the solution to the problem of artificial intelligence is to be found in the problem's disappearance [55, 6.521]. Nevertheless, new problems will arise. If machines based on artificial intelligence become more and more autonomous, are we then not in danger of losing control over them one day? Do we know how the most advanced algorithms do what they do? Precisely this question was recently addressed in an MIT Technology Review entitled *The Dark Secret at the Heart of AI* [56].

Today, learning machines are at the forefront of autonomous artificial intelligence. However, in what way do machines actually learn? How do they make decisions? Do they make the same mistakes as humans? These and similar questions necessarily remain unanswered, since an artificial learning system cannot explain to us why it did what it did. Owing to the overwhelming complexity of such systems, not even the engineer who designed the system will be able to find an answer.

These anticipated problems of control stand out particularly clearly in the context of learning machines, but they are in no way restricted to them. Already today, sophisticated computer programs are inscrutable in themselves. Such programs make computers operate like a black box, leaving open not only the question of the computer's inner states but also that of the correctness of their computations. Ultimately, the borderlines between simulated and experienced reality may disappear too. This applies wherever computers simulate complex processes in Nature and society (Sect. 7.10).

6.8 Modeling Semantic Information

Artificial neural networks are undoubtedly an essential stage toward a sophisticated digital intelligence. At the same time, they provide an impression of the fascinating perspectives of science at the interface of biology and information technology. However, concerning the further development of digital intelligence, an important question arises: Will computers one day also be able to distinguish between meaningful and meaningless information, or will they even autonomously generate meaningful information? This question leads right to the heart of artificial intelligence. It is the same question that is fundamental for life's origin and evolution (see Introduction).

Traditionally, the concept of meaning belongs to the domain of linguistics and philosophy of language. In linguistics, the meaning of signs and, in a broader sense, also the meaning of words, sentences and texts is denoted as "semantics". In ordinary language, other expressions are usual for the notion of meaning, such as sense, importance, content and so forth. It is an essential task of language philosophy to give such terms a precise interpretation by specifying them further.

An example of this is the theory of "logical" semantics, which goes back to the mathematician and philosopher Gottlob Frege. Within the framework of this theory, for example, a strict distinction is made between the terms "meaning" and "sense". According to Frege, the "meaning" of a sign or expression is the object to which it refers, while the "sense" is the manner in which it appears to us [57]. Thus, the terms "evening star" and "morning star" have the same meaning, because they refer to the same object, namely the planet Venus. However, they have a distinct sense, because the planet Venus is referred to in the term "morning star" in a different way from in the phrase "evening star".

The fact that the morning star is the same object as the evening star does not require any further explanation; yet the fact that the star observed at sunrise near the sun is the same object as the one that appears in the evening sky after sunset is by no means obvious. Rather, it is an insight that we owe to the astronomical observations of the Babylonians. Frege strove for such semantic clarifications, not only concerning proper names but also concerning sentences. His investigations ultimately led to profound insights into the truth value of linguistic utterances.

The concept of logical semantics, however, is too closely tailored to human language to be able to contribute to a general understanding of the dimension of meaning. Instead, we need a comprehensive approach that goes beyond information expressed in human language and can also be applied to algorithmic information (in connection with artificial intelligence) as well as genetic information (in connection with life per se). In other words, we are in search of a general approach to semantics that combines quite different forms of semantic information.

To implement this idea, we have first to bring the problem of semantics right down to the level of information theory. For this purpose, we consider the traditional model of sender and receiver, used by communication engineers. The scheme depicted in Fig. 6.20 represents in abstract form the information transfer from an information source (sender) to a receiver. If the information flow goes only in one direction, the meaning of the information transferred is specified exclusively by the interpretation which the receiver gives to the message. In bidirectional information flow, the meaning will be determined by the reciprocal communication between the receiver and the sender of the information. A biological example of a non-human communication is the information transfer between the genome and the cell's machinery, particularly in respect of its physical and chemical environment in general.

In interpersonal communication, however, the intentions of sender and receiver play a decisive role. Therefore, it makes a difference whether one is considering human or non-human communication. In interpersonal communication, the use of language exerts a great influence on the meaning of information. People who send out messages or exchange information with each other pursue, as a rule, practical intentions. They want, as the philosopher John Austin put it, "to do things with words" [58]. In language philosophy, this aspect of communication has become the subject of the theory of speech acts (Sect. 2.9). The communicative function of language, in particular its context-bound usage and the situation-related meaning of linguistic expressions, is a subject of the field of linguistic pragmatism.

Sender   Receiver

Fig. 6.20 Generalized scheme of sender and receiver. Schematic representation of the transfer of an item of information from a sender to a receiver and vice versa. If one considers only the information transfer from sender to receiver, the transfer process has the character of mere instruction. If one also takes into account the opposite direction of information flow, which establishes a feedback between sender and receiver, then the scheme represents communication

In the following pages, we will treat the sender of information as just a black box. In this case, it is unimportant whether the sender is pursuing any intentions with its message or not. Thus, we will only have to consider the reaction of the receiver. Let us furthermore suppose that the sender is transferring information as a sequence of signs or signals, for example, letters, symbols, speech sounds or similar (Fig. 6.21).

On the basis of this model, three aspects or dimensions of information can be distinguished (cf. [59, chapter 4]). The syntactic aspect refers to the manner in which the signs are arranged, and their relationship one to another. It characterizes the basal level of information. The semantic aspect includes the relationships among the signs and what they mean. The pragmatic aspect of information, in turn, comprises the relationships among the signs, what they mean, and the effects that they engender with the recipient.

Since the three dimensions of information stand in an ascending order of interdependency, one might come up with the idea of deriving, for example, the semantics of a piece of information from its syntax. However, we must abandon this idea at once, as there is no necessary connection between the syntax of a string of signs and the meaning associated with it, i.e., a connection that would make it possible to deduce the semantics from the syntax. As is known from human language, it is not possible to recover a mutilated or incomplete text by merely analyzing the syntax of the remaining parts of the text. Such a reconstruction can only succeed against the background of additional information.

In linguistics, the relationship between syntax and semantics is denoted as "arbitrary". That is to say that even though the syntax is the carrier of semantics, syntax does not *cause* semantics. Thus, semantics is an example of supervenience, a term by which the non-causal dependency between two

语言是自然界的一般原则
————————→
(1) String

(2) Evaluation

(3) Action(s)
————————→

Fig. 6.21 Scheme of information processing. The processing of a piece of information by a receiver involves three basic steps. (1) The receiver must identify incoming signals as a string of signs and analyze its syntactic structure. This step requires that the receiver has prior information about signs as such to be able to recognize the string as a piece of "potential" information. In the case shown here, the receiver needs prior knowledge allowing him to recognize the signs as Chinese characters (cf. the Chinese Room Argument, Sect. 6.7). (2) The receiver evaluates the contents of the string and associates meaning with it. Thereby, the semantic aspect of the information emerges. (3) The semantics of information, in turn, may induce a reaction of the receiver, in the broadest sense. It may modify the inner state of the receiver, its knowledge states, expectations and the like, and/or trigger interactions of the receiver with its environment. All possible reactions resulting from the receipt of a piece of information characterize its pragmatic aspect. The entanglement of syntactic, semantic and pragmatic information demonstrates that information is always context-bound and is therefore a relative entity

properties A and B are described (Sect. 3.8). However, the precise sense in which the relationship between syntax and semantics is considered to be arbitrary is a topic of lively discussions in linguistics. Maybe the link between semantics and syntax is only a convention established by the community using the language in question.

It can be objected that the terms "arbitrary" and "conventional" still imply causation, even if this is only given by the fact that the relationship between syntax and semantics has been fixed by the will of the language community. To avoid such inclarity, it would be better to denote the connection between syntax and semantics as contingent. Unlike the terms "arbitrary" und "conventional", the concept of contingency is neutral as regards causation; it merely expresses the fact that the relationship between syntax and semantics is not *necessarily* the way it is.

One could perhaps think that the contingent relation between syntax and semantics is a peculiarity of human language. However, this is not the case. It should be recalled that the same holds for the language of genes. Here, the semantics of genetic information is equivalent to the functional organization of living systems. Functionality, however, is a supervenient property of underlying structures. It is based on particular structures, but it cannot be derived from them (Sect. 5.4).

The semantics of linguistic expressions and the semantics of genetic information are comparable regarding their supervenient relation to their respective syntax. In both cases, there are no rules or laws that allow one to derive the contents of information from the information's syntactic form or its structure. Rather, syntax and semantics are different dimensions of information. There is no law-like connection between them, yet they are inextricably linked to each other.

Thus, the key to understanding semantic information lies with the recipient, who evaluates the information received. Before embarking upon an analysis of the relationship between sender and receiver, we must briefly review the concept of information as developed by the mathematician Claude E. Shannon at the end of the 1940s (Sect. 1.4). Shannon was guided by the idea that information has the task of eliminating uncertainties. Accordingly, he considered the information content of a message to be the higher, the less its expectation probability is on the side of the recipient.

Although Shannon's theory avowedly does not aim at the actual content of information, it nevertheless uncovers a general aspect of semantic information, namely the "novelty" value that it has for the recipient. Shannon was even able to give this aspect a precise mathematical form. Accordingly, the information content of a message is inversely proportional to its expectation value (details can be found elsewhere [59]). If, in the extreme case, the expectation probability of a message is one, then its novelty value for the receiver is zero.

Nevertheless, such a piece of information may well have some value, if it confirms an expectation on the part of the receiver. Confirmation, in turn, could be not only of theoretical but also of practical relevance for the receiver. In other words, the novelty value, given by the expectation probability of a message, is context-bound, as it depends on prior knowledge on the part of the receiver. Therefore, the confirmation of information may even trigger a learning effect at the receiver, changing expectation probabilities.

The question of how a piece of information affects the receiver falls within the scope of pragmatic relevance of information. Information may change the inner state(s) of the receiver and thereby also trigger—beside possible learning processes—actions and interactions of the receiver concerning its external environment.

At the end of the 1960s, the information theoretician Donald McKay outlined a model demonstrating how the inner states of a receiver might be modified by information. According to his model, the receiver functions like a "key-operated signal-box". With this comparison, McKay referred to the older one-track railway lines, where the engine-driver carried a key "which must be inserted in a special signalling lock at the station before his train can be cleared into the next section of the lines" [60, p. 25]. McKay argued that, similarly, the incoming information in the receiver (for example, the brain) serves as a key by which a particular selection from the range of possible configurations of signal-levers is made. If a different message comes, a different configuration is selected, by which "the recipient's states of conditional readiness for goal-directed activity" are modified. "Defined in this way," McKay elaborates, "meaning is clearly a relationship between message and recipient rather than a unique property of the message alone." [60, p. 24].

An example may illustrate this. Assume that person A informs person B that is raining. What will happen? Perhaps B does not intend to go out in the next few hours. Then B may not react at all to the information received from A. This, however, does not mean that the information from A does not influence B. If B is unexpectedly called away, B will look for an umbrella. Moreover, if a person C comes in from outside, B will probably ask if C has got wet. It is not the receiver's actual behavior that allows the significance of information to be seen, but rather the readiness of the receiver for goal-orientated action. What, in the end, the receiver actually does on the basis of the message received may then be denoted the "effectiveness" of a piece of information in the narrower sense.

The selective function of a piece of information narrows the recipient's action fields, the scope and options of actions. The actions themselves are triggered by the actual content of the information. In the particular case where the content is an order, a norm or similar, the information gives the recipient no options for action. This case may concern quite different forms of information, from the civil code or traffic regulations through to genetic information. Selectivity, for example, is essential for the nucleation of semantic information in living matter [61].

An essential feature of semantic information is, last but not least, its degree of complexity. The notion of complexity is here to be understood in the sense of algorithmic complexity theory, according to which the complexity of a message or a piece of information is tantamount to its incompressibility (Sect. 5.4). The less redundant a piece of information is, the greater is its information density. A greater information density allows one to encode a correspondingly greater amount of sophisticated, meaning-rich information. Accordingly, complexity is a precondition, which the syntax has to fulfil if it is to become a carrier of semantic information at all [61].

Novelty, confirmation, selectivity, effectiveness and complexity are general aspects of semantic information. They represent elements of a semantic code from which a recipient may build up its evaluation standard of semantic information (Fig. 6.22). In contradistinction to the usual understanding of the word "code", the semantic code does not provide any rules for the assignment of signs to another source of signs. Instead, the elements of the semantic code determine the value scale that a recipient applies to a piece of information that he wants to assess for its meaning (Fig. 6.23).

The semantic code is thus an objective as well as a subjective instrument for the evaluation of a piece of information. If all recipients are using the same standards of semantic information, the code represents an objective value scale. By attaching different weights to the evalution standards, the recipient expresses his subjective, i.e., individual and unique, understanding of the message's meaning.

Let us illustrate the idea of semantic code regarding language. Consider the message "Tomorrow it will rain in Berlin". Anyone receiving this message can evaluate it according to various criteria. For example, it may be assessed for its *novelty*. However, for a recipient who already heard the weather forecast, the value of the message will at best consist in its *confirmation*. Thus, for different receivers, the novelty value of a piece of information may have an entirely different weight. However, one might also ask for the *practical relevance* (*effectiveness*) of the message. Probably, this is equally large for all recipients who are exposed to the weather in Berlin on the following day. Conversely, it will have no (or only marginal) relevance for those recipients who are planning to spend that day somewhere else.

Comparable considerations can be applied to the *selectivity* of the message. Within the range of all conceivable weather forecasts, the above example, according to which it will rain in Berlin tomorrow, is highly selective—and thus considerably more meaningful than a message stating that "Tomorrow it will rain somewhere in the world." Yet even that information is dependent upon its recipient. For a recipient outside Berlin, both messages are presumably equally uninteresting and thus meaningless. On the other hand, the selective value of the message "Tomorrow it will rain in Berlin" increases if the recipient is going to be near Berlin the next day. The more detailed the message is, the less compressible it will be—conversely, the greater is its *complexity*.

Each element of the semantic code expresses a particular feature of semantic information. Therefore, one might suspect that this variety reveals a fundamental inconsistency of the concept presented here which aims at the general aspects of semantics. However, this impression is deceptive: the opposite is the case. The idea of the semantic code rather reflects the figure of thought deveoped by the philosopher Ernst Cassirer, according to which the particular and individual becomes manifest in its specific relations to the

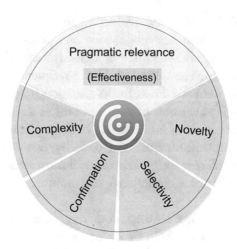

Fig. 6.22 Elements of the semantic code referring to linguistic information. The significance of a piece of information depends on the evaluation standard that the recipient applies to the information concerned. This standard, which may be termed the "semantic code", comprises a number of general aspects (elements) of semantic information. The elements of the semantic code do not refer to the *individual* content of information. Rather, they constitute the *general* frame of evaluation for this content. In other words, a recipient may assemble from these elements a value scale according to which he evaluates the meaning of a piece of information. Among the elements of the semantic code, pragmatic relevance occupies a superior position. In contrast to the other elements, pragmatic relevance has a fine structure reflecting all possible reactions of a recipient. The reactions that are finally induced by a piece of information may be denoted as its "effectiveness"

general (Sect. 4.2). In the present model, the general is given by the (general) elements of the semantic code, whereas the individual feature of semantics arises by the weighted relationship to its general features (Fig. 6.24).

It is remarkable that, in the brain, meaningful information is apparently constituted according to the same principle as described here. Experiments on the visual cortex of the cat seem to suggest such an interpretation. These investigations relate to the so-called "binding problem" ([62]). This problem concerns the question of how the brain re-connects decentralized, segregated information to coherent overall information. Experiments carried out so far indicate that brain cells stimulated by sensory stimuli combine to form a synchronously oscillating ensemble, whereby the individual information segments scattered over different areas of the brain are "bound", i.e., re-connected to a cognitive structure. In the following sections, we will apply the idea of integrative semantics to the problem of the origin and evolution of genetic information.

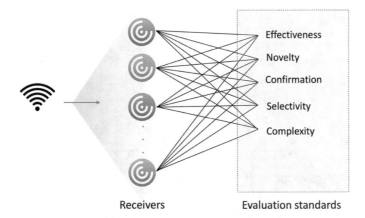

Fig. 6.23 Evaluation of information by various receivers. The evaluation of a piece of information demonstrating the principle of "integrative" semantics that is based on the elements of a semantic code. Each recipient evaluates information against the background of prior knowledge. The evaluation standard is general, as it is based on essential features of semantics such as (but not only) effectiveness, novelty, confirmation, selectivity and complexity. According to a recipient's prior knowledge, that recipient will give the general features a specific weight, thereby setting the frame for the evaluation of the significance of the information received. This is only another description of the fact that information is always context-dependent, i.e. receiver-related. In human information systems, the weighting is highly specific, depending on the recipient's prior knowledge, prejudices, desires, expectations and the like. Nevertheless, it is perfectly conceivable that one may develop appropriate methods to average the weighting of the semantic elements across all receivers

6.9   The Grammar of Genetic Language

The evolutionary biologist Theodosius Dobzhansky once said that nothing in biology makes sense except in the light of evolution [63]. Indeed, Darwin's theory of evolution has become the firm ground of modern biology. Without this theory, we would be completely unable to understand the origin of the incredible variety and purposefulness of life forms at all. Darwinian evolution undoubtedly provides a convincing explanation of the adaption of living beings to their environment, but does it also explain life's development toward higher complexity?

That this question is well justified, we have already demonstrated in the Introduction to this book. Now, we shall try to get to the bottom of the question. First, however, let us look briefly at the history of the theory of evolution. The idea of life's development was already outlined by the zoologist Jean-Baptiste de Lamarck half a century before Darwin [64]. However, Lamarck had a wrong idea about the mechanism of evolution. He believed

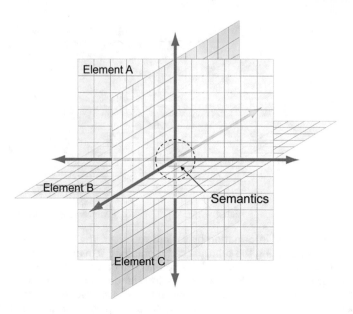

Fig. 6.24 Constitution of the meaning content of information. The meaning content of information arises by the entanglement of general aspects of semantics which build the receiver's evaluation standard. This schema has a general character. It is based on the emergence of the particular within the network of its general relationship (Chap. 4), sketched in Fig. 4.2. One can apply this schema to various aspects of reality, such as the causal interpretation of historical events (see Fig. 4.9) or the character of human time-consciousness (see Fig. 4.11)

that living beings passed on to their descendants particular abilities that they had acquired in the course of their lives, and that this led to continual adaptation and further development of the species.

Lamarck's theory of the inheritance of acquired characteristics was the first attempt, at the beginning of the nineteenth century, to explain living beings' incredible adaptation to their living conditions. However, there is no evidence whatsoever that the development of living beings is targeted, as Lamarck had in mind. It would also contradict the fundamentals of genetics, which of course Lamarck in his day could not have known. Thus, Lamarck engendered an idea of what could have led to the transformation of species in the Earth's history. However, with his explanation based on the idea that organisms' acquired abilities are inherited, Lamarck gave the world an unsustainable idea of the mechanism of evolution.

Only Darwin finally helped the idea of evolution to a breakthrough, because he was able to trace adaptation back to the principle of natural selection and thus give it a purely mechanistic explanation, without having to assert any kind of goal-directedness [65]. According to Darwin, the source of adaptation is the genetic variability of organisms, and the interaction between random genetic variation and natural selection eventually leads to an ever-better adaptation of organisms to their living conditions. Since genetic diversity is based on random processes such as mutation and genetic recombination, the whole process is strongly influenced by chance.

Although Darwin's idea of evolutionary adaptation possesses a high degree of plausibility and persuasiveness, the theory of evolution is still met in some circles by incomprehension to the point of utter rejection. This is primarily because his idea, like no other idea in the natural sciences, calls the religious creation narrative into question and instead assigns a creative role to chance and natural selection. However, even within science there is sometimes fierce debate about the Darwinian theory of evolution, in most cases centred around the issue of correct interpretation of the mechanism of evolution.

Indeed, there is still a need for clarification in this respect. Darwin's theory provides, strictly speaking, only the framework for our current understanding of evolution. It is, however, unclear how the frame is to be filled out in detail. This is by no means intended to question or to belittle the sophisticated theories of evolutionary biologists, especially in population genetics, that are based on Darwin's epochal discovery. It is a fact, however, that we have not yet really understood the actual process of the development of higher life forms and thus still lack an essential core of the idea of evolution.

The difficulties arise *inter alia* from the fact that natural selection does not automatically lead—as is often tacitly assumed—to a higher development of life. Evolution is more than mere selection. While selection is only a "blind" process, driven solely by the different reproductive behavior of organisms, evolution is to be understood as the development from primitive life forms to higher organisms. In other words, selection and evolution represent different aspects of biological development. The first refers to the mere fact of selection, which is a prerequisite of evolution. The second refers to the development of life forms. Thus these two factors differ, even though they are closely intertwined.

According to Darwin, selection directs the evolution of organisms toward optimum adaptation to their environment. Darwin had also clearly recognized the conditions under which natural selection becomes effective. Selection sets in when a population grows exponentially and its habitat or resources are limited. However, Darwin could only put forward plausible arguments for the process of higher development of life; he had no well-founded theoretical explanation for the increase in the complexity of life forms. Seen from this perspective, Darwin's theory is basically a theory of adaptation, but not really of evolution.

It is therefore understandable that in biology, controversial interpretations of evolution arise again and again. The extent to which the opinions differ is exemplified by the conflict between the evolutionary biologists Stephen Jay Gould [66] and Richard Dawkins [67]. Both admit natural selection as the driving force of evolution, and both support the notion of Darwinian evolution. Nevertheless, they arrive at very different assessments regarding the mechanism and course of evolution. While Gould and his colleague Niles Eldredge postulate a gradual development of life, in which long phases of punctuated equilibria are interrupted by explosive developmental thrusts, Dawkins sees a continuous evolution that only takes into account the "selfishness" of genes.

In another example, evolutionary biologists have been discussing for a long time the question of whether the image of the evolutionary tree provides an accurate representation of phylogenetic relationships at all, or whether these are not better represented by a shrub-like branching pattern. Another point of contention is whether, and if so then to what extent, the paths of evolution diverge or converge. While current views generally suppose that evolution is essentially a divergent process of development, the palaeontologist Simon Conway Morris came to the conclusion that there are convergent developments in evolution, and that these repeatedly led to similar forms of life [68].

Other theoreticians believe that the central importance imputed to selection in the evolutionary process must be revised. Thus, according to a theory developed by the population geneticist Motoo Kimura, a mutation can become fixed in a population solely through random fluctuations [69]. One speaks in this connection of a "genetic drift". Accordingly, a mutation does not even need a selective advantage over against other mutations. Compared with Darwinian evolution, it takes considerably longer for a "neutral" mutation to spread over an entire population. Nevertheless, the advocates of this theory are of the opinion that neutral selection, and the non-Darwinian evolution based on it, constitute the predominant causative factor in the development of organisms. Within evolutionary biology, the controversy has now raged for decades about the respective importances of Darwinian and non-Darwinian evolution in Nature.

All this shows that we are still far away from a comprehensive understanding of the evolution of life. Darwin's outstanding achievement has provided a mechanistic explanation for biological adaptation. However, it still seems to be a mystery why organisms have become more and more complex in the course of selective adaptation. Why did the development of life not stop after organisms had reached a certain degree of complexity? We will take up this question later on and offer an answer based on the model of language-driven evolution (see also Introduction).

Regardless of any questions concerning the mechanism of evolution, evolutionary theory must also be able to withstand epistemological objections. In contrast to the theories of physics, evolutionary theory allows only limited predictions. Yet the ability to predict is an essential criterion for distinguishing scientific from pseudo-scientific explanations. We recall that the philosopher Karl Popper only wanted to accept scientific theories that allow predictions that can also be refuted, principally by observation and experiment (Sect. 2.7). In the case of evolutionary theory, however, this criterion does not seem to be met. That is why Popper claimed that evolutionary theory is to be regarded as a mere metaphysical research program [70].

However, the molecular biologist Jacques Monod rightly pointed out that Darwin's theory of evolution contained some implicit predictions, which only came to light many years after Darwin's time [71]. In this context, Monod recalls an objection raised at the end of the nineteenth century by the famous physicist William Thomson (later Lord Kelvin) against the theory of evolution. Starting from the idea, current at that time, that the sun was basically a huge pile of coal from which the Earth receives its life-giving energy, Kelvin had calculated that the life-span of the sun was many orders of magnitude shorter than the time estimated by the evolutionists for the development of life. What Kelvin, however, could not have known was the fact that the sun consists of hydrogen, and its energy is generated by the nuclear fusion of hydrogen to give helium. Seen in this light, Darwin's theory had implicitly predicted the age of the Earth with greater accuracy than had the physicists of his day.

Other philosophers of science have criticised evolutionary explanation for being a tautology. This reproach refers to the central statement of evolutionary theory that the survival of a species is attributed to its adaptation to the environmental conditions ("fitness"). The critics assert that Darwin's explanation is an empty formula, as "fitness" cannot be determined independently and is therefore only detectable by becoming manifest in the fact of survival. Darwin's "survival of the fittest" would thus reduce to the tautology "survival of the survivor".

This harsh criticism is undoubtedly excessive. It derives its legitimacy solely from the limited possibilities of the theory of evolution to make predictions. Owing to the enormous complexity of living organisms, explanations of their adapted features are only possible in retrospect. Therefore, the impression may arise that adaptation itself can only be explained by the fact of survival. It was the achievement of Manfred Eigen to provide clarity concerning the essence and scope of the selection principle [29]. Eigen was able to demonstrate that natural selection is a principle which, under certain conditions, is effective even in non-living matter (see also Sect. 6.5). This insight has become the basis of the molecular theory of evolution, a theory that allows precise and experimentally verifiable prediction of the kinetics of selection among molecules [30].

According to our current state of knowledge, the emergence of life has been a quasi-continuous process leading from chemical evolution via the self-organization of biological macromolecules up to the development of a proto-cell. In the course of this development, the degree of organization of matter steadily increased. Despite this, material systems are still subject to the entropy law and tend to revert to states of lower order; therefore, the complex, orderly states of matter achieved by evolution must be continuously stabilized by a supply of free energy (Sect. 5.6).

Beside the physical prerequisites, the reproductive self-maintenance of a genetic system requires a large amount of information that co-ordinates and controls the system's structural and functional organization. For this reason, the emergence and evolution of biological macromolecules, which can store, process, and transmit information, has been essential for the development of living matter. Consequently genetic information, regulating the processes of life, must have a highly sophisticated content, i.e., semantics.

Let us take a closer look at the consequences that the concept of language will have for our understanding of biological evolution. For this purpose, we apply the sender–receiver model of communication engineering to genetics (Fig. 6.25). To make the model as simple as possible, we consider the very early stage of evolution, at which prebiotic nucleic acids still interacted directly with their physical and chemical environment. This is the phase in which the nucleation of semantic information must have occurred by natural selection and molecular evolution.

As plausible and straightforward as this picture may seem at first glance, the more challenging are the questions that it raises. As far as we know, there was no functional complexity, no semantic information, at the beginning of the evolution of life that could have provided a reference frame for progressive adaptation. In the beginning, the primordial Earth was a world of inanimate matter, natural laws, and contingent boundary conditions. So what caused molecules to accumulate the information required for the building of a living cell?

At first, we have to realize that information and communication presuppose language. Thus, we are required to develop a language concept that can also be applied to molecules such as nucleic acids. The linguist Noam Chomsky, for example, has defined language as a set (finite or infinite) of sentences, each finite in length and constructed out of a finite set of elements [72, p. 13ff.]. This abstract definition, typical in linguistic structuralism (Sect. 1.8), comprises all natural languages, in written or spoken form. Correspondingly, it also applies to (some) formalized systems of mathematics and—most important—even to nucleic acids and proteins. Nucleic acids, for example, consist of a finite alphabet of 4 elements, the nucleotides, which are arranged into a (finite or arbitrarily large) set of nucleotide sequences.

We can now take up once more the question of how genetic information could originate: Is evolution a perpetual motion machine capable of generating information from nothing? We have already come across this question in Sect. 5.6 and rejected the idea of creation *ex nihilo* as impossible. However, this answer challenges our understanding of evolution in a new way. It raises an issue that the Darwinian theory ignores. How complex, i.e., how information-rich, must the environment (i.e., the "receiver" of genetic information) be from the outset, if the enormous variety of complex life phenomena is to be able to emerge by selective adaptation?

This is the decisive question in the light of which the information-theoretical approach to life's origin and evolution can introduce completely new features and remove an explanatory gap in the Darwinian model of evolution. In other words, we are forced to search for a principle

Environment

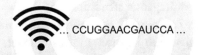

... CCUGGAACGAUCCA ...

Evaluation by natural selection

Fig. 6.25 Illustration of the genetic model of sender and receiver. From an information-theoretical point of view, evolution by adaptation may be regarded as a communication between the sender and the receiver of information (cf. Fig. 6.20). The sender is a genome, carrying information, while the receiver is the environment. It "evaluates" the content of the genetic information carrier according to the latter's ability to survive under the conditions given. In this respect, the environment represents an external information source that is not based on evolutionary adaption in the Darwinian sense but which could nevertheless have propelled prebiotic matter toward the nucleation of functional organization [73]. For this purpose, we must differentiate between natural selection on the one side and evolutionary adaptation on the other. We will see that natural selection alone can already lead to an evolution of the earliest structures of life.

To illustrate this process, let us consider the initial phase of the emergence of life, at a time when there was no genetic apparatus for translation RNA or DNA molecules into protein. At that time, natural selection could only act on the molecules' physical and chemical properties and the possible interactions between them. We know from theoretical considerations that selection criteria in such situations are based upon the molecules' reproduction rate, reproduction accuracy and decomposition rate [cf. 74]. Under prebiotic selection conditions, these parameters depend exclusively on the molecules' structural features. This has been verified by test-tube experiments with RNA molecules (cf. Sect. 6.5). Therefore, we will focus primarily on the so-called RNA world, which is considered the most probable prebiotic scenario in which life once emerged (cf. [75, 76]).

RNA molecules can adopt specific and fine-tuned structures by internal folding and thereby acquire catalytic properties that can even enable some of them to reproduce themselves. When such self-reproducing RNA structures are subjected to reproduction competition, selection between them takes effect, and this in turn is reflected at the level of their nucleotide sequence. An example has been discussed in connection with Fig. 6.8. Since selection will favor RNA structures that promote the self-reproduction of the RNA chains from which they are formed, those nucleotide sequences are selected that form and stabilize advantageous structures.

Pattern formation can already be considered a first step toward creating living matter's language (see below). In this respect, ribozymes are of particular interest. We remember that ribozymes are RNA molecules that have catalytic properties like those of enzymes (Sect. 6.5). In particular, ribozymes can catalyze their own reproduction [27]. Thus, the chemical properties of ribozymes already confer upon them the essential requirement for molecular self-organization, which is autocatalysis. Let us go into greater depth with this by looking at the so-called hammerhead ribozymes and their structural features.

The smallest hammerhead ribozyme that has been investigated in detail is a short double-stranded RNA motif, made up of two short nucleotide sequences (respectively 16 and 25 nucleotides long) that are held together by Watson–Crick base pairs (Fig. 6.26). One strand functions as an enzyme, the other one as the substrate on which the enzyme acts [77, 78]. The catalytic center of the RNA motif arises from the interaction of specific segments of both RNA strands. Such segments are essential for the molecule's self-reproducing function and thus force it to adopt a specific three-dimensional RNA motif structure. In all hammerhead ribozymes, the critical parts of the RNA motif are strongly conserved; in fact, this holds for all catalytic RNAs. Correspondingly, the primary nucleotide segments of catalytic RNAs may be considered as proto-words from which the evolution of genetic language could have started.

Thus, the mere fact of selection among self-reproducing RNA molecules already leads to the formation of syntactic structures in their nucleotide sequences, which may be considered a preliminary stage of genetic language formation. The nucleation of "proto-words" in an RNA sequence is illustrated by Fig. 6.27. It exemplifies the selective conservation of nucleotide patterns (red) to maintain the catalytic activity of an RNA structure. Moreover, the replication mechanism may also involve specific nucleotide segments that act as recognition signals, setting off RNA replication, and marking a first distinction between "foreign" and "own" (see for example [79]).

Apart from the features referring to RNA's nucleotide sequence, one must assume that the catalytic activity of prebiotic RNAs depended on a broad spectrum of external factors such as the presence of metal ions and proteins, and other physical conditions (temperature, ionic strength etc.). All these boundary conditions could have contributed to imprinting upon primordial nucleotide sequences a syntactic structure that later became the syntax of genetic language.

By now, the catalytic properties of RNA have been studied very thoroughly. It turned out that mechanisms of catalysis by RNA are more diverse than initially expected (cf. [82]). This has made catalytic RNAs an exciting research field that may be open not only for discovering novel catalytic RNA motifs, but also for a deeper understanding of the toolbox of genetic language (Table 6.4).

The proto-words of living matter's language do not have a uniform length. In contrast to the codons constituting the information content of genes, proto-words are not subjected to the genetic code's translation mechanism. Rather, they represent the syntactic pre-structure from which genetic language could evolve. The rules of its grammar are prescribed by the physical and chemical mechanism according to which the proto-words are anchored in the nucleotide sequence of prebiotic RNAs. We will see in Sect. 6.10 how cooperative interactions among RNAs bring about the connection of

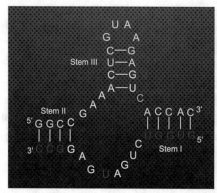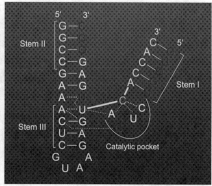

Fig. 6.26 Artificial all-RNA hammerhead ribozyme. Left: Sequence and predicted secondary structure of the artificial hammerhead RNA. In naturally occurring self-cleaving hammerhead RNAs, one of the ends is closed by a further loop. In the artificial construct, the molecule consists of two nucleotide strands, which are termed the enzyme strand (nucleotides: yellow or white) and the substrate strand (nucleotides: red or white). The active-site base C is highlighted in green. Conventional base-pairing is indicated by white lines. Right: The ribozyme's three-dimensional structural architecture, omitting for clarity the twisting of the double helices. Note that stems II and III are co-linear, forming a pseudocontinous long helix. The joint with stem I is bent around so as to bring the active-site base C close to the CUGA turn, forming the "catalytic pocket". The three-dimensional structure is reinforced by additional single hydrogen bonds (dotted lines) between the molecule's nucleotides (adapted from [78])

proto-words to give sentences of a proto-language, driving the evolution of living matter.

Analogous to human language, genetic grammar comprises the rules (laws) determining the building of words, word types, and sentence structures that constitute genetic language. Since there is no language without grammar, the nucleation of grammar must have preceded the actual development of genetic language. This means that genetic grammar has originated without any reference to possible semantics. The proto-language of living matter regulates certain physical and chemical process sequences. It can best be compared to the machine or programming languages that are used to simulate artificial or electronic life (cf. Sect. 6.6). However, natural selection and evolution finally raised the proto-language far beyond the level of mere mechanical operation. The key to a deeper understanding of this language must lie in the non-coding parts of the organism's genome.

Before we ask in Sect. 6.10 what the consequences of the above results are for the origin of life, let us point to another highly remarkable analogy between human and genetic language. For this purpose, we consider the

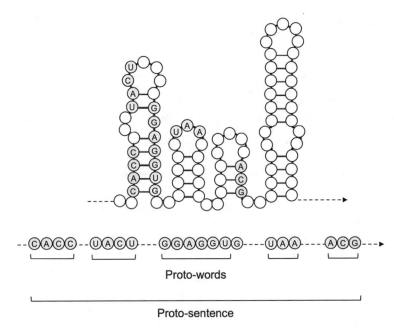

Fig. 6.27 Formation of a syntactic pattern in RNA by natural selection. Under prebiotic conditions, the reproduction rate, and thus the selection value, of self-reproducing RNAs depends only on their structural properties. Therefore, elements of the molecule's nucleotide sequence are selectively favored if they contribute to stabilizing an advantageous structure. The strongly conserved nucleotide segments of catalytic RNAs provide an example of this. There are also other conceivable mechanisms that could confer on the RNA specific structural properties favoring fast reproduction. We know from the virus Qβ, for instance, that particular nucleotide segments of their RNA serve as recognition signals for reproduction in their host cell [79]. The figure illustrates how the fixation of selectively favored nucleotide segments can be regarded as proto-words representing the pre-stage of a linguistic structure. Moreover, the three-dimensional folding of the secondary structure brings parts of the nucleotide sequence into spatial proximity to each other. This may lead to interactions between remote sections of the sequence, as is typical of the catalytic activity of biological macromolecules (see Fig. 6.26). The emergence of the hierarchy among proto-words gives them the structure of a proto-sentence, the beginning and end of which are shown in yellow

genome structure of the RNA virus Sars-Co-V2 (Fig. 6.28). Recent research has revealed that only part of the genome encodes the virus-specific proteins, which interact with the host's cell machinery to set off the rapid reproduction of virus particles. The non-coding parts of the genome provide regulatory information. This finding not only confirms the conjecture that the non-coding regions of the genome are significantly involved in gene

Table 6.4 Toolbox of genetic language. The organism's DNA contains many non-coding regions, i.e., genome segments that are not translated into proteins (for a review, see [80, 81]). RNAs that originate by transcription of DNA but are not translated into proteins are classified into short (small non-coding) RNAs (up to 100 nucleotides) and long non-coding RNAs (more than 100 nucleotides). Non-coding RNAs take on numerous regulatory tasks, at the levels of both transcription and translation. Part of the non-coding DNA is involved in gene expression. Some elements of non-coding DNA are responsible for turning the translation of genes on and off. Others control or modulate the translation of genes

Genome's non-coding elements	Function
Non-coding RNA	
[mRNA, rRNA, tRNA]	Involved in protein synthesis
[piRNA, miRNA, siRNA]	Regulation of translational activities
[long non-coding RNA (lncRNA)]	Highly diverse functions [80, 81]
Cis- and trans-regulatory elements	Modulation of gene transcription
Introns	Regulation/control of rRNA, tRNA and codons
Promotor	Regulation of gene transcription
Enhancer	Regulation of gene transcription
Superenhancer	Cell regulation
Inhibitor (Silencer)	Regulation of gene transcription
Insulator	Blocking of enhancer
Pseudogenes	Unknown
Transposon	Inducing or correcting mutations
Satellite RNA	Regulation of gene expression
	Stabilizing chromosome structure
Telomeres	Preservation of chromosomes' structural identity

expression, but it also points to a further basic analogy between genetic and human language.

To illustrate this, let us examine, for example, the sentence: "Everybody knows that a red rose is a transient beauty." This example demonstrates that only two words ("red" and "rose") have a direct relation to our external world, in that they denote a real, i.e., a physical object or property. All the other words in the sentence are either constructs of our mental world or words that serve as the support frame for grammatical sentence formation by putting words into grammatical order.

It is surprising that a high percentage of words in human language have no immediate relation to the external world. The same applies to genetic language. The genome's coding information, representing its reference to the outside world, makes up only a small fraction of the total genetic information. For example, only about 2% of the human genome's information is translated

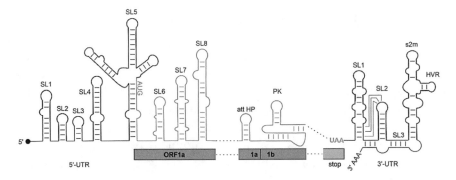

Fig. 6.28 Coding and non-coding regions of a viral RNA genome. Betacoronaviruses contain a long single-stranded RNA genome consisting of about 30,000 nucleotides. The secondary structure of conserved RNA elements of the virus Sars-CoV-2, shown here, was determined experimentally by NMR spectroscopy [83]. The coding region for viral proteins (orange) is embedded into two non-coding parts of the viral RNA (black), conserved within the betacoronavirus family. The non-coding parts are assumed to participate in regulating the production, replication, and translation of subgenomic mRNA and are thus taken to be crucial for viral propagation

into proteins. The remaining 98% is either not transcribed, or it is transcribed into RNA that is not translated into proteins. Thus, only a small part of our genetic information refers to the genome's "external" world by determining the phenotypic appearance of living matter. In contrast, the predominant proportion serves to give genes' information a "linguistic" expression.

The task-sharing between non-coding and coding information is a general feature of language and, as such, a downright necessity in genetic language. Earlier, the genome's non-coding parts were mistakenly referred to as "junk DNA". However, the realization that language is a natural principle of Nature demands an entirely different interpretation of the genome's seemingly superfluous information. It instead supports the conjecture that it is precisely in the non-coding information that the organization of genetic language is laid down. The analogy between human and genetic language is, in fact, more than just a metaphorical way of speaking. It has a vital guiding function for biology.

6.10 Language-Driven Evolution

We have seen that the grammar of genetic language is a consequence of physical and chemical laws. It consists of the chemical rules that induce the formation of specific nucleotide patterns ("proto-words") in autocatalytic RNA structures when these are subjected to natural selection. Furthermore, we have seen that the chemical mechanisms may also lead to a specific sequence of proto-words having a rudimentary sentence structure ("proto-sentence"). In short, genetic grammar can be considered a part of the chemical theory of autocatalysis.

In nearly the exact same words, Noam Chomsky described the grammar of human language "by an analogy to a part of chemical theory concerned with the structurally possible compounds. This theory might be said to generate all physically possible compounds just as grammar generates all grammatically 'possible' utterances. It would serve as a theoretical basis for techniques of qualitative analysis and synthesis of specific compounds, just as one might rely on a grammar in the investigation of such special problems as the analysis and synthesis of utterances." [72, p. 48]

Chomsky's further remarks on this issue likewise fit the grammar of genetic language: "A grammar of language L is essentially a theory of L. Any scientific theory is based on a finite number of observations, and it seeks to relate the observed phenomena by constructing general laws in terms of hypothetical constructs such as (in physics, for example) 'mass' and 'electron'. Similary, a language (…) is based on a finite corpus of utterances (observations), and it will contain certain grammatical rules (laws) stated in phomemes, phrases, etc. (…). These rules express structural relations among the sentences of the corpus and the indefinite number of sentences generated by the grammar beyond the corpus (preditions)." [15, p. 49]

Chomsky was guided by the thought that human languages' grammar is a universal feature of human language competence, having its roots in the innate structures of the human brain (Sect. 1.3). He further argued that grammatical structures could be justified without having to refer to the semantics of a language. Consequently, the universal grammar that Chomsky has in mind must somehow be encoded in the human genome. By uncovering the physical and chemical origin of the genetic grammar, we have now taken Chomsky's idea down to the level of prebiotic macromolecules. Thus, our description of the chemical mechanism from which, by natural optimization, a hierarchically organized syntactic structure can emerge must be considered the first step toward a comprehensive theory of the language of living matter.

Future research will be essential if we are to uncover the rules of genetic grammar by structural analysis of the genome's non-coding regions. Computer-aided searching for possible regularities will help look for a hint of the genome's grammatical organization. Nowadays, advanced algorithms such as SEQUITUR are available to identify regularities in a sequence of discrete symbols, offering insights into its possible lexical structure [84]. Such methods appear to be a promising way to uncover the regularities of genetic grammar.

So far, we have considered genetic grammar separately from the semantic dimension of language, just as Chomsky intended to do for human language. However, we will now ask how proto-semantics could originate in prebiotic matter. The mere selection of syntactic structures that follow grammatical rules is not yet sufficient to give such structures a meaning. Chomsky demonstrates this with an example from human language [72, p. 15]. It refers to the following sentences:

(1) Furiously sleep ideas green colorless.
(2) Colorless green ideas sleep furiously.

Both sentences are nonsensical. However, while sentence (1) is not grammatical, sentence (2) is grammatical. Thus, a grammatical sentence is not necessarily a meaningful sentence. This is why it seems justified to look first at the grammar of genetic language, i.e., detached from any semantic reference.

Nevertheless, the question remains as to how syntactic structures can acquire a semantic dimension. It is evident that this question concerns above all the evolution of proto-sentences, which must be considered the potential carriers of genetic information. The problem is all the more difficult because in the prebiotic phase of evolution there was no biosynthesis machinery that could have examined the RNA's structure for possible information content. In the RNA world, the only benchmark of selection was provided by the RNAs' inherent reproduction dynamics.

Prebiotic RNA molecules, however, did not exist in isolation. They must have interacted with the many chemical substances in their surroundings (proteins, metal ions, etc.). It is highly conceivable that the interactions between catalytic RNAs and other substances led to the emergence of reaction networks in which the molecules stabilized each other, thereby enhancing their combined selection value. Such cooperative reaction cycles could have developed, for example, from a so-called quasispecies, which consists of an RNA master copy and its mutant distribution (Sect. 6.5) [85]. Even though experimental verification of this kind of self-organization has not yet been achieved, there is no doubt that the emergence of functional organization in a molecular system requires, first and foremost, cooperative behavior of its components. However, the coupling of the selection values turns the RNA ensemble *as a whole* into the object of natural selection.

A model of this kind of molecular self-organization is the so-called hypercycle, developed by Manfred Eigen and Peter Schuster [29, 86, 87]. The model was initially thought to explain the origin of a protocell, because it starts from the existence of a translation mechanism for genetic information (Fig. 6.29). However, its basic idea is general and can just as well be applied to a self-reproducing reaction cycle that only consists of self-reproducing catalytic RNAs. Above all, the hypercycle explains how molecular systems could overcome a fundamental information barrier on the way to the evolution of higher complexity. Thus, the hypercycle is so to speak the needle's eye at the threshold from non-living to living matter.

The need for an information-integrating mechanism is immediately apparent: The more errors occur in the replication of a potential (or actual) information carrier, the higher is the number of variants that are subjected to evaluation in the selection process, and the faster evolutionary progress can occur. On the other hand, however, an accumulation of errors leads to the destabilization of information. At a critical error rate, the generation of information turns into an uncontrolled process of error accumulation, which finally leads to the collapse and complete dissolution of the information level attained so far by evolution.

The theoretical and experimental analysis of the error threshold has revealed that not more than around one hundred genetic symbols (nucleotides) can be reproduced reliably without the aid of sophisticated copying devices. Yet we know, naturally, that the genetic blueprint for an optimized reproduction apparatus such as that of the living cell requires far more than a hundred nucleotides. For this reason, an organizational principle is needed that makes possible the accumulation and transfer of a sufficient amount of information for the construction of such a reproduction apparatus. The hypercycle is a possible model of this.

Let us now consider the evolution of molecular language. We have argued that the grammar of this language is anchored in the physics and chemistry of the RNA world and that the formation of proto-words is already founded in the structural basis of catalytic RNAs. In this process, no adaptation to the environment is involved. It is solely the optimization of its catalytic function

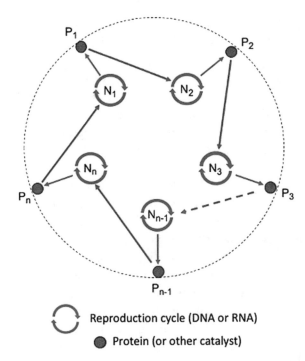

Fig. 6.29 Hypercycle: An information-integrating principle of early evolution. The prototype of cooperative interaction between biological macromolecules is the so-called hypercycle. It emerges by cyclic coupling of autocatalytic reproduction cycles of nucleic acids. The model depicted here already presupposes a primitive translation mechanism by which the information content of a nucleic acid N_{n-1} is translated into a protein P_{n-1} that catalyzes the next reproduction cycle N_n. The hypercycle model explains how molecular systems can organize themselves into a functional unit, thus overcoming an inherent information barrier on the way to the evolution of higher complexity. Coupling without translation machinery is also conceivable; ideal candidates for this are RNA molecules, as they themselves have catalytic properties. Hypercycles may therefore be of great importance for the nucleation of information in the RNA world

that contributes to the RNA's selective advantage, which, in turn, becomes manifest in specific segments of its nucleotide sequence. We have identified these segments as proto-words and proto-sentences from which genetic language could evolve.

The next step in the origin of life must have been the cooperative interaction of RNA molecules to overcome the information barrier on the way to complex sentence structures. Figure 6.30 illustrates this mechanism, using an example of human language. While at first natural selection only aimed at the self-reproductive properties of single RNA molecules, it now extends to organized RNA ensembles, evaluating their reproductive efficiency. Functional organization rests on principles such as cooperation, self-regulation, compartmentation, hierarchy formation, and others. These act like guard rails between which the semantics of genetic information could become manifest and evolve. In physical terms: they are boundary conditions channelling prebiotic development (Sect. 4.8). Adopting the terminology of Sect. 6.8, we may alternatively say that these principles represent elements of a semantic code constituting the "proto-semantics" of functional organization in living matter.

The self-organization of RNA molecules has a number of evolutionary advantages, not least in connection with the evolution of the language of living matter. For example, let us consider the genetic proto-words that emerge by the selective favoring of patterns in the nucleotide sequence of catalytic RNAs. Owing to the great diversity of possible catalytic RNAs, a correspondingly large number of proto-words of genetic language is possible. These words may be bound together, forming the proto-sentences of a genetic language, as is the case with hypercycles. In this way, a nearly unlimited number of (meaningless and meaningful) sentences can be generated by recombination, so that a vast reservoir of possible linguistic expressions is available to natural selection and evolution.

The model of language-driven evolution changes radically our traditional understanding of life's origin and evolution (Fig. 6.31). According to this model, the overall evolution process falls apart into a non-Darwinian evolution of the *syntax* and a Darwinian evolution of the *semantics* of genetic information. Both processes are based on natural selection. While the non-Darwinian development of syntactic structures is context-free, i.e., is only determined by criteria of functional organization, the Darwinian evolution of semantic information is adaptive and is, as such, strictly context-dependent. In the former, the structures of a proto-language emerge, whereas in the latter, the expression forms of this language must withstand confrontation with the organism's environment.

The fine structure of both processes reveals further interesting details regarding the evolution of life. However, before we summarize the main aspects, let us clarify two points. (1) Even though we refer here and there to the historical course of evolution, the principles guiding this process are general, i.e., independent of the *actual* course of evolution. (2) It is important to note that evolution only takes place in the presence of natural selection.

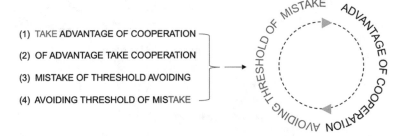

(1) TAKE ADVANTAGE OF COOPERATION

(2) OF ADVANTAGE TAKE COOPERATION

(3) MISTAKE OF THRESHOLD AVOIDING

(4) AVOIDING THRESHOLD OF MISTAKE

Fig. 6.30 Linguistic "hypercycle". The figure illustrates the generation of genetic information by cooperation between information carriers. Genetic words are represented by words of human language. Four phrases are considered to be in selection competition. The phrases (1) and (4) acquire a selective advantage when they become combined to build a more extensive information content. The benefit arises when, by cyclic coupling, the phrases are integrated into a self-reproducing unit in which the words stabilize each other's information content. In the same way, the prebiotic information barrier could have been overcome by the cyclic coupling of catalytic RNAs. This mechanism represents the epitome of functional organization. At first, the information carriers taking part in the organization attain semantic meaning only by the emerging organization, which itself may further differentiate and evolve by natural selection. As long as the semantics are restricted to the organization itself, it has a "proto-semantic" status. By the nucleation of a mechanism that leads to the translation of RNA into proteins, the proto-semantics become contextual and thereby the object of the Darwinian evolution of semantic information

Therefore, it is somewhat misleading if one refers, for example, to the chemical origin of biomolecules and their precursors already as "chemical evolution" (Sect. 6.6). Although the term has become commonplace, it only means that everything possible according to the laws of chemistry will necessarily originate under the corresponding reaction conditions. However, these conditions changed in the Earth's early history, suggesting the evolution of chemical substances. This, in turn, has led to the widespread opinion that the question of how life originated would already be answered if we were only sure that the essential components of life can originate under prebiotic conditions.

In contrast, the crucial question of the origin of life relates to the origin of genetic information [88]. How are we to explain the semantics of genetic information on the basis of physical and natural law? Since we cannot use the term "semantics" without referring to language, we must look for a natural principle that has the structure of language. This principle we have found in

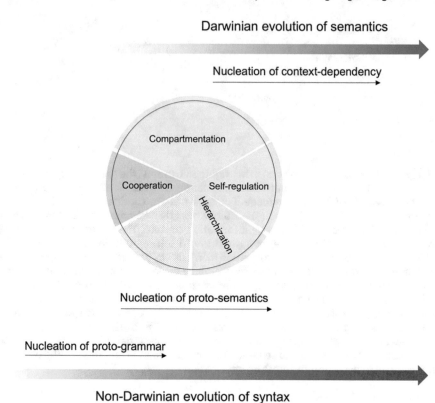

Fig. 6.31 Language-driven evolution of genetic information. Language-driven evolution has necessarily a dual character. It can be resolved into two intertwined but distinct processes: Non-Darwinian evolution of syntax and Darwinian evolution of semantics. Both processes are based on natural selection. (1) Non-Darwinian evolution is context-free. It does not require any semantic reference for developing syntactic structures from which a proto-grammar can emerge. The nucleation of proto-semantics may result from the spontaneous formation of cyclic couplings among RNA structures. The coupling includes cooperation, compartmentation, self-regulation, hierarchy formation and other principles characteristic of functional organization. Within the general frame of the functional organization, the evolution and diversification of the linguistic pre-structures took place. (2) After the invention of a genetic translation apparatus, the Darwinian evolution of semantics begins. All information—i.e., all forms of linguistic expressions generated by genetic grammar—are tested regarding their pragmatic relevance, i.e., their adaptive advantage. In contrast to non-Darwinian evolution, which is context-free, Darwinian evolution is strictly context-dependent. According to the language-based mechanism of evolution, the Darwinian evolution of life is nothing other than the evolutionary change of genetic language-use. Therefore, the non-Darwinian development of molecular language, starting from the nucleation of a proto-grammar, must be considered the primary source of the origin and diversity of life forms. This means that the evolution of life's genotype and phenotype are based on quite different mechanisms

molecular biology. It has its roots in the chemical structure of biological macromolecules in conjunction with natural selection. The backdrop to this is the prebiotic RNA world described above.

We have seen that RNA structures indeed have all the characteristics associated with language at the earliest stage of life's origin. They are built from a finite alphabet, the nucleotides, that allows the formation of a nearly unlimited number of different RNA structures. At first, selection could only be directed at the phenotypic properties of RNA structures themselves. However, natural selection and mutation are already sufficient for the emergence of syntactic patterns in RNA structures. Such patterns must be considered as being like rudimentary words and sentences of an emerging language in the RNA world and thus to be the preliminary stage in the nucleation of genetic information (Sect. 6.9). From this perspective, it is indeed justified to denote the generative principle of pattern formation as the "proto-grammar" of living matter. It seems that the main part of the nucleation of life took place at the level of syntactic structures. During this phase, the object of evolution was the structure of language itself, epitomizing the genetic organization of living matter.

The grammar of living matter's language is universal, but the language itself may have evolved, in a manner comparable to that of the development of human language. Different genetic languages would correspond to various forms of genetic organization and enable a genetic system to distinguish between "own" and "foreign". All words and sentences of genetic language compatible with a given organization are regarded as "own"; those that are incompatible are identified as "foreign". Such a "linguistic" distinction could, for example, be the basis for our immune system, as Niels Jerne already suspected when he delivered his Nobel lecture on "The generative grammar of the immune system" [89] (see also Introduction).

It can be assumed that only a fraction of the countless sequence alternatives that build sequence space (see below) will carry biologically relevant information. Accordingly, the space of biologically meaningful sequences is only a subspace of general sequence space. However, we have no idea how large the number of sequences containing semantic information actually is, particularly as the semantic component of information is a relative quantity that, owing to its context-dependence, changes according to changing context conditions.

Furthermore, we have argued that the proto-semantics of genetic information emerged by molecular self-organization of prebiotic RNAs, of which the emergence of hypercycles is an example. On the basis of the principle of cooperation, hypercycles can overcome the information barrier of early evolution by stabilizing their members against information loss. This stage of

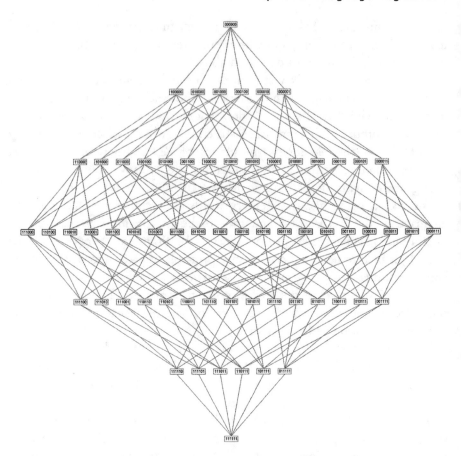

Fig. 6.32 Sequence space. All combinatorically possible alternatives of a given binary sequence constitute the so-called sequence space. The figure shows the sequence space of a series of six binary digits (from [90]). In total, there are 2^6 (=64) sequence alternatives and 6! (=720) pathways that reach from the sequence 000000 to the sequence 111111. Sequence space has turned out to be a useful tool to study the features of evolutionary optimization of biological macromolecules since, for example, the nucleotide sequence of a nucleic acid can be directly represented by a series of binary digits (see Sect. 5.4). The complexity of sequence space and with it the number of possible routes leading from one sequence to another increases dramatically with increasing sequence length. In general, the higher the dimensionality of sequence space, the higher is the probability that optimization by natural selection will not end in a local maximum but will find a way for further optimization. For this reason, evolutionary dynamics themselves exert a selection pressure, pushing nucleic acids toward greater chain lengths and thus to higher complexity

evolution, in which the nucleation of proto-semantics took place, must be considered the transition phase to Darwinian evolution.

However, from the proto-semantics, the substantial semantics of genetic information could only develop after the genetic translation apparatus had originated, giving the genotype a phenotypic expression. This is the point at which genetic information becomes context-dependent, and Darwinian evolution sets in. In Darwinian evolution, selection checks which of the numerous possible linguistic expressions, arising by mutation and recombination, fit the adaptive challenges of the organism's continually changing environment. Strictly speaking, the Darwinian model of evolutionary adaptation only describes the evolutionary change of the use of living matter's language. This means that the evolution of life has taken place mainly at the genotypic, rather than the phenotypic, level of organisms.

The idea of language-driven evolution makes possible a consistent description of life's origin and evolution that is based on physical and chemical law. Above all, it abnegates the criticism that evolutionary thought is merely a metaphysical research program, as asserted by the philosopher Karl Popper (Sect. 2.7). The concept of genetic language also answers the pending question of why the complexity of living matter increased at all. Why should the evolution of life have continued beyond the stage of simple hypercycles or proto-cells?

Obviously, there must be a driving force anchored in the evolutionary dynamics themselves that pushes genetic programs to become more and more complex. To address this problem, we consider the so-called sequence space, in which the evolution of genetic information takes place (Fig. 6.32). Sequence space is not a space that we can perceive with our senses. Rather, it is an abstract space, the coordinates of which are given by all possible sequence alternatives of a nucleic acid molecule. Since each position in a nucleotide chain can be occupied with one of the four basic building blocks, the dimensionality of the sequence space and the number of sequence alternatives increase with each building block that is added to the nucleotide chain.

Even for relatively short sequences, the corresponding sequence space has gigantic dimensions. A numerical example may illustrate this. The viruses stand at the theshhold of life and possess a genome that usually contains between one thousand and ten thousand nucleotides. A sequence of one thousend nucleotides, however, already has 10^{600} possible sequence alternatives. This is an incredibly large number! As the length of the genome increases, this number grows immeasurably.

The enormous number of sequence alternatives provides an impression of the potential diversity of genetic information. Since the expectation probability for a specific sequence decreases with an increase in the number of sequence alternatives, the dimension of a sequence space also tells us something about the degree of uniqueness of the sequences occurring in it. At the same time, the size of the sequence space provides information about the degree of freedom that evolution possesses at the various levels of complexity of living beings.

The immense dimensions of sequence space make the emergence of life seem highly enigmatic. The number of sequence alternatives for the genome of a simple virus already transcends all physical boundaries and pushes the probability of the random emergence of life in a single step to zero. Even the immense scope annd variety of the spatial, temporal and material conditions of the universe are by far inadequate to allow life to arise spontaneously by pure chance. A detailed analysis of this issue can be found elsewhere [74, 88].

Instead, the theory of the self-organization and evolution of life shows that natural selection is capable of performing a "search function" and thus of paving the way for evolution through the immense complexity of sequence space. In this way, within a physically possible time, a specific sequence can be selected, even if it is *a priori* highly improbable, from among an unmanageable number of alternatives.

How the processes of selection and optimization work together becomes better understandable when sequence space is transformed into a so-called "value space". This can be achieved by assigning a specific numerical value to each sequence, i.e. to each point in sequence space; this number specifies the selection value of the respective sequence. The result is a fitness landscape that resembles the mountain scenery described in Sect. 2.8 (see in particular Fig. 2.6). We remember that the search function anchored in the selection principle always leads evolution uphill in the fitness landscape. By this mechanism, evolution heads for the optimum point, that is, the point with the highest selection value. However, the evolutionary path is only determined in so far as it follows the general gradient of optimization. The exact path, in contrast, is open and depends on random "decisions" such as mutation and recombination. These ultimately impart to evolution the character of a historical process.

Owing to the enormous dimensions of genetic sequence space, biological evolution has only been able to open up a tiny fraction of the gigantic number of sequence alternatives. Beside the complexity of sequence space, another limitation comes into play that is characteristic of a historical process: the further optimization progresses, the more paths of evolution that were possible *before* are now excluded. Thus, evolution automatically leads to the perpetuation of the developmental path that it has taken. This limitation becomes the more noticeable, the further the optimization progresses. It makes evolution an irreversible process, which remains trapped in certain areas of sequence space as differentiation progresses.

Let us now return to the question raised above: Why does the complexity of living beings, and thus the dimensionality of genetic sequence space, increase in the course of evolution? The answer is provided by sequence space itself. Figure 6.32 shows how the number of possible pathways of biological evolution dramatically increases with the dimensionality of the sequence space. This raises the probability that local maxima in the fitness landscape are connected by ridgeways, saddle points and the like.

Such topological features have an enormous advantage, as they allow evolutionary optimization to progress from one local peak to the next, without the need to make the (forbidden) descent through an interjacent valley of the fitness landscape. Instead, evolutionary optimization can pass through the fitness landscape in just the same manner as we do in a real mountain hike. Here too, the experienced mountaineer will not descend into a valley to climb the next summit, but will stick to high-altitude trails in the mountain massif to get from one peak to the next.

The dependence of the efficiency of optimization on the dimensionality of sequence space has an interesting consequence for the evolution of genetic information. Since each extension of a nucleotide sequence by insertion(s), recombination and the like also raises the dimensionality of its sequence space, a large number of new evolutionary paths abruptly open up, making optimization considerably more effective. Thus, it is evolution dynamics themselves that lead evolution to an ever-higher level of complexity [73]. This is the sought-after explanation for the fact that the course of biological evolution has been accompanied by a steady increase in complexity.

The concept of sequence space has led us to fundamental insights into the mechanisms of evolution. Nevertheless, we still have no real idea of how much information, or what kind of information, is associated with the numerous sequence alternatives of a specific nucleic acid molecule. This is the point where the information theory of evolution reaches its limits. The reason for this is immediately clear: each possible sequence alternative of a biological information carrier represents a possible information content that only becomes manifest through evolution. This content is individual and unique. It cannot be specified either by theoretical terms or by any law-like algorithm.

Instead, the content reflects the context-dependence and historicity of its origin. It is contingent—that is, it is not necessarily the *way* it is. In this respect, the Darwinian evolution of semantics stands in contrast to the non-Darwinian evolution of syntax, which is not adaptive but develops only under the constraint of functional organization.

Thus, the application of information theory to biological evolution can explain the *existence* of information, but not its "suchness". It can reveal the constitutive principles leading to semantic information, but cannot grasp its unique content. For this reason, we shall need to explore sequence space experimentally, for example by using evolution machines (Sect. 6.5). It seems highly possible that, in this way, we may one day even advance to discovering entirely new features of living matter.

The technology of artificial evolution makes this possible, because it allows one to initiate a process of artificial evolution at any point in sequence space with a view to penetrating previously unknown regions of that space. This perspective conveys an idea of the huge range of possibilities for designing living matter that modern biology opens up to us. To exploit these possibilities to the full, we must learn to understand in detail the language of living matter. Only when we master its vocabulary and grammar will we be able to intervene constructively in the sequence space of genetic information.

References

1. Du Bois-Reymond E (1974) Vorträge über Philosophie und Gesellschaft (ed: Wollgast S). F. Meiner, Hamburg
2. Du Bois-Reymond E (1872) Über die Grenzen des Naturerkennens. Veit & Co., Leipzig
3. Boltzmann L (1979) Der Zweite Hauptsatz der mechanischen Wäremetheorie (1886). In: Broda E (ed) Ludwig Boltzmann: Populäre Schriften. Vieweg, Braunschweig
4. Haeckel E (1901) The Riddle of the Universe (transl: McCabe J). Watts & Co, London [Original: Die Welträthsel, 1899]
5. Hilbert D (1930) Naturerkennen und Logik. Naturwissenschaften 18:959–963
6. Eigen M (1967) Immeasurably fast reactions. Nobel lecture. Nobel foundation
7. Hell SW (2014) Nanoscopy with focused light. Nobel lecture. Nobel foundation
8. Neher E (1991) Ion channels for communication between and within cells. Nobel lecture. Nobel foundation
9. Sakmann B (1991) Elementary steps in synaptic transmission revealed by currents through single ion channels. Nobel lecture. Nobel foundation

10. Dulbecco R (1995) The Genome Project. Origins and Developments. In: Fischer EP, Klose S (eds) The Diagnostic Challenge: The Human Genome. Piper, Munich, pp 17–59
11. Venter JC, Adams MD, Myers E et al (2001) The Sequence of the Human Genome. Science 291(5507):1304–1351
12. Monod J (1971) Chance and Necessity: An Essay on the Natural Philosophy of Modern Biology (transl: Wainhouse A). Random House, New York [Original: Le hazard et la nécessité, 1970]
13. Doudna JA, Charpentier E (2014) Genome editing. The new frontier of genome engineering with CRISPR-Cas9. Science 346(6213):1258096
14. Cibelli J, Gurdon J, Wilmut I et al (2013) Principles of Cloning, 2nd edn. Academic Press, Cambridge/Mass
15. Küppers B-O (1995) The Context-Dependence of Biological Information. In: Kornwachs K, Jacoby K (eds) Information: New questions to a Multidisciplinary Concept. Akademie Verlag, Berlin, pp 135–145
16. Campbell KHS, McWhir J, Ritchie WA et al (1996) Sheep cloned by nuclear transfer from a cultured cell line. Nature 380:64–66
17. Liu Z, Cai YJ, Wang Y et al (2018) Cloning of Macaque monkeys by somatic cell nuclear transfer. Cell 172:881–887
18. Beyhan Z, Iager AE, Cibelli JB (2007) Interspecies Nuclear Transfer: Implications for Embryonic Stem Cell Biology. Cell Stem Cell 1(5):502–512
19. Sun YH, Zhu ZY (2014) Cross-species cloning: influence of cytoplasmic factors on development. J Physiol 592:2375–2379
20. Gurdon JB (1962) The developmental capacity of nuclei taken from intestinal epithelium cells of feeding tadpoles. J Embryol Exp Morphol 10:622–664
21. Yamanaka S (2017) Induction of pluripotent stem cells from mouse embryonic and adult fibroblast cultures by defined factors. Cell 126:663–676
22. Mills DR, Peterson RL, Spiegelman S (1967) An extracellular Darwinian experiment with a self-duplicating nucleic acid molecule. Proc Nat Acad Sci USA 58:217–224
23. Küppers B-O (1979) Towards an experimental analysis of molecular self-organization and precellular Darwinian evolution. Naturwissenschaften 66:228–243
24. Kramer FR, Mills DR, Cole PE et al (1974) Evolution in vitro: sequence and phenotype of a mutant RNA resistant to ethidium bromide. J Mol Biol 9:719–736
25. Rich A, Kim SH (1978) The three-dimensional structure of transfer RNA. Sci Am 238(1):52–62
26. Altmann S (1989) Enzymatic cleavage of RNA by RNA. Nobel lecture. Nobel Foundation
27. Cech TR (1986) A model for the RNA-catalyzed replication of RNA. Proc Natl Acad Sci USA 83:4360–4363

28. Bertalanffy LV von (1969) Robots, Men, and Minds: Psychology in the Modern World. George Braziller, New York
29. Eigen M (1971) Selforganisation of matter and the evolution of biological macromolecules. Naturwissenschaften 58:465–523
30. Küppers B-O (1983) Molecular Theory of Evolution: Outline of a Physico-Chemical Theory of the Origin of Life. Springer, Berlin/Heidelberg
31. Tjhung KF, Shokhirev MN, Horning DP et al (2020) An RNA polymerase ribozyme that synthesizes its own ancestor. Proc Natl Acad Sci USA 117 (6):2906–2913
32. Darwin F (ed) (1887) The Life and Letters of Charles Darwin, vol III. Murray, London
33. Miller SL (1953) A production of amino acids under possible primitive earth conditions. Science 117:528–529
34. Prusinkiewicz P, Lindenmayer A (1990) The Algorithmic Beauty of Plants. Springer, New York
35. Langton CG (ed) (1997). Artificial Life. Cambridge/Massachussetts.
36. Lindenmayer A (1968) Mathematical models for cellular interaction in development, Parts I and II. J Theor Biol 18:280–315
37. Gardner M (1970) Mathematical Games. Sci Am 223(4):120–123
38. Fraenkel-Conrad H, Williams RC (1955) Reconstitution of active tobacco mosaic virus from its inactive protein and nucleic acid components. Proc Natl Acad Sci USA 41(10):690–698
39. Noireaux V, Libchaber A (2004) A vesicle bioreactor as a step toward an artificial cell assembly. Proc Nat Acad Sci USA 101:17669–17674
40. Gibson DG, Glass JI, Lartigue C et al (2010) Creation of a bacterial cell controlled by a chemically synthesized genome. Science 329:52–56
41. Hutchison CA, Chuang R-Y, Noskov VN et al (2016) Design and synthesis of a minimal bacterial genome. Science 351(6280):1414
42. Cassirer E (1944) An Essay on Man: An Introduction to a Philosophy of Human Culture. Doubleday, Garden City/New York
43. Moor JH (ed) (2003) The Turing Test: The Elusive Standard of Artificial Intelligence. Springer, Dordrecht
44. Feynman RP (1960) There is plenty of room at the bottom. J Eng Sci 23 (5):22–36
45. Adleman LM (1994) Molecular computation of solutions to combinatorial problems. Science 266(5187):1021–1024
46. Adleman LM (1998) Computing with DNA. Sci Am 297(2):4–61
47. Benenson Y, Paz-Elizur T, Adar R et al (2001) Programmable and autonomous computing machine made of biomolecules. Nature 414:430–434
48. Unger R, Moult J (2006) Towards computing with proteins. Proteins 63:53–64
49. Shapiro E, Gil B (2008) RNA computing in a living cell. Science 322 (5900):387–388

50. Searle JR (1980) Minds, brains, and programs. Behav Brain Sci 3(3):417–457

51. Ebeling W, Rechenberg I, Schwefel H-P et al (eds) (1996) Parallel Problem Solving from Nature—PPSN IV. International Conference on Evolutionary Computation. Lect Notes Comput Sci vol 419

52. Ignatova Z, Martinez-Perez I, Zimmermann K-H (2010) DNA Computing Models. Springer, New York

53. Dennett DC (1980) The milk of human intentionality. Behav Brain Sci 3:428–430

54. Feynman RP (1967) The Character of Physical Law. The MIT Press, Cambridge/Mass

55. Wittgenstein L (1999) Tractatus logico-philosophicus. Dover publications, New York [Original: Logisch-Philosophische Abhandlung, 1921]

56. Knight W (2017) The Dark Secret at the Heart of AI. MIT Technol Rev 120(3)

57. Frege G (1980) On Sense and Reference (transl: Black M). In: Geach P, Black M (eds) Translations from the Philosophical Writings of Gottlob Frege. Blackwell, Oxford pp 56–78 [Original: Über Sinn und Bedeutung, 1892]

58. Austin JL (1962) How to Do Things with Words? Harvard University Press, Cambridge/Mass

59. Küppers B-O (2018) The Computability of the World: How Far Can Science Take Us? Springer International, Cham

60. McKay D (1969) Information, Mechanism and Meaning. Cambridge/Mass, MIT Press

61. Küppers B-O (2013) Elements of a Semantic Code. In: Küppers B-O, Hahn U, Artmann S (eds) Evolution of Semantic Systems. Springer, Berlin/Heidelberg, pp 67–85

62. Crick F, Koch C (1990) Towards a neurobiological theory of consciousness. Semin Neurosci 2:263–275

63. Dobzhansky T (1973) Nothing in biology makes sense except in the light of evolution. The Am Biol Teach 35:125–129

64. Lamarck JB (1914) Zoological Philosophy: An Exposition with Regard to the Natural History of Animals (transl: Elliot H). Macmillan & Co, London [Original: Philosophie zoologique, 1809]

65. Darwin C (1859) On the Origin of Species by Means of Natural Selection, or the Preservation of Favoured Races in the Struggle for Life. Murray, London

66. Gould SJ (2002) The Structure of Evolutionary Theory. Harvard University Press, Cambridge/Mass

67. Dawkins R (2006) The Selfish Gene. Oxford University Press, Oxford

68. Conway Morris S (2003) Life's Solution. Cambridge University Press, Cambridge

69. Kimura M (1983) The Neutral Theory of Molecular Evolution. Cambridge University Press, Cambridge

70. Popper KR (1974) Unended Quest: An Intellectual Autobiography. Fontana-Collins, London/Glasgow
71. Monod J (1975) On the Molecular Theory of Evolution (1973). In: Harré R (ed) Problems of Scientific Revolution: Progress and Obstacles to Progress in the Science. Oxford University Press, Oxford, pp 11–24
72. Chomsky N (1957) Syntactic Structures. Mouton, The Hague
73. Küppers B-O (2016) The Nucleation of Semantic Information in Prebiotic Matter. In: Domingo E, Schuster P (eds) Quasispecies: From Theory to Experimental Systems. Springer International, Cham, pp 67–85
74. Küppers B-O (1987) On the Prior Probability of the Existence of Life. In: Gigerenzer G, Krüger L, Morgan MS (eds) The Probabilistic Revolution 1800–1930, vol II. Probability in Modern Science. MIT Press, Cambridge/Mass, pp 355–369
75. Gilbert W (1986) The RNA world. Nature 319:618
76. Robertson MP, Joyce GF (2012) The origins of the RNA world. Cold Spring Harb Perspect Biol https://doi.org/10.1101/cshperspect.a003608
77. Pley HW, Flaherty KM, McKay DB (1994) Tree-dimensional structure of a hammerhead ribozyme. Nature 372:68–74
78. Scott WG, Murray JB, Arnold JRP et al (1996) Capturing the structure of a catalytic RNA intermediate: the hammerhead Ribozyme. Science 274:2065–2069
79. Küppers B-O, Sumper M (1975) Minimal requirements for template recognition by bacteriophage Q_β-replicase: approach to general RNA-dependent RNA synthesis. Proc Natl Acad Sci USA 72:2640–2643
80. Quinn JJ, Chang HY (2016) Unique features of long non-coding RNA biogenesis and function. Nav Rev Genet 17:47–62
81. Ransohoff JD, Wei Y, Khavari PA (2018) The functions and unique features of long intergenic non-coding RNA. Nat Rev Mol Cell Biol 19:143–157
82. Strobel SA, Cochrane JC (2007) RNA catalysis: ribozymes, ribosomes, and riboswiches. Curr Opin in Chem Biol 11:636–643
83. Wacker A, Weigand JE, Akabayov SR et al (2020) Secondary structure determination of conserved SARS-Co-2 RNA elements by NMR spectroscopy. Nucleic Acids Res 48(22):12415–12435
84. Nevill-Manning CG, Witten IH (1997) Identifying hierarchical structure in sequences: a linear-time algorithm. J Artif Intell Res 7:67–82
85. Domingo E, Schuster P (eds) (2016) Quasispecies: From Theory to Experimental Systems. Springer International, Cham
86. Eigen M, Schuster P (1978) The Hypercycle. Naturwissenschaften 65:7–41
87. Eigen M (1992) Steps Towards Life. Oxford University Press, Oxford [Original: Stufen zum Leben, 1987]
88. Küppers B-O (1990) Information and the Origin of Life (transl: Woolley P) MIT Press. Cambridge/Mass [Original: Der Ursprung biologischer Information, 1986]

89. Jerne NK (1984) The generative grammar of the immune system. Nobel lecture. Nobel foundation
90. Eigen M (2013) From Strange Simplicity to Complex Familiarity: A Treatise on Matter, Information, Life and Thought. Oxford University Press, Oxford

7

Epilog: Nature's Semantics

7.1 Unforeseen Progress

The mountain village Erice in the northwest of Sicily is a place with an eventful past. At the top of Monte San Guiliana, where the sacred site of a goddess of fertility was located in prehistoric times, an impregnable mountain fortress was erected in antiquity, first by the Elmiens and later by the Punics. Still today, the remnants of a Norman castle, which testifies to conquests in past centuries, can be found above the village. Because of the history of this place, it has almost symbolic significance that scientific meetings are held every year at the foot of Erice castle—meetings at which the latest research results in physics, chemistry and biology are discussed. Today, it is the natural sciences that have set out to conquer the world.

The summer school of 1971, which had the theme "Molecular and Developmental Biology", has remained vividly in my memory, for various reasons. I'm thinking here of unforgettable and almost literally "atmospheric" impressions, when in the evening hours the clouds were drawing in and swathing the old walls of the small mountain village with their fog. During the day, one had the feeling of looking down from an airplane at the nearby port city of Trapani, as if one were floating above the clouds. Not least, though, the scientific program of the summer school and the illustrious names of the lecturers gave me, as a fledgeling scientist, the feeling of a high-altitude flight. Among the participants were scientific celebrities such as the Nobel prize winners Francis Crick and Frederick Sanger, and the later laureates Paul Berg and Rita Levi-Montalcini.

© Springer Nature Switzerland AG 2022
B.-O. Küppers, *The Language of Living Matter*, The Frontiers Collection,
https://doi.org/10.1007/978-3-030-80319-3_7

Beside science, I was also occupied at the time with the question of whether outstanding scientists possess a particular disposition for virtue, morality and responsibility. I was especially interested in this question because in the early 1970s the consequences of science and technology were a dominant socio-political issue at German universities. In those years the development of science and technology was viewed very critically, not least by scientists themselves (Sect. 7.4).

However, my observations during the summer school in Erice only confirmed what I had already strongly suspected: as regards virtue and morals, scientists behave no differently from other social groups. This fact very quickly became clear when I had the chance to observe them not only in their academic surroundings, but also in the everyday situations of life—on excursions into the culturally rich Sicilian landscape, or in an exuberant atmosphere over a glass of wine. Scientists show the same strengths and weaknesses as all other people.

Accordingly, one encounters in the scientific community not only the elevated motives of scientific curiosity and the search for truth, but also the less noble symptoms of human competitive thinking: vanity and the craving for recognition. Even though such traits of character vary significantly from person to person, they make up an essential force that drives scientists to expose themselves to the harsh and challenging competition of ideas. This competition is additionally fostered in today's media-dominated society, which markets scientific success stories just as it does top sporting achievements. Occasionally, competition-based thinking even leads to serious misconduct, when individual researchers or groups of researchers deliberately disregard the basic rules of scientific knowledge, the supreme commandment of which is honesty. On the other side, the same competition also has a positive side-effect in so far as competition, aiming at external impact, is the best guarantee for public awareness of research and its control by society.

If it is true that the psychogram of human society is reflected in the scientific community, then it is unlikely that scientists will exhibit a moral behavior that goes beyond what is standard in society. No personal experience is needed to reach this insight. The construction of the atomic bomb is the irrevocable proof that scientists, like everyone else, act from a wide variety of motives and values. The belief that sharp analytical thinking, extensive specialist knowledge and a high level of education predestine a scientist to behave in an exemplary manner has long been refuted by the history of science. The correctness of this conclusion is unaltered by the fact that, nevertheless, scientists generally show a high degree of willingness to engage with moral issues.

The argument can be sharpened: scientists in general have no particular moral competence or outstanding moral insight. However, their awareness of such problems may be above average, because they may be particularly motivated by their intimate knowledge of their subject to raise moral issues and to contribute to their clarification. In short: Not moral behavior as such, but its rational justification, seems to require the ability to reflect scientifically. This is why scientists who are equally committed to morality sometimes entertain entirely contradictory views on the responsibility of the sciences. Moreover, scientists also differ in their personality structure. Some are risk-averse; others are willing to take risks. Some are conservative; others are progressive. Some are pessimists by nature; others are optimists. All these factors are decisive in determining a scientist's attitude to his life-world.

Against this background, the question arises as to whether there is such a thing as a general point of view that can be applied to the problem of responsibility. The phrase "to take on responsibility" implies that we want to satisfy our moral standards and are anxious to find an appropriate solution to the pressing problems of our time. However, this does not merely mean that we have recognized a problem as such and want to face the issue responsibly. More than that, it embodies a value judgment, insofar as we believe that we have identified the problem in a "true" form. Accordingly, taking on responsibility is always associated with a supposedly "correct" view of the issues. Therefore, it is not surprising that the problem of responsibility is just as complex as the problem of truth. In neither case can comprehensive answers, valid for all time, be found. Issues of responsibility have the same aporetic character as the truth problem (Chap. 2).

Before going on to discuss the responsibility of science in more detail, I would like to return briefly to the 1971 summer school. Another reason, perhaps the main reason, why this meeting made a lasting impression on me was that it came at a time that we may denote as the birth of modern genetic engineering. I remember that Paul Berg, who had arrived from Stanford, gave a lecture entitled "Tumor viruses. Suppression". In this lecture, he reported on recent experiments in his laboratory in which he and his colleagues had succeeded in merging the genetic material of a tumor virus (SV40) with a plasmid [1].

Plasmids are small ring-shaped DNA molecules that occur in the cytoplasm of bacterial cells (Fig. 7.1). They contain only a few genes; in these, among other things, resistance to antibiotics is inherited. Since plasmids can easily be introduced into cells, they are ideally suited for the transport of genetic material from cell to cell. For this purpose, however, it is necessary first of all to develop a method by which an external DNA segment can be inserted into

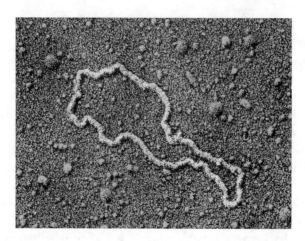

Fig. 7.1 DNA plasmid. False-colour transmission electron micrograph of a plasmid of the bacterium *Escherichia coli*. Plasmids are small, usually ring-shaped DNA molecules that occur in bacteria. They are separate from the main chromosome, replicate independently, and can be transferred from one cell to another carrying genes with them. This property of plasmids is one of the ways in which resistance to antibiotics spreads among bacteria. The plasmid shown here, designated pSC101, served as carrier molecule in the first experiments in genetic engineering (see text below). [Image: Stanley Cohen/Science Photo Library]

a plasmid. It was precisely such a procedure that Paul Berg presented in his lecture. However, what none of his listeners seemed to guess was that these experiments were to become the starting point of a far-reaching biotechnology, the consequences of which frighten many people.

Even so, strictly speaking, the path to modern genetic engineering had already been embarked upon in the 1960s. At that time, molecular biologists had already discovered important enzymes that would later prove to be useful tools in genetic engineering. The discovery of "restriction enzymes", which can be used to cut nucleic acids at defined locations, and the discovery of the so-called "ligases", with which the fragments can be reconnected, were also important steps on the road to genetic engineering [2]. Without these instruments, the development of recombinant DNA technique would never have been possible (Fig. 7.2).

Viewed in retrospect, it also becomes clear that the discoveries in molecular genetics almost inevitably had to lead to genetic engineering, and that Paul Berg was one of the researchers who, with their experiments in the early 1970s, opened the door to this. Nevertheless, I am quite sure that no one at

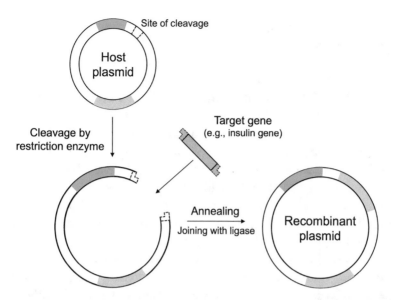

Fig. 7.2 Recombinant DNA technology. In the mid-1960s, molecular biologists started to investigate possible analogies between viruses that infect bacteria ("phages") and tumor viruses that infect mammalian cells. At the center of their interest was *inter alia* the question of how mammalian viruses can pick up genes and transfer them to other cells. Since plasmids can transfer specific properties to bacterial cells, it was an obvious idea to use plasmids as a vehicle for the transfer of a viral DNA to a bacterium. For this purpose, the DNA of the host plasmid must first be split with the help of special enzymes, called restriction endonucleases. In a second step, the ends of the DNA to be transferred are prepared so that they fit into the ends of the host plasmid. In a third step, the sticky ends are spliced. This procedure was the beginning of recombinant DNA technology

the summer school in Erice recognized at the time the dramatic implications of these experiments. In any case, no one in the auditorium expressed such an idea, and neither did anyone speak of any risks. Given the widespread critical attitude toward the possible consequences of science and technology at that time, such concerns would certainly have been raised if anyone had imagined them. Instead, the discussion was dominated by the perspectives which these extremely sophisticated experiments opened up for basic research.

Splicing the DNA of a tumor virus into a plasmid's DNA seemed unproblematic at first, because the DNA hybrid no longer fitted into the molecular packaging of the virus, so that the natural path of infection was interrupted. Moreover, there even seemed to be a way to use this effect to suppress the impact of dangerous viruses. This hope was reflected in the title of Berg's lecture: "Tumor viruses. Suppression".

Since plasmids can replicate in the bacterial cell, independently of the bacterium itself, the next step was to exploit the vector (carrier) properties of the plasmids to introduce the tumor virus into a bacterium and then to observe its proliferation in its new host cell. Such an experiment would undoubtedly have been an essential approach to elucidating the molecular basis of viral infection using a simple model organism, as biologists had successfully done with phages in the early years of molecular biology.

However, when Paul Berg and his colleagues drew up the plan for such experiments, the full extent of the risks of this new method was suddenly realized. The bacteria that Berg used to experiment with are naturally present among the human intestinal flora. Therefore, there was an acute risk that the experiments could get out of control and release a lethal pandemic.

Very soon, voices were heard that warned urgently against carrying out such experiments. However, the knowledge about the manipulability of genes was out there and could not be withheld; much less could it be withdrawn. Almost at the same time as Berg's research group, Stanley Cohen, Herbert Boyer and other researchers succeeded in assembling DNA segments from different organisms in the test-tube and incorporating them into the genome of a bacterium [3]. Shortly afterwards, a bacterial gene was transferred to baker's yeast, and a human gene to a bacterium. In the period that followed, the new methods, enabling organisms' genes to be newly combined across species boundaries, developed and spread almost like wildfire (see Sect. 6.4).

7.2   Future Visions: Organic Technology

In view of these breathtaking perspectives, it is not surprising that many people view current developments in biology with concern. It appears almost as if modern biotechnology is robbing the living world of its intrinsic value and degrading it to an abstract object of research and experiment. However, this is not the case. On the contrary, artificial evolution, described in Sect. 6.5 , does not lead to a devaluation, but to enrichment and intensification of the biosphere, as it may bring possible life to light.

The technological potential of living Nature was already recognized, centuries ago, by Leonardo da Vinci. In his notebook he wrote: "Human subtlety (…) will never devise an invention more beautiful, simpler or more direct than does nature, because in her inventions nothing is lacking, and nothing is superfluous." [4, p. 126] The universal genius of the Italian Renaissance was not only a gifted artist and designer, but also a skilled observer of Nature. For example, Leonardo studied with great care the wing feathers of a bird, because he wanted to use their construction principle for the design of a flying machine. It is noteworthy that Leonardo apparently had no intention of imitating the anatomical properties of the bird's wing in detail; he only wanted to explore its basics to transfer them to the mechanics of an apparatus for flying. Beside the ornithopter, shown in Fig. 7.3, he sketched numerous ideas for airborne objects such as fixed-wing gliders, rotorcrafts and parachutes.

In the same way, modern bionics does not see its task as that of reproducing the characteristics of organisms as accurately as possible. Instead, bionics is an analogy technique which makes use of the basic principles of animate Nature only to the extent that this helps in the solution of technical issues. In this respect, bionics is an example of an applied structural science that covers nearly all areas of modern technology (Sect. 4.9). The hybrid word "bionics", implying "biology" and "technics", already indicates this. Since living systems have been optimized in an evolutionary process lasting billions of years, and some of them have quite unusual material properties and

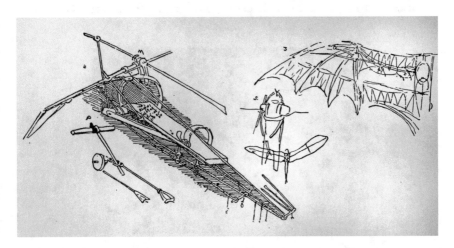

Fig. 7.3 Ornithopter (designed by Leornardo da Vinci). Leonardo was a pioneer of bionics. The drawing of a flying machine, shown here, documents one of Leonardo's attempts to solve technical problems by imitating Nature. [Image: World History Archive/Alamy Stock Photo]

forms of construction, they represent an inexhaustible reservoir of ideas for developers in technology.

Bionics focuses on the idea that evolution has succeeded in finding optimum solutions for almost every problem of biological adaptation and that, consequently, its inventions might also be used to solve complicated technical issues. Indeed, the list of natural examples on which technological solutions have already been modeled is correspondingly long (see [5]). For instance, the unusual adhesive power of a gecko, which can cling even to smooth vertical surfaces, has inspired bionics to develop nanostructures for glue-free adhesion. Another example is the leaf of the lotus flower, which served as a model for the development of superhydrophobic self-cleaning surfaces. The adhesive properties of the burdock, in turn, inspired the invention of "velcro", probably the best-known example of bionics.

These and many other examples demonstrate the manifold applications of bionics in materials research. No less spectacular are its applications in the field of robotics. Here, there seem to be no limits to the technical ingenuity that we can acquire by copying Nature. An example of this is shown in Fig. 7.4. The picture shows three artificial ants which can move and interact with each other by way of their electronic devices. They can even simulate cooperative behavior and solve complex tasks. One may realistically expect that future production systems will be based on intelligent components that allow robots to adapt flexibly to different production requirements by making

Fig. 7.4 Robotic ants. Just as real ants move fat beetles into their burrows, robotic ants can cooperate to move objects that they could never move alone. The borders between natural and artificial systems begin to disappear in such technological creations. [Image: courtesey of Festo SE&Co]

"intelligent" decisions on how best to carry out orders that they receive from higher control instances.

Bionics, however, does not only try to use the solutions that biological evolution has found; it also tries to take on the problem-solving strategy itself. This allows engineers to find technological solutions outside the natural range. By this strategy, extremely complicated technical problems such as the optimization of pipe bends, hot steam nozzles, contact lenses, wings and the like have already been solved (see [6]).

Another promising approach at the interface of biology and technology is the development of biohybrids consisting of inorganic and organic material [7]. Figure 7.5 shows such a biohybrid that is constructed from a metal, a transparent polymer and organic muscle tissue, in this case of the rat. This hybrid behaves like a "living" robot. It can swim, and it reacts to light pulses. This curious creature is not quite an animal, an not quite a machine, but a "medusoid". It supports the vision of constructing larger biohybrids that, one day, may also work in a natural environment.

With such considerations, we are entering the borderline area of science and technology, where reality and vision merge into each other. Visions, however, are indispensable for scientific progress. We must continually create new pictures of the future, because these motivate us to search for knowledge in a specific way. In this respect, vision has a leading function in the cognition process. Our visions help us to depart from our deep-rooted ways of thinking and thus create the preconditions for us to free ourselves from traditional images of reality.

The development of computer technology illustrates how quickly visions can become reality. When the era of modern automatic calculators began in the 1950s, the theoretical biologist Nils Barricelli designed the visionary model of an evolution that takes place within the computer "medium" [8]. Indeed, in the course of the worldwide networking of computers, a comprehensive digital evolution has already begun that far exceeds Barricelli's expectations. The historian of technology George Dyson even suspected that, at the end of digital development, a global intelligence could arise that filters all information flows and thus influences decision-making processes in economics, society and politics [9]. Today we do not seem to be far away from the fulfilment of that prophecy.

The computer age has also produced its own culture, the so-called "cyberculture", the guiding science of which is cybernetics. The main elements of cyberculture are bioelectronic space (cyberspace) and the

Fig. 7.5 Hybrid robots of inorganic and organic matter. Hybrids of inorganic and organic material are a preliminary stage on the way toward a cybernetic organism ("cyborg"). The objects shown here are biohybrids, a few millimetres in size, that look like stingrays. Powered by rat cardiac muscles, they swim in response to a pulsating light source. The experiment, carried out by the physicist Kevin Kit Partner and his colleagues, was designed to help us understand in more detail how muscular pumps work [7]. However, in addition, such experiments may provide insight into the relationships between sensory inputs and responding behavior. This could one day become a model of synthetic cognition. [Image: courtesey of Kevin Kit Partner]

bioelectronic organism (cyborg). The latter is a hybrid being whose physiological functions are supported by mechanical and electronic devices, similar to the artificial creature shown in Fig. 7.5. Cyberculture appears to be the cultural precursor of a future world in which man and machine, natural and artificial, animate and inanimate, will merge.

However, that is far from all. Even the boundary between mind and matter appears to be disappearing, to the extent that the neurosciences and information technologies move closer together. The idea of connecting a computer directly with nerve cells, which at first seemed completely unrealizable, has already taken shape at the horizon of the neurosciences. First experiments, carried out by Peter Fromherz, have indeed succeeded in establishing a closed circuit for signal transmission between a semiconductor chip and the nerve cells of a sludge worm (Fig. 7.6). Thus, a signal originating from the chip was transmitted to a nerve cell, from where it was then transferred to a neighbouring nerve cell, and the response to this signal was registered on the chip again [10]. Such "neurochips" are a significant step on the way to merging artificial and natural processing of information.

Of course, today's possibilities of neurotechnology are still a long way behind the capabilities of the human brain. Therefore, the idea that one day we will be able to build a neurocomputer that can take on the brain will remain an unfulfilled dream of the computer visionaries for a long time to come, even though scientific and technical progress is taking place much faster than expected.

Indeed, the human brain is so immeasurably complex that there are very high hurdles to its technical reproduction. Nevertheless, experiments such as that shown in Fig. 7.6 have shown that hybrids in which organic and technical components exchange information are in principle possible. This opens up exciting applications in the field of neurotechnology. An obvious

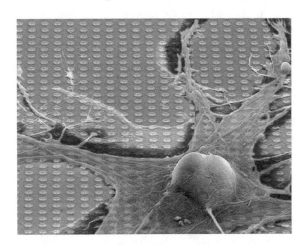

Fig. 7.6 Hybridisation of electronic and grey matter. The image shows the fusion of a computer chip (128 × 128 sensors) with the nerve cell of the sludge worm *Lymnaea stagnalis* [10]. A signal is transferred from the chip to the nerve cell and vice versa. The link between the chip and the ionic channels of the neurons was established by special brain proteins that glue the neurons onto the chip, so that the electrical signal of the neuron can pass to the chip and back. [Image: courtesey of Peter Fromherz]

application of this technique could be to use it for screening the pharmaceutical effects of drugs. Moreover, it is conceivable that neurochips will one day be able to take over the function of an artificial retina, so that blind people will be able to see it with the help of neuroprostheses. It is also conceivable that neurochips can be used to gain new insights into the perceptual functions of the brain, from which conclusions may one day be drawn for the construction of future neurocomputers.

This shows more than anything else the tremendous power of the cultural change that is currently emanating from the life sciences. This cultural upheaval is in some ways reminiscent of the beginning of the twentieth century. At that time, there was also a progress-oriented current of opinion, which focused exclusively on the future. This current, known as "Futurism", originated in art at that time, but it also influenced thinking in science, technology, society and politics. It is amazing to see how futuristic thinking seems to be repeating itself at the beginning of the twenty-first century. This time, the trigger is modern biology—combined with computer technology.

The advances regarding artificial life, artificial intelligence and artificial evolution force us to make a radical break with our traditional ways of thought. It is above all our conception of the essence of the natural that demands a new, futuristic interpretation. With the dissolution of the boundaries between animate and inanimate Nature by science and technology, the traditional understanding of the natural and the artificial must also change. What we have hitherto regarded as "natural" has for centuries been shaped by our view of "animate" Nature as the epitome of a wise and planned creation. Modern science, however, has revealed that living matter originated from inanimate matter through physical self-organization and that the corresponding processes can be set artificially in motion by employing modern biotechnology. For this reason, life can no longer be regarded as the result of a unique act of creation, something that the human race is obliged to preserve. Instead, we are called upon to shape Nature in a future-oriented manner. Needless to say, this task demands a high degree of responsibilty.

7.3   Control by Renunciation?

Scientific discoveries are neither predictable nor plannable. Usually, their technological implications only become visible when the various scientific findings necessary for the development of a technology converge. In the case of genetic engineering, this is all the more remarkable as its tools were already present in Nature and were well known to be used by living organisms. For example: without them, sexual reproduction, which leads to the recombination of the female and male genomes, would be impossible. In this process, the maternal and paternal genetic material is taken apart, individual genes are exchanged with each other and are then put together again to form a functional genetic blueprint for the offspring.

Natural genetic engineering was indispensable for the evolution of higher organisms, including mankind. Therefore, in retrospect, the development of artificial genetic engineering should have appeared to be the most natural thing in the world. Nevertheless, it took the creative power of man to assemble knowledge about the genetic procedures of Nature, initially scattered in a multitude of individual insights, into a single edifice of knowledge capable of being translated into technological innovation. The same applies to the innovative ideas of physics. For example, essential elements of relativity theory were already known before Einstein put them together to form a new theory that revolutionized physics.

Nevertheless, it is an essential feature of human knowledge that its development cannot be planned and certainly cannot be targeted. Hilbert's failed attempt to systematize all mathematical knowledge by applying a formalized procedure showed this clearly (Sect. 5.2). The advancement of knowledge is unpredictable, simply because acquiring it depends on unpredictable factors such as intuition and indeed sometimes on sheer good luck.

In this connection, I remember a notable discussion between the philosopher Karl Popper and the biologist Konrad Lorenz, which took place in 1983 at the European Forum in Alpbach and dealt with the question of how we acquire knowledge. Popper put forward his deductive model of cognition, while Lorenz took an inductive point of view (cf. Chap. 2). Popper insisted that progress of knowledge takes place when we form hypotheses that lead to predictions about the world which withstand all attempts to refute them. Lorenz, in contrast, defended his view that we gain new ideas by generalizing from individual observations and that new ideas arise through our noticing interconnections between our observations. Popper argued from the perspective of the epistemologist, Lorenz from that of the behavioral biologist.

Independently of how deductive and inductive cognition are weighted, an essential part in the process of discovery is played by circumstances under which a discovery is made. Regarding this point, Popper and Lorenz were in agreement. A certain amount of foreknowledge, the perception of relevant facts, intuition, the deciding flash of thought—all this and more must come together for a fruitful hypothesis to emerge and for our findings to develop further.

Last, but not least, the working environment of a scientist also has a considerable influence on the course of research. After all, scientific knowledge is not usually acquired by people working in seclusion, but rather under the influence of an exchange of ideas between scientists. Several people or research groups are always involved, either directly or indirectly, in any process of scientific discovery. This interdependence means that scientific research is today highly collaborative and globalized.

Thus scientific knowledge cannot just be forgotten or wiped out. Consequently, we are forced to deal constructively with our findings. The dramatist Friedrich Dürrenmatt staged this insight impressively in the play *The Physicists*. His main character, the ingenious physicist Möbius, goes so far as to seek refuge in a lunatic asylum to bar access to his ground-breaking findings—in vain, as it turned out. At the end of the play, Möbius must resignedly state: "Whatever has once been thought can never be taken back." [11, p. 122]

It is quite conceivable that Dürrenmatt, with the name of his main character, Möbius, was alluding to the so-called Möbius strip, named after the mathematician August Ferdinand Möbius. The Möbius strip is a circular ribbon, twisted so as to make it one-sided, conveying the idea of inescapability. In any case, the message of the play is clear: whether we like it or not, we have to live with our insights. There is no point in searching for ways in which we can hide knowledge. Instead, we must learn to be proactive in dealing with the knowledge that we have, and in controlling its possible consequences.

An obvious idea for avoiding the dangers of scientific knowledge would be to refrain from further research. Paul Berg took this path when he stopped experimenting with the SV40 virus. Moreover, he also co-organized the famous Asilomar Conference, at which leading geneticists discussed the necessity and possibility of controlling and limiting their research. Later, Asilomar became a globally visible sign that scientists are not only aware of their responsibility, but also that they are willing to take it on in word and deed [12].

However, how exactly can and should science shoulder its responsibilities? Is, in the end, self-regulation by self-restriction the *ultima ratio* of responsible science? Moreover, would this kind of self-imposed responsibility be at all practicable? If the reports from the Asilomar conference are credible, doubts soon arose about the idea of self-control. According to the geneticist Philippe Kourilsky [13], a great deal of perplexity and helplessness emerged in the course of the Asilomar conference, because it was not even possible to reach agreement among the experts on the fundamental question of how to assess the risks of genetic research. Under these circumstances, one must ask how a behavioral codex governing future research can ever be reached, given the uncertainty that always accompanies scientific knowledge.

Kourilsky asked this question, too: "How can reality, Nature and the extent of possible risks be clarified without experimenting, without taking a minimum of risk? If I can trust my memory and my perception, the confusion after the three-day debate was greater than ever, and a consensus was not in sight. Many participants were reluctant to offer the journalists present a spectacle of powerlessness." [13, p. 144; author's transl.] In the end, a joint resolution was adopted, but this was mostly because the participants feared that politicians, under the pressure of an insecure and frightened public, could seize control from scientists and make far-reaching politically based decisions against genetic research. Kourilsky provides detailed information on how, finally, the mass media took over opinion leadership by staging and supporting a campaign against research in genetics. Thus, the positive impetus that was intended to come from Asilomar was ultimately reversed, and the public reaction to Asilomar became a prime example of how responsibility can be dealt with irresponsibly.

From Asilomar, we learn that new knowledge can only be evaluated and controlled within the overall framework of existing, comprehensive and advancing knowledge. This framework can only be provided by the methodically secured knowledge acquired in science itself, and not by some transcendent wisdom based on beliefs, religious convictions, emotions or the like.

For example, if we want to assess the risks of genetic research with the highest possible reliability, we need to investigate the properties of genes, their stability, their transfer from organism to organism, their information content under changing environmental conditions and more. In short, it is necessary to conduct comprehensive basic research on genetically modified organisms. This is the only way to assess reliably the possible dangers associated with, for example, transgenic organisms. The same holds for the possible opportunities offered by genetic research. The question whether new genetic techniques— therapeutic cloning, stem-cell research, cell therapy and the like—may keep their medical promises and lead one day to cures for serious disorders such as Parkinson's or Alzheimer's disease, can only be assessed on the sound foundation of basic research.

A responsible discussion on the risks and chances of science and technology is only possible in the context of a body of knowledge that is continually renewing and refining itself. Therefore, all efforts to stop basic research in high-risk areas are irresponsible, sometimes highly so. Where, after all, are we to find the knowledge that is necessary for rational risk assessment, if the required research is restricted from the outset? This was another question that the Asilomar conference failed to answer. It eventually led the discussion into a dead end and sealed the participants' "powerlessness" to which Kourilsky refers in his book.

The dependence of knowledge upon knowledge means that the idea of a self-restriction of research, as initially debated at the Asilomar conference, is ultimately an illusion. The idea that the risks of genetic research could be averted by a worldwide agreement binding on all genetic researchers, allowing only experiments that have received prior approval, turned out to be unrealistic, as a consensus on the risks of genetic research could not be reached, even amongst the scientists themselves. The origin of these difficulties lies not only in differing assessments of scientific facts, but it is also related to the hypothetical character and the incompleteness of any scientific knowledge.

Because renouncing basic research is not a solution to the problem of responsibility, and because indeed such a moratorium on research would ultimately be directed against a responsible approach to science, the burden of decision is frequently (and gladly) referred back to the individual scientist. According to a widespread argument, every scientist is his own supervisory authority. He has the freedom to control his activities and to restrain his quest for knowledge. For this reason, it is stated, every scientist has to develop ethical standards for himself in interaction with the scientific community and, if necessary, to abstain voluntarily from certain research activities. Following a model from antiquity, even a declaration based upon the Hippocratic oath has been proposed, according to which every scientist should commit himself to use his art only for the benefit, and never to the detriment, of humankind.

However, the appeal to the individual responsibility of scientists fails to do justice to the problem. The independent handling of scientific knowledge requires that a scientist be able to make his decisions rationally and on the basis of the most comprehensive knowledge background available. However, he is not allowed to ascribe final certainty to any conclusion that he reaches on his own. In other words: Not only the search for truth, but also truthfulness in dealing with scientific knowledge, is part of a scientist's professional ethics. This integrity, however, obliges him to question every item of knowledge and, if necessary, to replace it by a new one that approximates more closely to the ideal of truth. In this respect, as the biologist Hans Mohr has emphasized, every scientist has already "made a preliminary moral decision in favor of knowledge" [14 p. 4; author's transl.].

The truths of our scientific knowledge about the world have only a hypothetical character. If they nevertheless can claim a high degree of validity, this is solely because they are methodically secured. This "method" consists of confronting scientific findings with new observations and hypotheses time and again, and thus continually questioning them. The real driving force of the quest for knowledge is, therefore, the critical handling of knowledge—which, however, itself implies an obligation to acquire more knowledge, because, according to the research logic of the philosopher Karl Popper, the validity of a scientific hypothesis can only be maintained as long as it proves itself under continued attempts to refute it. This means that the logic of gaining scientific knowledge in fact calls for unrestricted basic research, since the method of falsification as an instrument of truth control can fulfil its critical function only at the meta-level of more extensive knowledge.

The insight that the validity of scientific knowledge can only be guaranteed in the context of more comprehensive knowledge is a conclusion that can be reiterated indefinitely. For this reason, a scientist who feels committed to the search for truth and who is willing to take on responsibility in science cannot at the same time be committed to a self-imposed restriction of his research. Such a commitment would lead to a contradiction. Any scientist complying with such a demand would be violating the principles of science. Not even the risk assessment, on which a possible relinquishment of further research could be based, can be regarded as robust, as all the findings used in risk analysis are of a provisional nature and may at a later date prove to be incorrect. To sum up these arguments: a scientist can only follow the Hippocratic oath and use his knowledge for the benefit of humankind if he gives first priority to an unrestricted search for knowledge.

7.4 The Phantom of Instrumental Reason

The freedom of research is based on the idea of pure and objective knowledge that opens up for humankind the possibility of designing the future. However, the term "pure" knowledge, understood as knowledge for its own sake, is hugely controversial. There are philosophers who see in the concept of pure knowledge only an idealized thought, far removed from reality. They justify their criticism by asserting that the goal of scientific discovery always serves a particular interest. This argument is put forward principally by philosophers who consider the process of human cognition to be more or less dependent on social conditions.

There is no doubt that human cognition and human action are interrelated. Equally indisputable is the fact that interests guide human actions. As a result, the temptation to short-circuit this connection, and to link knowledge and human interest directly to each other, is quite apparent. The social philosophy of Jürgen Habermas is an example of this. His philosophy goes so far as to ascribe to social conditions a constitutive function for human knowledge, comparable with Kant's philosophy of Apriorism (Sect. 2.5). In his writings, Habermas generally confronts science and technology with the accusation that their real motive is not at all the search for pure and objective knowledge, but rather the interest-led mastering of Man and Nature (see [15, 16]).

With his criticism of science and technology, Habermas follows the so-called "Critical Theory" of the "Frankfurt School" of social sciences, founded in the 1960s by Theodor Adorno and Max Horkheimer. The philosophers of the Frankfurt school saw it as their task to continue the tradition of Western criticism of reason against the background of Karl Marx's critique of society and Sigmund Freud's psychoanalysis.

The direction of thrust of the criticism of science and technology coming out of the Frankfurt School was given by Herbert Marcuse, who succinctly pronounced: "The liberating force of technology—the instrumentalization of things—turns into a fetter of liberation; the instrumentalization of man." [17, p. 131] In the same vein, Habermas warned against the "dangers of the reduction of reason to technical rationality and the reduction of society to the dimension of technical control" [16, Appendix, note 14]. This criticism culminated in the assertion that the self-imposed degrading of human reason to instrumental reason would inevitably consolidate the existing "relationships of domination" of Man over Nature.

Adopting this sociopolitical maxim, Habermas placed science and technology under the general suspicion of being an ideology [15]. This may seem strange, because one would probably give immediate credit to science for being resistant to ideologies of all kinds, on account of its critical understanding of truth. In contrast, "ideologies" are usually understood as cultural currents that doggedly hold on to their preconceived maxims. According to Habermas, however, an ideology is something like a "false consciousness" at the level of collective action that, for reasons of self-justification, propounds fake motives for its actions instead of declaring its real motive [15]. In such false consciousness, Habermas argues, science and technology purport to strive for pure and objective knowledge while, in truth, they serve the interest of instrumental reason.

The image of science outlined by Habermas indeed stands in stark contrast to the perception that science has of itself. Thus, scientists are convinced that their findings are objective because they follow strict methodological rules which are free of interests of any kind. For Habermas, however, the methodological constancy of science is precisely the expression of its false consciousness. The rigorous methods of science, Habermas claims, have a "protective function", since they ensure progress in science within a frame that is better not discussed. With this excuse, according to Habermas, the crucial aspect—the link between knowledge and interest—is deliberately ignored by science.

It is in the nature of things that Habermas' criticism of science is directed primarily against the positivist cognition ideal of science. We recall that, according to positivism, the sciences are confronted with a world of pre-structured facts, and it is their task to describe and explain these by rules, principles and laws (Sect. 4.2). This idea Habermas rejects flat out as being uncritical. He argues that there are no directly given, objective facts at all. Instead, the alleged "objectivity" is given within the context of the human interests that steer our acquisition of knowledge.

Under the influence of positivism, according to Habermas, the philosophy of knowledge has degenerated to a mere theory of science, which is only interested in the rules describing how scientific theories are constructed and examined, but which has completely cut itself off from the question of the conditions of cognition. Therefore, Habermas demands, a critical philosophy of knowledge must return to Kant's issue of the constitution of possible objects of knowledge—a perspective that would be denied by positivism *eo ipso* because it refers exclusively to that which is found.

In contrast to Kant, however, Habermas does not seek the foundations of knowledge in the cognitive faculty of man, but in the social conditions under which the process of cognition takes place. The *a priori* conditions under which reality is structured and objectified thus shift from the guidelines of intuition and intellect (Kant) to knowledge-guiding interests (Habermas). In this way, Kant's Apriorism of cognition disappears under the cover of sociopolitical interests.

This is not the place to analyze the philosophy of Habermas in detail. To do this, we would have to dismantle and to analyze the conglomerate of pre-Socratic, Platonian, idealistic, hermeneutic and many other motifs that Habermas took over from the history of philosophy, and we would then need to relate these to the motives that led him to his philosophy. Thus, we would be forced to follow his maxim and ask which social and political interests guided his philosophy in particular and Critical Theory in general. Such a task would go well beyond the scope of this book.

Instead, let us emphasize one point that reveals Habermas' motive in attacking the positivist sciences. His remark about the essence of theory gives a decisive hint: "Through the soul's likening itself to the ordered motion of the cosmos, theory enters the conduct of life. In *ethos* theory molds life to its form and is reflected in the conduct of those who subject themselves to its discipline." and "This concept of theory and of life in theory, has defined philosophy since its beginnings." [16, p. 298] Thus, Habermas is attempting to reanimate an ethical and social understanding of theory that goes back to pre-Socratic philosophy, again by his social doctrine. According to the old idea, the norms of the (reasonable) world and the norms of reasonable action form a unity; in short, what is and what ought to be are related to each other. Positivistic science, according to Habermas, has abandoned traditional theoretical thinking, which was connected with the claim of practical effect, in favor of the illusion of pure theory, which is directed solely at describing what exists.

What Habermas is calling for here is nothing less than a return from a "descriptive" to a "prescriptive" understanding of theory, which he claims is justified in the social Apriorism at which his social doctrine is aimed. At the same time, this change also entails a shift in the objectification of reality, in so far as the "fact objectivism" of conventional science is now replaced by the "normative objectivism" founded in social theory.

Against this background, Habermas sets out to destroy the "illusion of objectivism", in order to counteract the "practical consequences of a restricted, scientistic consciousness of the sciences" [16, p. 312]. Consequently, the concept of "objective" fact becomes the central target of his attack on the positivist self-image of modern science. Habermas' argument culminates in the claim that there are no facts at all that are free of values and interests. Instead, he concludes, the so-called "objective facts" found by science are always permeated by interests that had guided the cognition process.

Habermas is right insofar as facts, regarded as primary reality, are an idealization that cannot be accessed even by positivistic science. Every perception is selective, and every observation directed at some object is based on a prior theoretical understanding. Therefore, statements of facts indeed have their specific place in the context of background knowledge. This context is also influenced by the interests of the recognizing subject as well as the circumstances of perception and observation. In short, there are no context-free facts.

These arguments, however, do not in any way lead us to conclude that the objectivism of science is merely an illusion. In science, the objectivization of facts takes place at an entirely different level, namely in the relationship of facts to each other. Isolated facts are irrelevant to science, because they have no substantial relation to reality at all. It is rather the fabric of facts in which facts become manifest as significant parts of objective reality. A mere collection of isolated facts says just as little about the structure of reality as the mere accumulation of notes results in a coherent piece of music.

Relationships between facts are the rock-bottom foundation of the exact sciences. For example, the search for causal relationships, which is the basis for law-based explanations, is equivalent to the search for temporal asymmetries between facts (cf. [18, p. 13 ff.]). Facts themselves, in turn, enter the laws only in the form of initial and boundary conditions. They narrow down the set of possible solutions of the differential equations describing natural laws, but they do not violate the objectivity of the natural event. On the contrary, the concept of natural law would lead to absurdity if it only had solutions of technical usability, i.e., ones determined by human interests.

The assertion of Habermas, according to which empirical facts as such only become manifest through prior organization of our experience within the functional circle of instrumental action is stated without proof and lacks any substantiation. His arguments against positivism are consistently vague and uncritical, as they do not go into the deep structure of science. In fact, they miss the mark at the start, because modern science has long since abandoned the primitive housing of laboratory science assigned to it by Habermas. The structural sciences, for example, are immune against his criticism, as they do not refer to facts at all. Instead, they approach only the abstract, structural and relational aspects of reality, as mathematics also does (Sect. 4.9).

Ultimately, Habermas' criticism of positivist science turns out to be itself an ideology. It purports to destroy the dangerous "objectivistic appearance" of science, yet it erects a new and far more hazardous objectivism on its ruins, namely the objectivism of norms, oriented toward the antique concept of theory. Following this tradition, norms are given priority over the facts.

The norms of proper life, Habermas believes, must be raised like a lost treasure. As we have already seen with his concept of theory, he uses a thought figure from Greek antiquity as a template for his philosophy. In this case, it is the Socratic method of "maieutics" (midwives' art), intended to bring to light the true and objectively binding norms of right living. This method was first employed by Plato in his famous dialogs. In these dialogs, each named after a key person being questioned by Socrates, a certain subject matter is analyzed by Socrates' asking and his interlocutor's responding to questions, from which the truth emerges by degrees.

Habermas has formalized this idea in the framework of modern language philosophy and extended it to the so-called "ideal discourse community." According to this, a community should be able to achieve a reasonable consensus on the norms of human action, provided that it adheres to the rules of rational communication (for details see Sect. 2.9). However, any rational discourse about the implications of science and technology is doomed to failure if it is based on an ideologically predetermined judgment about science. How is one to gain and justify norms of proper life in modern society, if one denies the factual objectivism of science and thereby destroys the epistemic foundation of science and technology?

The philosophy of Habermas thus begins to look like a reflex reaction, delayed for centuries, to Bacon's program of striving for the rule of Man over Nature (Sect. 1.2). It prohibits a favorable view of science and technology since the term "rule", in the context of the Critical Theory, has negative connotations from the outset, as a synonym for suppression and exploitation. The forward-looking, trend-setting model of mastering Nature by science and technology, a model committed to humanity and enlightenment, as envisioned by Bacon and Descartes, is thus thwarted. The idea of ruling over Nature, which was intended to liberate humankind from the hardships of life, is instead demonized and reinterpreted as the phantom of instrumentalized reason acting against humans.

In society, the attack on the positivism of science has promoted a pervasive anti-intellectualism that is generally directed against progress and innovation. It pursues two goals. On a theoretical level, it it supposed to liberate our world-understanding from the alleged "illusion" of scientific objectivism. On a practical level, it is supposed to take the ground away from under the feet of scientism, which wants to bind human action to the guidelines of a positivistic image of reality. In making such attacks, Habermas undermines not only confidence in the objectivity of scientific knowledge, but also the credibility and authority of the sciences per se. Of course, this is the declared intention of his criticism, with which he supposedly counteracts the alleged indoctrination of society by instrumental reason.

A philosophy that traces human knowledge back to cognition-guiding interests inevitably shares the boat with an intellectually fashionable trend that has been labelled "postmodern philosophy" (Sects. 3.10 and 4.1). Both of these contest the objectivism of science, and both claim that scientific knowledge is bound to social or, respectively, cultural conditions. Since such conditions are contingent by their very nature, scientific knowledge must necessarily be considered contingent as well. Even if the "postmodern" understanding of science follows its own absurd logic, it becomes clear how short the path is from the ideology of interests to the postmodern maxim of arbitrariness ("anything goes"), when science and technology are put into the straitjacket of a socio-philosophical doctrine (Sect. 3.10).

Of course, every epistemological position can and must be critically questioned. Positivism is not exempted from this. Nevertheless, there is no doubt that positivism best meets the requirements of critical thinking guided by experience. The philosophy of Habermas, in contrast, has no more than a discourse theory of truth with which to oppose positivism, one that leaves the decision on the claims to the validity of statements and norms of action solely to the discourse about these issues. It turns truth into a self-fulfilling prophecy. However, the advancement of scientific knowledge never emerged from a democratically organized discourse community as such; rather, it arose and arises from the confrontation of both the individual and the community with their objectively given, outer world. The link to the individual's inner world is provided by perception, observation, experience and abstraction. These cognitive elements constitute the positivist access to the world and thus the foundations of science.

This is without prejudice to the fact that the process of cognition also takes place within a certain social and cultural frame. However, that does not make human cognition a process that must necessarily be subjected to the procedures of an ideal discourse community. On the contrary, the exact opposite seems to be the case. Progress in knowledge is only possible under conditions in which the "collective" reason and the individual's reason temporarily go their separate ways. This, in turn, means that the individual breaks out of the traditional structures of thought and thus places itself outside the world of common thought. The actual motor of creative thinking is the break with conventional ways of thinking, not classification in the conventional tradition. Therefore, scientific creativity is first and foremost a property of the individual and not of some discourse community that refers to the ideas and ideals of times past.

The criticism of scientism that surfaces in the social theory of Habermas is not new. It has accompanied science since its unstoppable rise in the nineteenth century. Yet the criticism has always had the same goal, namely, to deprecate modern scientism, because scientism supposedly reduces human reason to mere scientific rationality and thus promotes the instrumentalization of human thinking.

As a consequence of the ongoing criticism of scientism, a general devaluation of scientific-analytical thinking can be observed in large parts of society, which philosophers of science and cultural theorists already pointed out decades ago. In the 1930s, Richard von Mises and Johan Huizinga saw themselves constrained to warn of a systematic philosophical and practical "anti-intellectualism" which in their view was to be encountered in many circles, running to a downright hostility to science. Von Mises termed the attitude of mind that was directed against the positivist sciences "negativism", because it was "directed toward restricting and deprecating the role of reason and analytical thinking in our efforts at comprehension of our environment" [19, p. 58].Huizinga even castigated anti-intellectualism as a "spiritual illness" and the "shadow of tomorrow" [20].

Indeed, anti-intellectualism, mixed with social and political ideologies, has long since cast its shadows over today's knowledge society [21]. It has not only undermined, at least in part, the acceptance of science and technology; it has also become the biggest obstacle to the responsible handling of science in society. It therefore seems all the more grotesque when Habermas denotes science and technology as ideology and thus makes the victim of anti-intellectualism, namely science and technology, the perpetrator.

Questioning sciences' claim to objectivity, Habermas attempts to tear down the borderline between science and ideology which runs alongside the positivist attitude of science. This borderline represents, as it were, the watershed between science and ideology. If one destroys the foundations of positivism, science becomes open for ideologies of all kinds. A prime example of this is the pseudo-scientific doctrine of the agronomist Trofim Lysenko in the 1930s. Lysenko rejected the generally approved concepts of genetics and heredity and replaced them by assuming that environment is the decisive factor in biological development, particularly in the growth of plants. In the Stalinist period of the Soviet Union, this idea fell onto fertile ideological soil. Supported by the mass media, it became an official scientific doctrine of the Soviet political system for several decades. However, the inadequacies, mis-interpretations, counterfeits and ideological motives running through Lysenko's theory became evident by and by. Under pressure from prominent scientists, the doctrine of Lysenkoism was finally abandoned in the 1960s. It had not only set back genetic research, but also agricultural development, in his country for a generations.

No ideology can override the fundamentals of genetics, atomic physics, or any other scientific concept. All attempts to do this are bound to fail. The humanities, in contrast, as long as they do not feel committed to the positivist ideal of science, are inherently susceptible to ideology. For example, within the hermeneutic humanities, the phenomena of reality are not only inter-preted in different ways, but are already perceived as fundamentally different from the objects of cognition of positivistic science. The hermeneutic method eludes the objectivism of science by propagating the inseparable unity of man, Nature, society and history with regard to the cognition process (Sect. 3.4). The hermeneutic approach to reality is also the basis on which Habermas developed his theory of society, in which—following the ancient Greeks —"be and ought" form a unity.

In reality, however, the alleged "critical" philosophy is nothing more than an ideology that opens the floodgates for anti-intellectualism. The faster scientific progress is, the stronger the ideologically fomented resistance against science and technology becomes in society. Contemporary anti-intellectualism is reflected in the same slogans that von Mises already criticized a century ago: "Besides the slogan 'synthesis, not analysis' there are the favorite formulas 'soul against reason' or 'spirit versus intellect', or 'living intuition instead of dead formalism', and expressions such as 'holism', 'reverence for life', 'intuitive comprehension of the world', 'Wesensschau', etc. Sometimes, one hears such statements as 'This may be logically correct, but it does not stand up against the facts of life,' or 'The sterility of thinking has to give way to a stronger feeling of life,' and other similar things. Phrases of this kind are built up into entire philosophical systems. They dominate the relevant columns in newspapers and literary magazines and at times provide the formation of organized groups." [19, p. 58] From today's point of view, there is nothing to add.

7.5 Only Knowledge Can Control Knowledge

"The sole purpose of science is to honor the human spirit." With these powerful words, the eminent mathematician Carl Gustav Jacobi described the ethos of the emerging sciences of the nineteenth century [22, p. 454 f.]. This idea, however, which has become the leitmotif of modern science, was rejected by the philosophers of the Frankfurt School as being uncritical because, in their view, science is not committed to cognition at all, but merely pursues the business of instrumental reason, which focuses exclusively on the mastery of Man and Nature.

One cannot deny that in the natural sciences the path to the technical mastery of Nature is already embarked upon with every experiment. By experimenting with Nature under controlled conditions, we extort a specific reaction from Nature. In other words: By performing laboratory experiments, we bring about particular events. This is a form of skill denoted in Greek antiquity by the word *téchne* (Sect. 1.2). Therefore, the controlled experiment always provides information on how to effect or bring about something. In this sense, experimentation already implies the possession of potential knowledge concerning the technical control of Nature.

However, this aspect of the experimental sciences must not lead to the coarsened view that the experimental method solely serves the purpose of gaining control over Man and Nature. One can only come up with such a thought if one grants practical knowledge a primacy over theoretical knowledge, that is, if one subordinates the theoretical knowledge entirely to practical issues. Even if the sciences initially arose from observations and practice in daily life, one cannot simply generalize from this fact and claim that it is valid over more than two thousand years of science's history. On the contrary, if science had exclusively tailored the goals and methods of its search to the acquisition of practical knowledge, then it would have fettered itself in its quest for cognition and thus narrowed down the range of obtainable insights.

Only on condition that the search for scientific cognition is not pursued from the outset with the goal of possible applications is research free to tread new and innovative paths. Knowledge originates by recombination, i.e., by pooling and the new combination of findings, ones that have been acquired independently of each other without any particular intended purpose (Sect. 2.2). It goes without saying that, in this way, practical knowledge may also arise, opening up new possibilities for the mastery of Nature. Nevertheless: Knowledge can only unfold freely, in all directions, under the umbrella of pure, non-utilitarian and open-ended basic research. This holds not only for theoretical knowledge but also for practical knowledge.

Let us deepen the idea of "pure" or value-free knowledge by appealing to the concept of information. We have to look again at the three aspects of this concept that we termed the syntactic, semantic and pragmatic dimensions of information. As explained in Sect. 6.8, there is an ascending mutual dependence of the three dimensions, which leads to an entanglement of the pragmatic aspect with the syntactic and the semantic ones. Thus, the content of information, its semantics, cannot merely be reduced to its pragmatic dimension, i.e., to the action content. There is a reciprocal dependence between semantics and pragmatics in so far as the content of information changes under changing circumstances.

Nevertheless, there must also be an invariant aspect of the changing assessments of a piece of information. Otherwise, the information would have no identity at all. This invariant aspect of information is given by its syntax. In a figurative sense, the syntax of a piece of information represents something like the value-neutral aspect of information that we may call "pure" information. Pure information has no actual, but only potential, content. It only gains its semantic aspect from the circumstances under which it becomes operational and thereby pragmatically relevant.

These considerations can be transferred directly to the structures of human knowledge. The changing value that an item of knowledge can have requires a value-neutral substrate, i.e., an invariant aspect of knowledge. If there were no such thing as "pure" knowledge, the talk of the ambivalence of scientific knowledge would make no sense at all.

The structural sciences, above all mathematics, are sciences whose cognition goals are the ontologically and pragmatically neutral elements of pure knowledge. We remember (Sect. 4.9): An inherent characteristic of the structural sciences is that they abstract from all physical or material descriptions of reality and describe reality solely by mathematical terms and symbols and their transformations. Thus, the laws of the structural sciences are derived from real laws by eliminating all descriptive and empirical constants and retaining only the logical and mathematical ones. In this respect, the structural sciences represent, like mathematics, a logical framework for a possible science. Similarly, the basic concepts of the structural sciences are all-embracing abstractions of reality, with the help of which the characteristics of complex phenomena in Nature and society can equally be described. The structural concept of language, which is at the center of this book, is an impressive example of this. Thus, regarding structural sciences, we can certainly speak of the fact that there is pure knowledge and that the guiding interest of these sciences is only the search for cognition itself.

One could dismiss the controversy about the possibility of pure cognition as an academic, entirely theoretical discussion that immediately fades into the background when knowledge becomes pragmatically relevant. However, the question then immediately arises as to how we should deal with this knowledge. Of course, it would be nice if there were an objective system of values that are equally recognized by all people. However, as von Mises has already pointed out, this idea is illusory, because "the existence of an objective scale of moral values, inborn in all men, is not supported by any observation" [19, p. 370]. On the contrary: "The acceptance or rejection of a normative system by a group occurs through a collective act of resolution, in which every individual acts under the influence of the various impacts by society." [19, p. 370]

Even if the collective decision-making process is designed to follow the rules of an ideal discourse community, it ultimately would not lead to an objective value system, binding for all people. At best, the discourse could guarantee the rationality and the democratic procedure of decision-making. However, no discourse community has access to absolute truth, since the idea of truth is, like all other human ideas, subject to the relativism of cognition (Chap. 2). Consequently, all values—and with them, all value systems—are relative.

In any effort to justify rationally the existence of specific values and value systems, one has always to realize that values are culturally conditioned and depend on temporal, social and individual living conditions. This fact must be considered, if science, or more precisely, the scientist, is confronted with responsibility. Like all other people, the scientist with his value ideals, is embedded in a particular life-world that is first and foremost shaped by his own circumstances of life.

During the Second World War, the physicists who took part in the "Manhattan Project" and thus in the development of weapons of mass destruction are a monitory example. Neither the persons leading the project, such as Robert Oppenheimer or Edward Teller, nor Albert Einstein, who supported the construction of the atomic bomb in a letter to US president Roosevelt, were unscrupulous people. Instead, as ambiguous as this may seem at first sight, the reasons they adduced for their participation in the Manhattan Project were above all moral ones. This illustrates how strongly value decisions and moral actions are context-bound and, therefore, situation-dependent.

In its perversion of recognized fundamental values, we encounter value relativism in its sharpest form. The death penalty, the targeted killing of people to avert danger or the "just" preventive war are all examples of how, within the framework of a community based on the rule of law and committed to humanity, fundamental values are turned into their opposite by employing rational arguments. Sometimes, science and technology are also affected by this. The destruction of experimental facilities by radical opponents of genetic engineering or animal-welfare activists is usually based on a logic that places the alleged right to self-assertion above a democratically based legal system.

At the same time, these examples make it clear that value relativism inevitably establishes a ranked scale of values. This, in turn, has a regulatory function for man's community life and thus appears to be indispensable. Conversely, however, it cannot be concluded from this that there is an absolute and objective order of values in the world that is independent of our actions. Such ideas arise mostly from metaphysical or religious world explanations, which try to establish the norms of human behavior once and for all, according to a moral understanding that has its origin in the deep past. We need to look ahead. We shall have to be prepared for the fact that "to an ever-increasing extent scientific knowledge, i.e., knowledge formulated in a connectible manner will control life and the conduct of men" [19, p. 370].

After all we have heard so far, it does not seem to make much sense to demand from scientists any self-regulation or self-restriction of their research.

As the Asilomar Conference made clear, the demand for such self-control is pointless, as a consensus about the extent of control and how it is to be exercised could not even be achieved among the scientists themselves. Already at that time, it became apparent that the assumption of responsibility itself presupposes extensive knowledge, which for its part requires more research, not less. This applies generally: Responsibility for knowledge can only be taken on against the background of other knowledge that is more substantial than the knowledge to be controlled. This is the "knowledge spiral" underlying any responsible actions by humans.

Thus, the demand that some risky research projects should be discontinued, or banned, is only a feeble attempt to evade responsibility in dealing with scientific knowledge. If one were to yield to such demands, as the biophysicist Manfred Eigen pointed out, "already today, practically nothing more would be allowed to be researched, and we would also have to forget most of what we already know" [23, p. 34; author's transl.].

The demand to restrict man's pursuit of cognition in a preventive way is just as absurd as an attempt to ban fishes from swimming or birds from flying. We come across the same insight again and again: The controlled handling of scientific discoveries requires new scientific findings that are richer and more sophisticated than those that have to be mastered. On the other hand, the increase in scientific knowledge always leads to new problems and challenges that can only be reliably assessed and controlled by renewed knowledge. Thus, scientific progress and its ambivalent perspectives drive the constantly rotating spiral of human knowledge.

However, if only knowledge can exercise control over scientific cognitions and their applications, it must be ensured that knowledge is allowed to unfold and develop freely. On this premise, the contemporary ethics of science have the primary task of advocating the unrestricted freedom of scientific research. It must be clear that, for knowledge control, a continuous search for knowledge is unavoidable. It must be the supreme imperative of a future-oriented society. Only basic research can provide the foundation for a comprehensive assessment of science and technology, one that makes possible an effective self-checking of knowledge.

Ethics are founded on the core idea that moral norms can be derived from the generalization or universalization of human rules of behavior. Since the basic patterns of human behavior have undoubtedly been shaped by evolution, ethics must already demand, for its own sake, unrestricted freedom of research. It is to be expected that the universals of moral behavior such as truthfulness, fairness, solidarity and the like will find a more profound explanation in the evolutionary history of man. It is evident that only basic research will be able to substantiate this hypothesis.

The biological roots of organisms' social behavior are the subject of sociobiology [24]. Sociobiology is confronted *inter alia* with the difficult question of how, under the competitive conditions of biological evolution, the justified self-interest of living beings can be reconciled with the collective interest of their congeners. Put generally: How can conflicting values be harmonized for the optimum development of living beings? For example, if one wants to understand how a consistent pattern of social behavior could develop by evolution, one has to investigate the behavior patterns of living beings in the interplay of opposing evolutionary strategies, such as cooperativity and competition, in a kind of cost–benefit analysis. However, this requires experimental and theoretical investigations up to evolutionary experiments with genetic material in the test-tube.

Thus, it cannot be the goal of the ethics of science to limit or standardize basic research. On the contrary, the ethics of science can only follow the cognition process, namely as a subordinate authority to keep a check on the scope of actions made possible by scientific discoveries. In other words, responsibility in and for science cannot be perceived in a metaphysical world of absolute and unchangeable values, but only in the world of concrete human actions, the value structure of which must be continually adapted to the changing state of knowledge of the sciences.

Contemporary ethics of science must not rely on any super-historical values, which are given *a priori*; it must be anchored in the scientific process itself and subject to constant control by the progress of knowledge and the reality of life. Only a code of ethics that considers the findings of science and thus ultimately also the relativity and openness of human knowledge will be able to monitor adequately the scope of human actions resulting from scientific progress.

The parallels between biological evolution and the evolution of human knowledge cannot be overlooked. The mechanism of natural evolution also applies to the development of knowledge: the larger the range of new findings that enter into competition with traditional knowledge structures, the more effective the progress of knowledge is. Seen in this way, any attempt to place the scientific process of the acquisition of knowledge under normative constraints, which determine what is allowed to be researched and what is not, is counterproductive as regards optimizing human knowledge and the possibilities of controlling it.

To emphasize once more: We need first and foremost ethics of knowledge that advocate the optimization of human knowledge. Only in this way can man's scope for action be expanded to optimize the design of scientific and technical life. We do not need a universal ethical draft that prescribes the conditions of right living for all time. Rather, it is the prudent testing of new possibilities for designing reality, guided by the current state of knowledge, that will open up new paths to the future of mankind.

7.6 The Nature of Ecological Balance

It would seem to be a matter of course that scientific knowledge has to be included in solving the problems raised by science and technology. Reality, however, is different. With the anti-intellectual attitude that has spread within the modern knowledge society, trust in science is increasingly fading. This development is fueled not least by those philosophers who, from a wide variety of motives, question the objectivity and the achievements of science and attempt to drive a wedge between human reason and scientific rationality.

The extent of the backwardness and deep conservativism of the anti-intellectual mentality is very clearly illustrated by the social and political disputes over the most primordial state of Nature, as the source of all life, which it is imperative to preserve. Just as in the religious creation narrative, the balance of Nature is seen as an expression of a perfect harmony, into which human action has to fit.

The hypostasis of Nature is a legacy of Greek antiquity and has shaped the Western image of Nature over two millennia. In Aristotelian philosophy, it was the idea of a poetic and creative Nature that led to Nature's personification (Sect. 1.1). Later on, in German idealism, it was the idea of "Nature as a subject" that was set against the idea of "Nature as an object" (Sect. 2.6). Romanticism, finally, stylized Nature as "Mother Nature", the life-giving and nurturing ground of man's life-world. The idea that Nature acts like a reasonable person makes the harmony between man and Nature seem almost enforced. To the extent, however, that preservation of the existing order of Nature is becoming an overarching norm for human action, the path to a scientific and rational redefinition of the relationship between man and Nature is, or will become, blocked.

To return to basics: What is the real character of Nature, what are Nature's semantics? Specifically: What is really implied by the demand that the environmental balance be maintained at all costs? This demand is often combined with the assertion that maintaining biodiversity would have a stabilizing effect on ecosystems. The generally accepted rule of thumb, according to which a species-rich ecosystem is stabler than a species-poor system, at first seems entirely plausible. An ecosystem can react more flexibly to disturbances from inside and outside, the more complex its structure is. A species-rich system would be expected to possess a robust self-regulating force and thus high stability. Yet, on the other side of the coin, it is also possible that an ecosystem's complexity might increase its susceptibility to disturbances. This may explain why a species-poor ecosystem, such as the Alpine region, is less susceptible to interference, and is ultimately stabler, than a species-rich ecosystem such as a tropical rain forest.

There are indeed two contrasting views on the question of how biodiversity affects the balance of an ecosystem. Yet each may be right in its own way. In fact, it is characteristic of life's dynamics that antagonistic principles can work together in a well-balanced relationship. This is particularly evident in biological evolution, where the interaction between competition and cooperation, reproduction accuracy and error-proneness, are indispensable conditions for evolutionary progress (for more details see [25]).

Let us pursue this point by looking at the actual character of an ecological equilibrium. However, even the attempt to provide an adequate definition of the concept of an ecosystem already presents difficulties. The relevant textbooks typically answer that an ecosystem is an interactive system, involving both living beings and their inorganic environment, and that it largely regulates itself. The word "self-regulation" is decisive in this context. It expresses the fact that fluctuations in the population sizes of individual species, or the energy flow and the turnover of matter, are dampened by certain compensatory mechanisms, such that the overall system remains to a large extent in a steady state of equilibrium. However, "to a large extent" indicates that self-regulation in fact describes an ideal situation that can only approximately be realized in a real ecosystem.

Let us take a closer look at the facts. From a physical point of view, ecosystems, like every living being too, are open systems. They continuously exchange energy and matter with their environment. Accordingly, the ecological equilibrium is not static, but dynamic. The latter is defined as an equilibrium state in which all inflows and outflows are balanced, so that many parameters of the system remain constant. Thus, for example, the population numbers, and the flows of energy and/or of substances can be stationary in an ecosystem. Therefore, the equilibrium can take on entirely different forms. The unambiguous characterization of an ecological balance thus always requires a statement of the components to which the stationary state refers.

The spatial delimitation of an ecosystem is likewise difficult. Where are the boundaries of an ecosystem, if one does not want to consider the entire earth as the extreme case, an overarching ecological system? This difficulty is also reflected in the definition given above, according to which an ecosystem is a system that "to a large extent" regulates itself. Indeed, as mentioned, the vagueness of this phrase conveys clearly the fact that the concept of ecological equilibrium is an idealization that is only approximately fulfilled by a real ecosystem. This caveat generally applies, even though it is quite possible to classify ecosystems by further criteria, giving the idea of an ecosystem a certain internal cohesion—spatial or functional characteristics, for example. On the other hand, if the concept of ecological equilibrium is a mere idealization, how can one justify the demand for the maintenance of ecological balance? Or, conversely: how strongly may an ecosystem deviate from the ideal state of equilibrium before it collapses irreversibly?

A general consideration may help to clarify this fundamental issue. Let us examine the possible reactions of an ecosystem to fluctuations, i.e., deviations from the sizes of populations within it. Such fluctuations are either generated within the system by random processes or they are induced by external disturbances. In principle, there are three reaction patterns that describe how a system can respond to such fluctuations: The system can dampen fluctuations, amplify them, or behave indifferently toward fluctuations (Fig. 7.7).

Obviously, complete control of fluctuations, and thus the maintenance of an exact equilibrium state, would mean the end of all innovation processes in an ecosystem. Only variations, like the appearance of new species, can lead to those innovations that enable the system to evolve. The opposite behavior, in which every fluctuation of the system is amplified, would inevitably lead to the degradation and finally to the complete collapse of the ecological order. The intermediate case of a system that behaves completely indifferently to fluctuations does not correspond to ecological reality either, as such a mechanism would lead to utterly irregular behavior, in which no ordered state could exist permanently.

Systems that have several mechanisms at their disposal for reacting to fluctuations appear to react with particular flexibility, which enables them to

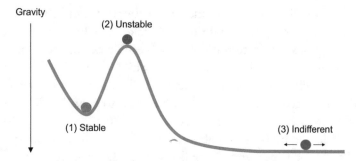

Fig. 7.7 Different balances. Illustration of various equilibrium concepts by a ball that is only under the influence of gravity. (1) Stable equilibrium: after a small deviation from the equilibrium position; the ball always rolls back to its initial position. (2) Unstable equilibrium: after a small deviation from the equilibrium position, the ball no longer returns to that position, but moves further and further away. (3) Indifferent ("metastable") equilibrium: after deviating from the equilibrium position, the ball remains at rest. The concept of stability defined here can easily be transferred to ecological equilibrium, where movement of the ball would correspond to fluctuations in the system's parameters, for example the population numbers

withstand disturbances. For example, in an ecosystem, the dampening of fluctuations (stability) and the amplification of fluctuations (instability) are equally important. This concerns both the systems' long-term behavior in response to environmental changes (external disturbances) and their evolutionary response, induced by internal disturbances. On the one hand, the system must be able to stabilize temporarily an advantageous state of order. This can be achieved by regulating fluctuations with negative feedback. On the other hand, the system must also be able to integrate innovations. This is done by amplifying and fixing advantageous fluctuations through positive feedback.

Therefore, in an ecosystem, innovations can only be exploited through non-equilibrium processes that lead to irreversible changes in the system. This is precisely the way in which a system develops an evolutionary behavior that creates new forms of ecological order. Ecosystems are a further example of the fact that life's evolution is based on antagonistic principles: in this case, on stability *and* instability, i.e., preservation *and* development.

In contrast, the popular notion of ecological balance is more in keeping with the image of chemical equilibrium, which suggests stability and immutability. Thus, to avoid such wrong associations, it would be better to replace the term "ecological equilibrium" by that of "ecological order". The notion of order fits better to biological reality, as it is not bound one-sidedly to the aspect of stability; instead, it also takes care of the issue of creative change engendered by the principle of trial and error.

The fact that natural ecosystems are closer to stability than instability indicates that those systems have already been optimized, so that advantageous fluctuations become increasingly rare, that is, the number of further possible advantageous fluctuations is much smaller than the number of possible disadvantageous ones. Strictly speaking, the ecological order is a "metastable" state. Whenever favorable changes occur, the system's original balance will collapse and be replaced by a new one that is better adapted to the new conditions. Thus, instability is a constitutive characteristic of an ecosystem, and it is indispensable for its evolutionary development.

Profound ecological changes have occurred time and again in the course of life's evolution. The most impressive example is the dramatic transformation of the earth's atmosphere in the earliest phase of biological development. According to all we know about this phase, the first microorganisms, archaebacteria and eubacteria emerged in an oxygen-free, i.e., reducing atmosphere. "Oxygen-free" means that all the oxygen present was bound in water or carbon dioxide. It was only in the further course of evolution that the blue-green algae developed; these obtain their energy from photosynthesis, thereby releasing oxygen as a waste product. Since the photosynthesizing blue-green algae were much more energy-efficient than their contemporary organisms, they multiplied vastly, and the reducing atmosphere was gradually transformed into an oxidizing atmosphere. The result was a mass death of anaerobic organisms that were unable to cope with the new living conditions.

The popular idea of an everlasting balance of Nature that must be preserved is the consequence of a wrong understanding of life's evolution. In reality, Nature has always been subject to constant change, often with dramatic impacts on the organisms and their environment. On the evolutionary time-scale, Nature must be seen as gigantic machinery in which life forms are continually destroyed because they have to give way to new, better-adapted forms of life.

From this point of view, the traditional image of a rationally acting, life-giving and life-preserving Nature seems untenable. The extinction of living beings by natural selection is just as much a hallmark of "Mother Nature" as the individual organism's pre-programmed death. There is only a handful of species that have survived extended periods of biological evolution. Only biological families and subfamilies are "long-lived", but not the corresponding genera and species. Ignoring this fact is tantamount to denying the history of Nature. The demand for the preservation of ecological diversity becomes even more contradictory if it is (purportedly) justified by referring to the natural history of life and its importance for man's living conditions.

In view of the destructive concomitants of natural evolution, it appears not only understandable, but also imperative, to concede to mankind the right to put an end to Nature's autonomy, by intervening in its effect structure and adapting Nature to his needs. Man's ability to master Nature has conferred upon the human species an enormous selection advantage over all other living beings. Man is not a failure of Nature, but a hugely successful product of evolution, gaining selective advantage through the superiority of his actions.

For Nature surrounding man, on the other hand, there is no right to a life of its own. Instead, Nature is, for man, a mere environment that he must use, as does any other living being, for his own benefit. However, this does not mean that man is allowed to exercise arbitrariness in dealing with his natural environment. Instead, there are good reasons why mankind should deal with its natural surroundings circumspectly and why man should leave an inhabitable climate for later generations, thus guaranteeing the survival of the human species itself. However, it is precisely for this reason that man must break away from the "unnatural naturalness" propagated by dogmatic conservationists.

Scientific knowledge is not based on human wishful thinking or ideologies. Rather, science has to clarify how living beings can coexist with all the other inhabitants of an ecosystem that is characterized by conflicting properties such as competition, symbiosis, displacement etc. Biological evolution has found admirable ways of combining its different driving forces, which are often in conflict with each other, into a creative ensemble. This must be the guiding principle that defines the limits of human intervention in the environment, and on which we must base our dealings with Nature.

7.7 Does Nature Pursue Purposes?

The appeal to take possession of Nature, and to design it responsibly, necessarily conflicts with the conservative understanding according to which Nature must be preserved in its "original form", whatever this might have been. There are even philosophers who ascribe to Nature a purposeful behavior, directed toward its own preservation. A prominent example of these is Hans Jonas. In his "philosophy of the organic", Jonas poses a profound question, one that Aristotle (Sect. 1.1), Immanuel Kant and others have already discussed extensively: Is there, in living Nature, an ultimate purpose or a final cause directing organic events?

In contrast to Kant, who considered this question to be undecidable and therefore used the concept of purpose merely in the form of an "as-if-it-were consideration", Jonas resolutely argues in favor of an ontological teleology according to which all organisms follow a fundamental goal, namely that of self-preservation. The assumption that there is an objective self-purpose in Nature serves Jonas at the same time for the justification of a fundamental value endowed by Nature on all living beings, namely, the right to existence. Pursuing this thought, Jonas developed his much-respected ethic of responsibility, according to which "the plenitude of life, evolved in aeons of creative toil and now delivered into our hands, has a claim to our care in its own right. A kind of metaphysical responsibility beyond self-interest has devolved on us with the magnitude of our powers relative to this tenuous film of life, that is, since man has become dangerous not only to himself but to the whole biosphere." [26, p. 136]

There are reasons why it is problematic to infer from being to values. Philosophy denotes this issue as the "is–ought gap". In the present case, however, there is no convincing argument that would allow one to conclude compellingly that the existence of organisms implies a duty to conserve them. This *non sequitur* rests upon a fallacy, one that is typical of metaphysical ontologies that do not take account of reality. Accordingly, Jonas' assertion is in direct contradiction with the recognition by modern science that the evolution of living beings unavoidably entails the extinction of other living beings. In that light, any normatively enforced preservation of a species would contradict the essence of Nature.

Living beings can only develop in the free interplay of variation and selection. This, however, presupposes the possibility that a species can be replaced at any time by a better-adapted species. Therefore, the conservation of existence cannot be interpreted in the sense of a purposeful and deterministic behavior of Nature, but only as being the result of a (statistically substantiated) selective advantage in the continuous struggle for existence.

Although Nature is exposed to the blind mechanism of natural selection, Jonas nevertheless assumes a purpose-driven behavior of Nature as the basis of his philosophy. Such a philosophy could hardly be further from reality. The teleological misconception is typical of a hypostatized image of Nature, in which Nature takes on the role of an active participant. It must indeed be admitted that evolutionary biologists sometimes use teleologically colored language when describing the natural processes of adaptation. However, the teleological manner of speaking in biology is explained solely by the fact that the concept of adaptation requires some kind of template to which the evolving system can adapt. That is why the idea of adaptation allows one to use the terminology of (apparent) purposefulness. Nonetheless, one cannot conclude from the (apparently) target-oriented optimization process that it serves an overarching goal. In fact, evolutionary adaption can be well explained without recourse to a final purpose or an ultimate cause.

The conclusion that any change implies an objective goes back to Aristotle, who asserted that every movement follows a direction and therefore necessarily runs toward a goal (Sect. 1.1). In the light of modern science, however, this view can no longer be held. Evolutionary adaptations may look as though they are pursuing an ultimate goal or plan. In reality, however, only the *gradient* of evolution is predetermined, and not some target (Sects. 2.8 and 6.10).

Modern biology knows how to avoid the teleological misconception: it distinguishes carefully between goal-directedness (teleology) and expediency (teleonomy). The philosophy of Jonas, in contrast, is based squarely on the idea that Nature pursues the overriding goal of its own preservation. For him, the maintenance of existence is elevated to the status of an overarching norm, to which all living things are subjected. It obliges humankind to preserve not only Nature but also itself. The commandment to preserve Nature is absolutely obligatory for humankind—even if man could in fact live under other conditions that he himself designed.

What is more, if one day humankind were to decide, in free self-determination, not to reproduce any more, then according to the ethics conceived by Jonas that would be utterly irresponsible. This, however, would be a ludicrous scenario. There is no rational justification for such an ethic committed rigorously to the idea of human self-preservation at any price. Rather, there are good reasons to demand that each generation of humans leave behind a worthwhile world for future generations.

The diversity of life forms that we know could only have come into being because the principle of conservation of existence is *not* universal. Much rather, the evolution of life presupposes the extinction of less well-adapted forms of life. As a result, the path of evolution is also studded with countless species that have been extinguished in favor of subsequent species. In other words: Nature's "creatorship," which is based on natural selection, has never concerned itself with the "fullness of life" that it has produced. Instead, at any time, the fullness of life only existed temporarily, because it always had to give way to new forms of life. A concept of existence that does not include the "vital" change of the living world necessarily remains an empty one.

Without the death of the individual and the extinction of species, there would have been no evolution. Put differently: life's existence is based just as much on the destruction of existing life as on its reproductive preservation.

Only in the interaction of these principles could life have arisen from inorganic matter in the first place and have developed into the diversity of forms that constitutes the richness of animated Nature today, and whose actual state Jonas considers it mandatory to preserve.

Obviously, Jonas is misjudging the semantics of Nature, the dichotomous character of evolution, in which conflicting principles such as preservation and destruction, construction and decay, stability and instability, precision and error-proneness, competition and cooperation etc. act together. The fact that mutually opposed principles can nevertheless unite into a creative optimization process is ultimately what makes up the fascination of natural evolution.

Let us summarize: Life on Earth has developed from the constant conflict between the preservation and the destruction of living beings. If this irresolvable connection is reduced to the demand that Nature should be preserved in pristine form, this leads to a static and completely unrealistic understanding of Nature. If Nature itself were to follow that principle, then this would violate the fundamental value of life's existence and thus be self-contradictory. A philosophy that justifies the demand for the preservation of Nature by appealing to the permanent existence of Nature in its present state has an anemic image of Nature. This image is abstract and empty, and it leaves open the crucial question of Nature's semantics.

Science has a sobering answer to this question. In the narrow sense, Nature is the animate Nature that originated from the self-organization of matter in an environment that initially was inanimate. It has further evolved by natural selection, the main feature of which is competition and displacement among organisms. Nature does not have goals or values that are oriented "toward the whole". Quite on the contrary: Nature resembles a blind predator–prey system, the members of which compete mercilessly with each other. From this point of view, all living beings must be regarded as selfish reproduction machines, which have not the slightest interest in Nature's historically developed causal network. Even the phenomenon of co-operativity, as observed in Nature, only serves organisms in the sense that they gain an advantage over their competitors.

The popular demand that man free himself from his culturally entrenched anthropocentrism and abandon his claim to autonomy over Nature is not justified by scientific knowledge. On the contrary, we cannot avoid putting man at the center of controlled Nature. Only in this way can today's knowledge society regain the freedom of its actions—a freedom that normative ethics of preservation, religious dogmatism and backward-looking, romantically tinted views of Nature all threaten to deprive it of.

7.8 Nature as a Cultural Task

Rigorous ethics directed to preserving Nature are dangerous because they obscure our actual task, which is to design Nature in a future-oriented way. The only option we have is to delay humanity's inevitable end for as long as possible. This goal can only be achieved through science and technology, and not by an ideology based on a deluded view of Nature that ignores scientific cognition and rationality.

We shall only be able to meet the challenges of the future if we break away from a static and emotionally charged concept of Nature and adopt the value-neutral idea of Nature as an environment. Nature is man's environment and nothing else. Man must design his natural habitat so that it is in a dynamic state of equilibrium controlled by ourselves. However, this task presupposes that we understand man and Nature as an interwoven fabric of effects in which the natural and the artificial, Nature and culture, present themselves on an equal footing as manifestations of one and the same reality.

With their understanding of reality, Far Eastern cultures are already very close to this goal. There, unlike in the tradition of Western thought, Nature is not seen as a gift from a Creator that takes on a unique position because of its origin. Instead, reality is understood as a unity in which the natural and the artificial, the living and the non-living are on the same level. The harmonization of appearances perceived as being in contrast to one another is in Eastern cultures the basis for the syncretism of tradition and modernity, of man and machine, of the natural and the artificial (see Sect. 1.7).

Wonderful examples of the harmony of Nature and culture are provided by Japanese gardens. In each of these, everything is arranged by human hand to form a unique work of art (Fig. 7.8). Yet, nevertheless, everything in the garden seems to be completely natural, because designed Nature "is not something different from the world, but condensed Nature", a "garden of paradise", a "vividly captured landscape" [27]. The mysterious symbolism which Zen garden art radiates allows such gardens moreover to become "philosophical landscapes" (cf. Fig. 1.10). Culture emerges here to the extent that the differences between animate and inanimate, natural and artificial, dissolve.

In the same sense, the philosopher Hans Blumenberg has pointed out that "culture is generally the expansion of the diversity of forms brought about by favoring processes of dissolution of what is naturally given—up to the fine arts, where this has been concealed for a long time" [28; author's transl.]. Thus, in Far Eastern cultures, we find an anticipation of what will be

Fig. 7.8 Nature as a cultural task (a Japanese garden in the prefecture Shimane). Japanese gardens are an epitome of the harmony of Nature and culture (Sect. 1.7). They are a model of how to design landscape and save the immanent proportions of Nature at the same time. We must look for corresponding solutions in the age of science and technology to steer and to transform our environment in a forward-looking way. [Photo: KazT, Shutterstock]

inescapable for humanity in the twenty-first century: the abandonment of historically grown Nature in favor of a naturalness designed by humans. Following this objective, we have, with the biologist Hubert Markel, to understand "Nature as a cultural task" [29]. This task corresponds precisely to the original meaning of the Latin word *cultura*, initially understood as the maintenance and processing of the landscape, including everything that man in the broadest sense creates himself—in contrast to Nature, which he has not designed or changed.

It is a mistake to believe that there can be a Nature without human influence and intervention. Instead, we have to take advantage of science and technology to compensate for the disruptions of the ecological balance that human actions bring about. One possibility, which has already been practiced for more than a hundred years, is the assisted colonization of new habitats by threatened species. Another strategy designed to combat loss of biodiversity is conservation translocation, improving the focal species' status, restoration of the functions or processes of natural ecosystems, and the reinforcement of populations by releasing species into a community of con-species (see, for example [30]).

The task of restoring species outside their indigenous habitat or introducing ecological replacements for extinct forms confronts us with a number of serious challenges. One of these is that of identifying the habitat for species and delineating sharply enough the appropriate ecosystem, including food sources, predators and parasites. Furthermore, all aspects of the abiotic environment have to be considered, essential for the continuation of the species in question and the variation of its habitat over space and time that may occur following climate changes.

It has been objected that species restoration by assisted colonization will create, in the end, more problems than it solves [31]. The release of alien species into a new habitat may lead to a decrease in its regulation by natural enemies, which in turn may provide the introduced species with a selective advantage over competing indigenous species ("enemy release hypothesis") [32]. Yet a mere preservation strategy cannot solve the ecological crisis (Sect. 7.6). Considering the full range of possible future climates, assisted migration will probably be indispensable for many species [33]. Otherwise, the catastrophe would be allowed to run its course.

Moreover, there is evidence that environmental changes in conjunction with species invasions are contributing to a widespread emergence of novel, self-sustaining ecosystems that have no historical precedents [34]. Recognition of this calls into question the traditional practices of the conservation of native species and their historical continuity. Instead, we may have to realize that attempts to conserve or restore Nature in its historically developed form may be doomed to failure because disrupted ecosystems tend to reorganize themselves into novel ecosystems.

Consequently, entirely new approaches may be needed to develop a natural environment for humans that fits human interventions into Nature and is adapted to a continuously changing world. This is a cultural task of unprecedented magnitude, and it can only be mastered by scientific progress and technological innovations.

At present, synthetic biology and assisted evolution are the most advanced tools for redesigning Nature. Synthetic biology is a multidisciplinary research field that comprises quite different techniques and applications directed toward the creation, modification and recombination of biological material (Chap. 6). Assisted evolution is a technique by which one tries to accelerate naturally occurring evolutionary processes to support the building of new ecological balances or even new ecosystems.

An example of this new approach to ecological problems is enhancing coral reef resilience by assisted evolution [35]. As is well known, global climate change over the past decades has entailed severe changes in the temperature and chemistry of the oceans and a shift in sea level, which in turn has had a profound impact on the stability of marine ecosystems. The rapid mass bleaching of coral reefs is the most clearly visible sign of this development (Fig. 7.9). For example, in 2016, the largest coral reef in the world, the Great

Barrier Reef, lost 30% of its corals, and in 2017 a further 20%, indicating that the reef does not regenerate fast enough to balance the decay of its diversity caused by rising water temperature [36]. The Great Barrier Reef is a case study of an ecosystem on which external constraints are imposed that cannot be automatically balanced by natural processes (Sect. 7.6).

Human-assisted evolution is an auspicious way to accelerate naturally occurring processes of adaptation. In combination with synthetic biology, the application of assisted evolution represents a paradigm shift regarding the future orientated management of Nature. The instruments of this technique include the managed movement of individuals with favorable traits into populations to reduce local maladaptation to environmental change (assisted gene flow), the manipulation of microbial symbionts and the enhanced supply of larvae, adapted by interspecific hybridization (for details see [35]).

The Great Barrier Reef is home to an estimated 400 of the world's 1200 coral species. It is evident that any release of biotechnologically manipulated corals could outgrow the present diversity of species and thus change, or even destroy, the existing species balance of the Great Barrier Reef. Beside the need to intensify research into corals, a deeper understanding of the inherent proportions of Nature is required to judge the possible risks of assisted evolution.

Given the challenges that lie ahead, the famous statement by the sophist Protagoras handed down from antiquity, "Man is the measure of all things", takes on central significance. This sentence, directed initially against the metaphysical doctrine of the unchangeable, truly existing being, obliges us today to shape what exists and to make scientific knowledge the measure of all things, namely of that "which is, which will be and which will not be" [37, 80B1; author's transl.]. Only the consistent implementation of this maxim will release man from the constraints of a two-thousand-year-old attitude according to which Nature must be conserved in its "primary" form. Only this cultural upheaval will enable man to understand himself as autonomously acting being that perceives its ability to design its natural living conditions with a "sense of proportion".

7.9 Nature's Intrinsic Proportions

Remarkably, in the history of culture, there have always been phases during which man has developed a particular sensitivity to the intrinsic proportions of Nature. An example from early Greek antiquity is the Pythagorean theory of order. For the Pythagoreans, the primal ground of being was not matter,

Fig. 7.9 Ecological damage to the Great Barrier Reef. The Great Barrier Reef's coral bleaching is an example of the degradation of a substantial ecosystem by global warming. Corals live in symbiosis with algae, which produce energy by photosynthesis. With increasing water temperature, corals release more and more algae, which leads to the death of the coral polyps and becomes visible in the loss of the coral's color. Restoration measures focus *inter alia* on breeding corals with increased thermal tolerance. The attempt to enhance the resilience of coral reefs through assisted evolution is a crucial step in avoiding irreversible damage to ecosystems. [Photo: Chris Jones, Great Barrier Reef Marine Park Authority]

but form; numbers were regarded as an expression of form. At the same time, however, they understood numbers not as abstract, but as sensual entities. The numerical ratios, the Pythagoreans believed, determined the essence and order of things and the harmony of the cosmos (Chap. 2).

However, it was not until the thirteenth century that the mathematician Leonardo Fibonacci discovered an intrinsic measure of Nature that is still of interest today. Fibonacci was interested in the processes of natural reproduction. Among other things, he investigated the rate at which a pair of rabbits multiply. By abstracting from the natural process, he derived a general rule of reproduction describing the progressing size of the rabbit population. It can be displayed by the series of numbers beginning with the digits 0 and 1. Each further number results from the sum of the two immediately preceding numbers: 0, 1, 1, 2, 3, 5, 8, 13, 21, 34, 55, ….

Behind the Fibonacci series is an exponential law. However, the law's basis is not Euler's number e (= 2.718...), but the reciprocal numerical proportion of the so-called Golden Section (or Golden Ratio) g (= 1.618...). The Golden Section is a geometric measure of the continual subdivision of a line into two (A and B; Fig. 7.10). The proportions of the Golden Section appear whenever the ratio between the lengths of the whole line ($A + B$) and the larger

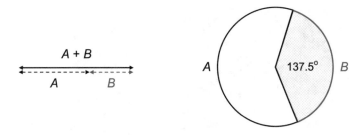

Fig. 7.10   The proportions of the Golden Section. Left: A line divided into the sections *A* and *B* fulfils the proportions of the Golden Section if the ratio of *A* + *B* to *A* is the same as the ratio of *A* to *B*. Right: If a full circle is partitioned according to the Golden Section, the larger angle that results is 222.5° and the smaller one is 137.5°. The latter is known as the Golden Angle

section (*A*) is the same as the ratio between the length of *A* and that of the smaller section *B*. This ratio is the irrational number *g*. The Golden Section, to which Johannes Kepler attributed divine properties, has long been regarded as an ideal of beauty and harmony in occidental art. Accordingly, the Golden Section's proportions are found in many epochs in art history: in ancient Greek architecture, in artworks of the Renaissance and modern art, and in the paintings of cubism, for example.

The most fascinating thing for us is Kepler's observation that the arrangement of the leaves of numerous plants has a direct relationship with the Fibonacci numbers and the Golden Ratio. Wherever one finds spiral arrangements—be it in the distribution of leaves on a branch, the arrangement of branches on a tree or the distribution of seeds in a fruit capsule—one inevitably comes across the Fibonacci numbers and the proportions of the Golden Section (Fig. 7.11).

The Golden Section emerges particularly clearly in the well-proportioned shell of the cephalopod *Nautilus pompilius* (Fig. 7.12). Its geometric form is the logarithmic spiral, whose mathematical intersecting lines are all divided according to the proportions of the Golden Section and whose linear dimensions form a Fibonacci series. The shell's logarithmic spiral form has the advantage that it enables the shell to grow which retaining the same shape all the time. Each increase in length is balanced by a proportional increase in radius, so that the individual chambers can become larger, while still keeping their shape. Such spiral shapes are frequently observed in Nature, especially where proportional growth is required [38].

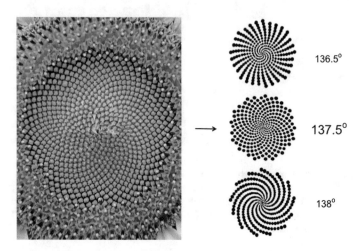

Fig. 7.11 Spiral arrangement of the sunflower's seeds. Head of a sunflower showing 34 right-handed and 55 left-handed spirals. These two numbers are consecutive members of a Fibonacci series. The angle of divergence corresponds fairly exactly to that of the Golden Section (137.5°), as can be seen in the computer simulation of the spiral patterns for various angles of divergence (sketches on the right)

In many cases, as in the arrangement of a sunflower's petals, the logarithmic spiral form is striking. Therefore, it did not need the instruments of modern science to discover and describe this phenomenon. Kepler, and before him Leonardo da Vinci, had already correctly observed the connections. However, it is only in recent times that physico-chemical models have been developed for this purpose (see [39]).

It goes without saying that the proportions of the Golden Section do not answer the global question of what standards we should apply in dealing with the natural environment without running the risk of crossing the limits of instability. Nevertheless, the fact that Nature indeed has intrinsic proportions encourages us to search for further yardsticks that may be hidden in Nature and determine its stability.

Regarding this problem, research has already yielded significant results. In the 1920s, the chemist Alfred J. Lotka and the mathematician and physicist Vito Volterra investigated independently of one another the population dynamics in predator–prey systems. From their models, three rules could be derived which provide details about the long-term behavior of two populations in such a relationship. However, the Lotka–Volterra rules can only be applied on the assumption that other biotic or abiotic environmental factors do not influence the population dynamics.

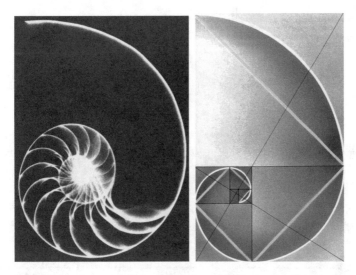

Fig. 7.12 Shell of *Nautilus pompilius*, a cephalopod living in warm ocean waters. X-ray image showing the regular subdivision of the shell into chambers (left). The shell's basic geometry makes up a so-called logarithmic spiral, the exact form of which is shown on the right. This idealized form is obtained by recursively dividing a Golden Rectangle (that is, a rectangle with dimensions corresponding to A and B in Fig. 7.10) into a square and a smaller Golden Rectangle. If the diagonally opposite points of the consecutive squares are joined up, the result is a Golden Spiral. The logarithmic spiral frequently occurs in Nature, presumably because it allows growth to occur while the basic form is retained. Every increase in length is accompanied by a proportional rise in radius so, that a structure can grow without changing its shape [38]

The mathematical equations from which the Lotka–Volterra rules were derived apply to idealized conditions of population growth. Real systems, however, are much more complicated than those considered in the mathematical model. Even so, such empirical investigations as have been possible so far have revealed relatively good compliance with the Lotka–Volterra rules (Fig. 7.13).

For idealized conditions, the first rule states that the numbers of individuals in the predator population and the prey population fluctuate periodically, whereby the growth curves are slightly offset against each other in time. The population trajectory of preys runs ahead. The second rule states that under stable environmental conditions, the average size of each population will remain constant in the long term. The third rule, finally, formulates the consequences of an external disruption of the predator–prey relationship. If both populations are reduced at the same time, in proportion to their

Pelts (thousand)

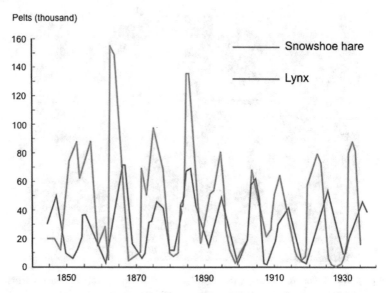

Fig. 7.13 Predator–prey relationship between hare and lynx. The numbers of pelts of lynxes and snowshoe hares delivered by the Hudson Bay Company from 1845 to 1935. The diagram reveals the main features of the Lotka–Volterra model. If the prey animals (snowshoe hares) are present in large numbers, the predators (lynxes) will reproduce strongly because of the favorable food conditions—assuming that no other limit is set to their population density, for example by the environment. However, the increase in lynx numbers leads to a sharp decline in the population of snowshoe hares and, subsequently, to a decline in the predator population as well. In consequence, the population numbers of predators and prey oscillate, whereby the rises and falls in the prey's growth curve precedes corresponding rises and falls in that of the predators. Since the fluctuations in the population numbers of predators and prey are coupled with each other through negative feedback, they regulate and stabilize each other. The population numbers of predators and prey vary around their average values, so that the system is in a stationary state when averaged over an extended period

respective population size, then afterwards the average size of the prey population will increase in the short term while, in contrast, the predator population will—assuming it to be dependent upon this single source of nutrition —decline still further as a consequence of the temporary decrease in the availability of its food. Thus in this case the predators will initially undergo a twofold decline, because both their number and the number of their prey are reduced. Among other consequences, the third rule confirms the expectation that the predator population will risk extinction if the reduction in the prey population becomes too large. Pest-control measures using insecticides provide a case in point.

The impetus for the development of modern population dynamics came from "An Essay on the Principle of Population" by the economist Thomas Malthus, published in 1798 [40]. In this essay Malthus considered the consequences of growth under the constraints of limited resources; the essay had a significant influence on science in the nineteenth century. The most prominent example of this is Darwin's theory of evolution by natural selection, which was decisively influenced by the work of Malthus.

At the beginning of the nineteenth century, the mathematician Pierre-François Verhulst derived for the first time a general equation for limited growth [41]. The so-called logistic equation has since been applied to numerous problems, going far beyond the domain of ecology. In our time, the logistic equation could finally be analyzed with the help of modern computers in hitherto unknown detail. Such investigations have led to entirely new insights into "Verhulst dynamics". In particular, they have uncovered a universal boundary between ordered and chaotic behavior in complex systems.

It is worth taking a closer look at Verhulst dynamics, because the boundary between order and chaos is an essential aspect of human interactions with Nature. A scenario for Verhulst dynamics is a growing population that is supplied with food, but whose habitat is limited. Under these conditions, the population will grow until its size reaches a critical value, given by the capacity of its ecological niche. Thus, the population's growth rate depends on its food supply and the size of its niche defining the growth parameter r.

The logistic equation is a nonlinear equation which—depending on the growth parameter r—may have drastic consequences for the growth process. Without going into the mathematical details, one can extract from a computer simulation of Verhulst dynamics an important piece of information: There are certain areas of the growth parameter where the population dynamics show irregular behavior, with numerous bifurcation points, and in the end this turns into utterly chaotic behavior (Fig. 7.14).

Even more surprising was the discovery by the physicist Mitchell Feigenbaum that the value of the growth parameter at which the transition from order to chaos occurs is universal and applies to all possible scenarios described by the logistic equation (for details see [42]). These insights into the dynamics of complex systems, which we owe to chaos research, also throw light on Nature's intrinsic proportions. These proportions, however, are hidden in the depths of Nature. They cannot be observed directly and—in contrast to the Golden Section—they do not give any place for a

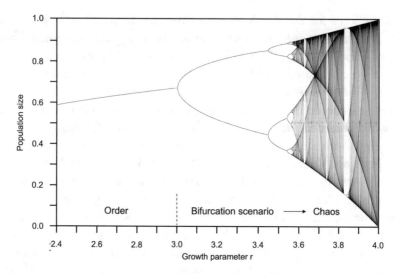

Fig. 7.14 Bifurcation diagram of the logistic equation illustrating the path from regular behavior (order) into chaos. The figure shows a computer simulation of the growth of a population with variable growth rate (Verhulst process). It is supposed that the population size is limited, for example, by its habitat. The population is normalized such that its maximum value is 1. The image shows the population size depending on the growth parameter r (in the area from $r = 2.4$ to $r = 4.0$). By varying the growth parameter (along the abscissa), the population size shows complex behavior. At growth parameter values above 2.4, the growth stagnates at a population slightly above 0.6. If r becomes larger than 3, a remarkable phenomenon occurs. The population now oscillates between two sizes ("fixed points"). The splitting of population size ("bifurcation") is repeated when the growth parameter passes the value $r = 3.448...$, and the population now oscillates between four sizes. The next bifurcation takes place at $r = 3.544...$. Further bifurcations occur at ever shorter intervals until r reaches the limiting value of 4.669.... At that point and thereafter, chaos is seen. The bifurcation tree dissolves entirely into a random pattern of fixed points. No more predictions are possible (adapted from [42])

contemplative admiration of Nature. Nevertheless, such cognitions of the physical nature of order and chaos are of great importance for ecological problems, as they also provide information about the universal limits of Nature's stability.

7.10 Science in the Age of Global Change

Given the ecological tasks that lie ahead, it becomes clear that the search for the right proportions regarding our dealing with Nature will be central for life in a finite world. The historically developed Nature herself will be our best teacher in this respect. Nature's intrinsic proportions, which reflect Nature's semantics, point to the limits at which stability turns into instability, order into chaos. We must learn to act along this critical borderline with sound knowledge based on science and technology in order to maintain the dynamic order of our natural environment in a rapidly changing world. However, regarding the problems caused by global changes, the sciences are reaching their limits. Empirical sciences can no longer draw upon their classical experimental methods, by which Nature is subjected to controlled experimental conditions. Instead, experimental access to Nature is increasingly being replaced by computer simulations.

That is not to say that complex phenomena are in principle excluded from experimentation. Sometimes, it is even possible to test experimentally scientific hypotheses that refer to aspects outside physically accessible reality. For example, already at the beginning of the twentieth century, laboratory experiments were reported that were designed to clear up questions regarding the structure of sunspots and the flocculi of the sun's chromosphere [43]. However, such experiments only represented a terrestrial simulation, intended to explain astrophysical phenomena that were not directly accessible to experimental investigation.

Perhaps the most important difference between classical experiments and computer simulations only becomes visible with the increasing complexity of the phenomena under investigation. As outlined in Sect. 4.7, any experiment requires the initial and boundary conditions to be decided upon in advance. By variation of these conditions, the behavior of the experimental system is then tested. However, if a system depends on numerous further parameters, then the question of whether the selection of experimental conditions was adequate becomes increasingly pressing. A well-known example of this is the study of meteorological phenomena.

Moreover, here one has also to take account of the fact that numerous physical laws are often interacting with each other. In contrast to simple systems, complex systems largely elude simplification through abstraction, isolation and idealization. This is the reason why scientific explanations and predictions become more and more imprecise with increasing scope and complexity of the phenomena being studied.

At the same time, computer simulations inevitably move to the forefront of research. In contrast to traditional experiments, computer simulations do not extend our empirical knowledge. However, they do enable us to explore the various solutions that can result from a specific set of initial conditions, boundary conditions and parameters with which we are modeling a complex system, such as the earth's atmosphere. In this sense, computer simulations primarily have a heuristic function, increasing our general knowledge about reality's complexity. Modern computers can do this because the speed of computation has increased by many orders of magnitude in recent decades. Thus, computers can be used as a mathematical microscope to make visible the fine structure of the solution to a set of nonlinear equations. The Mandelbrot set is an excellent example of this (Sect. 3.2). Another example is that of Verhulst dynamics (Sect. 7.9): without modern computers, the cognitive wealth of the logistic equation could not have been revealed.

Despite the enormous progress that scientific research owes to computer technology, the question of the reliability of knowledge gained by computer simulations arises. Are computers really doing what we think they do? At first glance, this question might seem a little odd, because the machines are strictly following an algorithm imposed by the user. Nevertheless, the question of computers' reliability is entirely reasonable. This had already become apparent in the 1960s, when the meteorologist Edward Lorenz noticed that computer simulations, even if they are based on deterministic algorithms, may nonetheless show utterly unexpected behavior. It turned out that some algorithms depend so sensitively on the input values that even the smallest deviations in the initial conditions of the computation can lead to entirely different results (Sect. 4.7). It should be noted in passing that the question of the reliability of computer simulations arises *a fortiori* in connection with autonomous computer systems that can organize their computing operations themselves (Sect. 6.7).

Lorenz had rediscovered a physical phenomenon that in fact had already been known for half a century through the mathematical work of Henri Poincaré. Its fundamental significance, however, was only recognized accidentally, when computer simulations were performed. This phenomenon, known as deterministic chaos, can fundamentally limit the predictability of a physical process, even if deterministic laws govern the process (Sect. 4.7). The Verhulst dynamics demonstrate impressively how the smallest variations in the growth parameter repeatedly lead to new bifurcations, finally turning order into chaos (cf. Fig. 7.14). One has to keep such phenomena in mind when attempting to explain and forecast complex dynamical processes such as the climate change on Earth.

We need to look at global warming in more detail, as it addresses science's limitations in a particular way. On the one hand, science is supposed to investigate the causes of global warming, the future development of the earth's atmosphere and the impact that climate change will have on ecology and environment. On the other, because of the importance of these results for society, science must rebuff all attempts by environmental organizations and political institutions to appropriate them with a view to reinterpreting scientific cognitions as absolute truths in order to enforce a radical economic and ecological reconstruction of modern industrial society.

Regarding the natural history of the earth, global warming is not new. There are many indications that there have been enormous temperature changes in the past in which warm and glacial periods alternated (Fig. 7.15). Compared with these changes, which are estimated to have been up to 20 degrees Celsius, the increase of 0.18 degrees Celsius in the average temperature during the past ten years seems to be almost negligibly small. However, above all, it is the rapidity of the current rise in temperature that is alarming. It appears that the increase will be orders of magnitude faster than ever before in the earth's history, in which the temperature changes extended over millions of years.

This impression, however, could be wrong. All that we know about the earth's climate changes in the geological past is based on estimates, and these refer to extended geological periods. We do not have any information about short-term fluctuations in temperature within these periods. However, such fluctuations must have occurred again and again, as they are a characteristic feature of thermodynamically open systems. They will have been inevitable, just as entropy fluctuations around thermodynamic equilibrium are unavoidable (Sect. 5.5). Therefore, it must be regarded as certain that changes in temperature took place in the early life of the earth, ones comparable to the increase in global temperature observed in recent decades.

Rapid changes in temperature, as such, cannot yet be called a distinctive feature of our age. The earth's atmosphere is thermodynamically an open system, exchanging energy with its surroundings: interstellar space and the earth's lithosphere, biosphere and oceans. Therefore, changes in the atmosphere's temperature always have natural causes that are independent of human actions, that is, they are caused solely by general geophysical, geochemical and biochemical factors (Table 7.1).

With the beginning of industrialization, however, the climate on Earth also fell under human influence. In particular, the fossil-fuel emissions of carbon dioxide (CO_2) contribute to the so-called greenhouse effect, which causes global warming of the atmosphere. Beside the increase in atmospheric CO_2, a

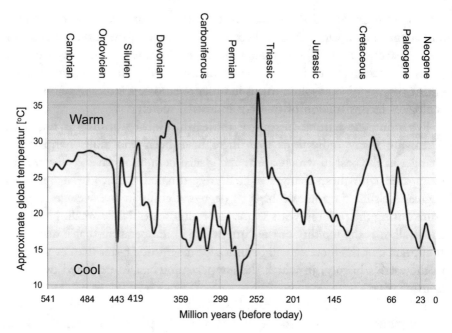

Fig. 7.15 Reconstruction of past temperatures on Earth. Presumed temperature near the earth's surface during the past 500 million years (estimation based on [44]). The changes shown are long-term changes, ranging over millions of years. We do not know the fine structure of these changes on a smaller time-scale (such as decades). Thus, a direct comparison between the past (long-term) and the present (short-term) warming fails to allow any conclusion about the cause of the current warming. Even a rapid rise in global temperature could have taken place very often in the earth's history, as small fluctuations are not detectable in the long-term curve on the scale of geological reconstruction. For a climatic reconstruction of the most recent period of earth's history (Quarternary) see [45]. [Image: adapted from Wikimedia Commons]

substantial rise in methane, another greenhouse gas, is observed. For this, however, no convincing explanation is available at present. Thus, the crucial question is not whether there is a human influence on the climate, but how significantly industrial CO_2 emission really is for the earth's climate change.

Let us examine this in a little more depth. The concentration of CO_2 in the atmosphere is regulated by the so-called carbon cycle, which links the atmosphere, the biosphere, the lithosphere and the hydrosphere together into a balanced carbon-transfer system (Fig. 7.16). Consequently, any disturbance of the carbon cycle will also lead directly or indirectly to a disturbance of the global ecological balance.

Table 7.1 The variety of factors influencing climate change. The most significant causes that increase (↑) and/or lower (↓) the terrestrial atmospheric's temperature. Since all these factors are interlocking, it seems impossible to give a precise weight to each of them in inducing climate change. This holds, in particular, for the greenhouse gases such as carbon dioxide and methane. They are thought to have a strong impact in global warming. Industrial CO_2 emissions may play a role, as may also the increase in methane, the detailed cause of which is unknown at present

Source	Influence on temperature of Earth's atmosphere
Sun	Varying radiation intensity (Solar constant; sun's activity) ↑↓
Quasi-periodic variations of the Earth's orbital parameters (Milanković cycles)	Long-periodic change of the solar radiation reaching the Earth. Warm and cold periods of the Earth's surface (↑↓)
Plate tectonics	Multi-causal influences on earth's climate ↑↓
Volcanism	Effects on temperature ↑↓
Large igneous provinces	Large-volume outlet of magmatic rocks↑↓
Greenhouse gases	Methane ↑ ; Carbon dioxide ↑

There are numerous points at which the carbon cycle could be influenced. Human interventions into Nature are only one of these. Consider, for example, the industrial emission of CO_2. This is just one factor within the carbon cycle, and yet it already demonstrates how significant the uncertainties regarding climate change are. The amount of CO_2 emitted worldwide per year is only known from estimations that are far less precise that the values that would be required for a sound scientific analysis. Other factors—for example, stratospheric ozone concentrations and changes in land use—would also have to be included, as well as the rise in atmospheric methane concentration, which is at least as problematic as that of carbon dioxide.

Projections of climate change are mostly based on simple extrapolations which suggest an at least apparent correlation between the increase in atmospheric temperature and industrial CO_2 emission. However, one cannot simply separate one cause from the network of all causes that together determine climate change. The entanglement of causes and their weighting is a notoriously difficult problem of historical development (Sect. 4.6). This applies not least to the history of Nature. Even if one had exact numerical knowledge of all parameters and all the laws that govern climate change, there

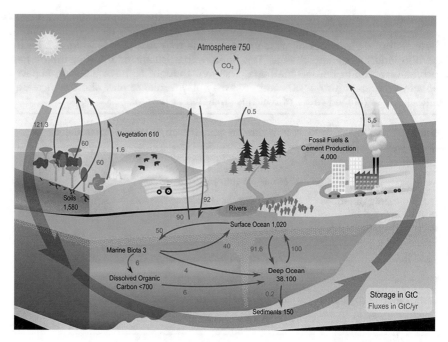

Fig. 7.16 Terrestrial carbon cycle. Carbon is the most crucial of the elements on which life is based. The diagram shows the carbon exchange between the atmosphere, the biosphere, the lithosphere and the oceans. Any change in one carbon reservoir will lead to a change in all the others. Beside physical, chemical and biological factors, anthropogenic factors also lead to disturbances of the carbon cycle's balance. However, the magnitude of global warming, the importance of human influence upon it and the various hypotheses concerning it remain controversial issues. [Image: NASA]

would remain a gap of knowledge resulting from the limits of calculability inherent in computer simulations (see Sects. 4.7. and 7.9).

Climate researchers are very well aware of the uncertainties in climate modeling that arise from incomplete knowledge of external factors and in consequence of their internal variability. Thus, a model may respond differently to the same external influence, owing to the properties of the atmosphere, the ocean and the coupled ocean–atmosphere system [46].

Consequently, there are numerous models of the earth's climate change which examine quite different possible causes as a basis. However, all these models only demonstrate the variability of computer simulations, depending on their input information. Global problems have no simple solution. This applies just as much to the computer simulations themselves as to the reach of the scientific forecasts that are based on them.

Another issue that has been the subject of attempts to make computer-based predictions is the world's population. Here again, the complexity of the mechanisms involved makes exact prediction impossible. Before 2014 most calculations showed the global population peaking at nine billion by 2070 and then easing to 8.4 billion by 2100. In 2014, however, the U.N. had to revise these numbers steeply upward, to 9.6 billion by 2050 and 10.9 billion by 2100. Owing to the substantial volatility of growth parameters, forecasts of global and regional growth are always uncertain, as shown in this case by comparing the old and new projections (Fig. 7.17). Predictions of population, like those of climate, cannot be made with sufficient accuracy to be of any real value. Instead, one has to be content with rough forecasts, which must be perpetually adjusted to reflect the actual development. The differences between current and earlier estimates are sometimes considerable, demonstrating the major uncertainties inherent in predicting global changes.

In fact, global changes possess all the features of historical processes. This includes, in particular, their multicausal background, which can only partially be grasped. Forecasts are nevertheless important if one is to recognize trends and risks of global developments. However, these forecasts must not be confused with solid truths. The more parameters and variables enter a model, and the larger the periods are to which prognoses refer, the more significant the uncertainties become. This applies to all forecasts of global change.

With the increasing complexity of phenomena, the empirical sciences reach their limits, because they can draw neither upon experiments nor upon science's primary methods of simplification, abstraction and idealization. At the same time, the symmetry of explanation and prediction, which is a trademark of the scientific understanding of simple phenomena, breaks down once a certain degree of complexity has been reached.

However, scientific models that rely on estimations, extrapolations, indirect observations, average values, reconstructions and so forth are entirely dependent on computer simulations. They can lead into a virtual reality in which there is no sharp borderline between computational results and empirical truth. Moreover, scientific truths are not decided by some social consensus, as the consensus theory of truth (Chap. 2) would have it, nor by some undefined consensus of research, but only by the intrinsic control procedures of science, applied over time. One has to bear this in mind if one wishes to derive norms for human behavior based on the scientific forecasts that refer to global changes, such as climate change or the disbalance of Nature's ecology.

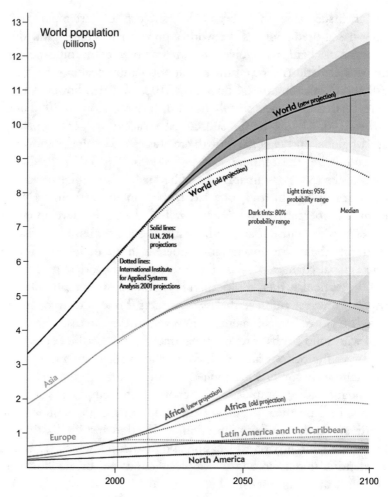

Fig. 7.17 Forecasts of world population growth. The diagram shows projections for the populations in different world regions and overall. The curves reveal significant differences in growth expectations (adapted from [47], based on [48] and [49])

There can be no doubt that we should do everything possible to minimize the impact of human actions on our natural environment. Fresh scientific insights and technological innovations may support this aim, as demonstrated by the example of the human-assisted regeneration and evolution of threatened coral reefs (Sect. 7.8). To combat climate change, even methods of "geoengineering" have been discussed [50]. In fact, there is already a project in the test phase in which tiny particles are sprayed into the stratosphere that are expected to reflect part of the sun's heat back into space; this, it is believed, may help in countering climate change.

Society is, however, also bound to preserve and improve humans' social and economic living conditions. Given the undampened growth of the world's population, we must expect a steady increase in global energy demand in the coming decades. However, even now there is a severe imbalance in the use of energy resources. Therefore, the goal of the decarbonization of the atmosphere must go hand in hand with structural changes in economics that are subordinated to the guiding principle of economic effectiveness regarding the multitude of alternative changes. The demand for a radical reconstruction of modern industrial society could call into question the world population's energy supply, and it thus hides the risk of another danger, namely, a massive disturbance of the worldwide economic balance.

The only guideline for dealing with global and multi-causal problems is cautious trial and error, with a sense of the correct proportions of the problem and with a willingness to go new and unforeseen ways. Life's evolution epitomizes a balanced interplay of opposed dynamical properties such as stability and instability, competition and cooperativity, accuracy and error-proneness. Exploring the intrinsic proportions of Nature will be the key to understanding Nature's semantics and for dealing with Nature responsibly.

References

1. Jackson DA, Symons RH, Berg P (1972) Biochemical method for inserting new genetic information into DNA of Simian virus 40. Proc Natl Acad Sci USA 69:2904–2909
2. Arber W (1979) Promotion and limitation of genetic exchange. Experientia 35:287–293
3. Cohen SN, Chang ACY, Boyer HW et al (1973) Construction of biologically functional bacterial plasmids in vitro. Proc Natl Acad Sci USA 70(11):3240–3244
4. Richter JP (2016) The Notebooks of Leonardo Da Vinci, vol 2. Dover Publications, Mineola
5. Nachtigall W, Wisser A (2015) Bionics by Examples: 250 Scenarios from Classical to Modern Times. Springer, Berlin/Heidelberg
6. Ebeling W, Rechenberg I, Schwefel H-P et al (eds) (1996) Parallel Problem Solving from Nature—PPSN IV. International Conference on Evolutionary Computation. Lect Notes Comput Sci, vol 419

7. Nawroth JC, Lee H, Feinberg AW et al (2012) A tissue-engineered jellyfish with biomimetic propulsion. Nat Biotechnol 30:792–797

8. Barricelli N (1957) Symbiogenetic evolution processes realized by artificial methods. Methodos 9:143–182

9. Dyson GB (1997) Darwin Among the Machines: The Evolution of Global Intelligence. Addison-Wesley, New York

10. Zeck G, Fromherz P (2001) Noninvasive neuroelectronic interfacing with synaptically connected snail neurons immobilized on a semiconductor chip. Proc Natl Acad Sci USA 98(18):10457–10462

11. Dürrenmatt F (2006) Selected Writings, vol 1, Plays (transl: Agee J). The University of Chicago Press, Chicago [Original: Die Physiker 1962]

12. Berg P, Baltimore D, Brenner S (1975) Asilomar conference on recombinant DNA molecules. Science 188(4192):991–994

13. Kourilsky P (1987) Les Artisans de L'Hérédité. Odile Jacob, Paris

14. Mohr H (1987) Natur und Moral. Wissenschaftliche Buchgesellschaft, Darmstadt

15. Habermas, J (1970) Technology and Science as "Ideology". In Toward a Rational Society: Student Protest, Science, and Politics (transl: Shapiro JJ). Beacon Press, Boston, pp 81–122 [Original: Technik und Wissenschaft als Ideologie, 1969]

16. Habermas J (1987) Knowledge and Human Interests (transl: Shapiro JJ). Polity Press, Cambridge [Original: Erkenntnis und Interesse, 1968]

17. Marcuse H (1964) One Dimensional Man. Sphere Books, London

18. Simon HA (1997) Models of Bounded Rationality: Empirically Grounded Economic Reason, vol 3. MIT Press, Cambridge/Mass

19. Mises R von (1968) Positivism: A Study in Human Understanding. Dover Publications, New York [Original: Kleines Lehrbuch des Positivismus, 1939]

20. Huizinga J (1936) In the Shadow of Tomorrow: A Diagnosis of the Spiritual Ills of Our time. W.W. Norton & Co, New York [Original: In de schaduwen van morgen, 1935]

21. Harrington A (1996) Reenchanted Science. Princeton University Press, Princeton

22. Jacobi CGJ (1881) Letter to Legendre from July 2, 1830. In: Borchardt CW (ed) Jacobi: Gesammelte Werke, Bd. I. G. Reimer, Berlin

23. Eigen M (1989) Perspektiven der Wissenschaft: Jenseits von Ideologien und Wunschdenken. DVA, Stuttgart

24. Wilson EO (1975) Sociobiology. Havard University Press, Cambridge/Mass

25. Küppers B-O (2016) The Nucleation of Semantic Information in Prebiotic Matter. In: Domingo E, Schuster P (eds) Quasispecies: From Theory to Experimental Systems. Springer International, Cham, pp 67–85

26. Jonas H (1985) The Imperative of Responsibility: In Search of an Ethics for the Technological Age. The University of Chicago Press, Chicago

27. Itoh T (1984) The Gardens of Japan. Kodansha International, Tokyo
28. Blumenberg H (2001) Die erste Frage an den Menschen. All der biologische Reichtum des Lebens verlangt eine Ökonomie seiner Erklärung. Frankfurter Allgemeine Zeitung (2. Juni 2001), Frankfurt, pp I-II
29. Markl H (1986) Natur als Kulturaufgabe. DVA, Stuttgart
30. Seddon PJ, Griffiths C, Soorae PS et al (2014) Reversing defaunation: restoring species in a changing world. Science 345(6195):406–412
31. Elton CS (2020) The Ecology of Invasions by Animals and Plants (with contributions of Ricciardy A, Simberloff D). Springer Nature, Cham
32. Keane RM, Crawley MJ (2002) Exotic plant invasions and the enemy release hypothesis. Trends Ecol Evol 17(4):164–170
33. Hällfors MH, Aikio S, Fronzek S et al (2016) Assessing the need and potential of assisted migration using species distribution models. Biol Conserv 196:60–68
34. Hobbs RJ, Higgs ES, Hall CM (eds) (2013) Novel Ecosystems: Intervening in the New Ecological World Order. Wiley-Blackwell, Oxford
35. Van Oppen MJH, Oliver JK, Putnam HM et al (2015) Building coral reef resilience through assisted evolution. Proc Natl Acad Sci USA 112(8):2307–2313
36. Hughes TP, Kerry JT, Baird AH et al (2018) Global warming transforms coral reef assemblages. Nature 556:492–496
37. Diels H, Kranz W (1974) Die Fragmente der Vorsokratiker, Bd. 1–3. Weidmann, Berlin
38. Thompson DW (1942) On Growth and Form. New Edition. Cambridge University Press, Cambridge
39. Richter PH, Schranner R (1978) Leaf arrangement. Geometry, morphogenesis, and classification. Naturwissenschaften 65:319–327
40. Malthus TR (2008) An Essay on the Principle of Population. Oxford University Press, Oxford
41. Verhulst P-F (1845) Recherches mathématiques sur la loi d'accroissement de la population. In: Nouveaux Mémoires de l'Académie Royale des Sciences et Belles Lettres de Bruxelles, vol 18. M. Hayez, Brüssel, pp 14–54
42. Peitgen H-O, Richter PH (1986) The Beauty of Fractals. Springer, Berlin/Heidelberg
43. Hale GE, Luckey GP (1915) Some vortex experiments bearing on the nature of sun-spots and flocculi. Proc Natl Acad Sci USA 1(6):385–389
44. Scotese CR (2002) PALEOMAP Project. (www.scotese.com)
45. Bradley R (2015) Paleoclimatology: Reconstructing Climates of the Quaternary, 2nd edn. Academic Press, New York
46. Deser C, Phillips A, Bourdette V et al (2012) Uncertainty in climate change projections: the role of internal variability. Clim Dyn 38:527–546
47. Fischetti M (2014) Up, up and away. Sci Am 311(6):1000
48. Gerland P, Raftery A, Ševčíková H et al (2014) World population stabilization unlikely this century. Science 346(6206):234–237

49. Lutz W, Sanderson W, Scherbov S (2001) The end of world population growth. Nature 412(6846):543–545
50. Keith W, Weisenstein DK, Dykema JA, Keutsch FN (2016) Stratospheric solar geoengineering without ozone loss. Proc Natl Acad Sci USA 113(52):14910–14914

Author Index

© Springer Nature Switzerland AG 2022
B.-O. Küppers, *The Language of Living Matter*, The Frontiers Collection,
https://doi.org/10.1007/978-3-030-80319-3

Subject Index

Printed in the United States
by Baker & Taylor Publisher Services